THERAPEUTIC POTENTIAL OF MARINE BIOTOXINS

From Poison to Pharmacy

THERAPEUTIC POTENTIAL OF MARINE BIOTOXINS

From Poison to Pharmacy

Santhanam Ramesh, PhD
Ramasamy Santhanam, PhD
Subbiah Balasundari, PhD

AAP | APPLE ACADEMIC PRESS

First edition published 2026

Apple Academic Press Inc.
1265 Goldenrod Circle, NE,
Palm Bay, FL 32905 USA

760 Laurentian Drive, Unit 19,
Burlington, ON L7N 0A4, Canada

CRC Press
2385 NW Executive Center Drive,
Suite 320, Boca Raton FL 33431

4 Park Square, Milton Park,
Abingdon, Oxon, OX14 4RN, UK

Library and Archives Canada Cataloguing in Publication

Title: Therapeutic potential of marine biotoxins : from poison to pharmacy / Santhanam Ramesh, PhD, Ramasamy Santhanam, PhD, Subbiah Balasundari, PhD.
Names: Ramesh, Santhanam, 1979- author | Santhanam, Ramasamy, 1946- author | Balasundari, Subbiah, author
Description: First edition. | Includes bibliographical references and index.
Identifiers: Canadiana (print) 20240537882 | Canadiana (ebook) 20240537920 | ISBN 9781774918883 (hardcover) | ISBN 9781774918890 (softcover) | ISBN 9781003610991 (ebook)
Subjects: LCSH: Toxic marine algae—Therapeutic use. | LCSH: Microalgae—Therapeutic use.
Classification: LCC QK568.T67 R36 2025 | DDC 579.8—dc23

Library of Congress Cataloging-in-Publication Data

Names: Ramesh, Santhanam, 1979- author. | Santhanam, Ramasamy, 1946- author. | Balasundari, Subbiah, author.

Title: Therapeutic potential of marine biotoxins : from poison to pharmacy / Santhanam Ramesh, Ramasamy Santhanam, Subbiah Balasundari.

Description: First edition. | Palm Bay, FL : Apple Academic Press, 2025. | Includes bibliographical references and index. | Summary: "Marine biotoxins are poisonous compounds produced mainly by certain kinds of microalgae. These microalgae and their toxins may adversely impact the marine environment and fisheries and even humans directly or indirectly. What is little known is that these toxins and poisons also have therapeutic potential that can be effective against various diseases. The new title, Therapeutic Potential of Marine Biotoxins: From Poison to Pharmacy, explores the state of the art of the therapeutic potential of marine biotoxins, emphasizing the present gaps in the existing knowledge on marine biotoxins. The book covers the types and chemistry of marine biotoxins; marine phyla yielding biotoxins; the impact of marine biotoxins on the environment, on fisheries, and on human health; and marine phycotoxins, invertebrate toxins, and fish toxins and their therapeutic potential as pharmacological agents for several disorders, including cancer, diabetes, hypertension, dyslipidemia, and obesity. The book provides the names of marine taxonomic groups and further explores the therapeutic potential of certain groups, including marine fishes, marine algae, and more. The impact of marine biotoxins on human health is explored using case informative studies. As one of the very few books on this topic, this book will be of great value as a standard reference for pharmaceutical researchers as well as researchers and faculty and students of various disciplines such as biomedical sciences, pharmaceutical sciences, marine biology, and fisheries science. It will also serve as a valuable reference for libraries of colleges and universities and as a guide for the pharmaceutical industries involved in the development of new marine biotoxin-derived drugs"-- Provided by publisher.

Identifiers: LCCN 2024057736 (print) | LCCN 2024057737 (ebook) | ISBN 9781774918883 (hardcover) | ISBN 9781774918890 (paperback) | ISBN 9781003610991 (ebook)

Subjects: MESH: Marine Toxins--therapeutic use | Marine Toxins--chemistry

Classification: LCC QP632.M37 (print) | LCC QP632.M37 (ebook) | NLM QW 630.5.M3 | DDC 615.9/45--dc23/eng/20250206

LC record available at https://lccn.loc.gov/2024057736
LC ebook record available at https://lccn.loc.gov/2024057737

ISBN: 978-1-774918-88-3 (hbk)
ISBN: 978-1-003610-99-1 (ebk)
DOI: 10.1201/9781003610991

About the Authors

Santhanam Ramesh, PhD
Vice-Principal, Karuna College of Pharmacy, Kerala University of Health Sciences, Palakkad, Kerala, India

Santhanam Ramesh, PhD, is Vice-Principal at the Karuna College of Pharmacy, Kerala University of Health Sciences, Palakkad, Kerala, India. He has 15 years of teaching and research experience in pharmaceutical sciences, and his fields of specialization include natural products, pharmaceutical nanotechnology, biomaterials, and marine wastes of pharmaceutical importance. Dr. Ramesh is also a visiting professor in the Department of Pharmacology, North-Caucasian State Humanitarian and Technological Academy (North-Caucasian State University), Cherkessk, Russia. To his credit, he has 13 books published with internationally reputed publishers. He is a member of the International Society for Pharmacoeconomics and Outcome Research, the British Society for Nanomedicine, and the Academic Pharmacy Group of Royal Pharmaceutical Society, London. He earned his PhD from Jawaharlal Nehru Technological University, Hyderabad, India.

Ramasamy Santhanam, PhD
Former Dean, Fisheries College and Research Institute (FC &RI), Tamil Nadu Dr. J. Jayalalithaa Fisheries University), Thoothukudi, India

Ramasamy Santhanam, PhD, is Ex-Dean, Fisheries College and Research Institute, Tamil Nadu Veterinary and Animal Sciences University, Thoothukudi, India. He has about 50 years of teaching and research experience in marine/ fisheries sciences. He has been serving as a fisheries expert for various government and nongovernment organizations in India and abroad. To his credit, Dr. Santhanam has 45 books on various aspects of marine life. He was a member of the American Fisheries Society, United States; World Aquaculture Society, United States; Global Fisheries Ecosystem Management Network (GFEMN), United States; and the IUCN's Commission on Ecosystem Management, Switzerland.

Subbiah Balasundari, PhD
Dean, Dr. M.G.R Fisheries College and Research Institute, Tamil Nadu Dr. J. Jayalalithaa Fisheries University, Ponneri, , Tamil Nadu, India

Subbiah Balasundari, PhD, is Dean at the Dr. M.G.R. Fisheries College and Research Institute at the Tamil Nadu Dr. J. Jayalalithaa Fisheries University, Ponneri, Tamil Nadu, India. She has 26 years of teaching and research experience in fish processing. To her credit, she has four textbooks on fisheries science and 20 research papers. She has developed a number of contemporary value-added fish products and has disseminated the technologies to fish processing industries of India. She has completed six projects through state- and national-level funding agencies and has established various demonstration units in fisheries enterprises for the benefit of stakeholders. She is a member of the World Aquaculture Society, Asian Fisheries Society (Indian Branch), Agricultural Scientific Tamil Society, and Society of Fisheries Technologists.

Contents

Foreword

I am very pleased to write this foreword for the book *Therapeutic Potential of Marine Biotoxins: From Poison to Pharmacy* by Santhanam Ramesh, Ramasamy Santhanam, and Subbiah Balasundari. Biotoxins produced by marine microalgae are known to cause paralytic shellfish poisoning (PSP), neurotoxic shellfish poisoning (NSP), diarrheic shellfish poisoning (DSP), and amnesic shellfish poisoning (ASP) in humans when they ingest biotoxin-contaminated shellfish. Despite these unfavorable aspects, the unique structure and diverse functionality of these biotoxins make them valuable and interesting therapeutic compounds. Several toxicological studies have unfolded the potential of these marine-microalgae-derived toxins as promising pharmacological effectors.

This interdisciplinary book, prepared by experts in both pharmacy and fisheries, represents an incredible resource that deals with aspects such as marine biotoxin types and their chemistry, biotoxin-producing marine taxonomic groups, source organisms of marine biotoxin-associated poisonings, marine biotoxins and human health (case studies), and therapeutic potential of phycotoxins and zootoxins derived from the different groups of marine biota.

Dr. Santhanam Ramesh and his team have superbly succeeded in creating the most useful overview of the hitherto untouched aspect, namely, therapeutic potential of marine biotoxins, a treatise that will be most welcomed by students and researchers of the faculties of pharmacy, fisheries science, and marine biology. Further, the libraries of colleges and universities will find this volume worth having.

I strongly hope and expect that this title will amply provide an effective learning experience and reference source for the professionals of fisheries and pharmacy. I congratulate the team of authors for their timely contribution.

Dr. B. Ahilan, PhD
Dean, Fisheries College and Research Institute,
Tamil Nadu Dr. J. Jayalalithaa Fisheries University,
Tamil Nadu, India

Preface

Marine biotoxins are poisonous compounds produced mainly by certain kinds of microalgae such as diatoms (phytoplankton), blue-green algae (cyanobacteria), and dinoflagellates. During the formation of "blooms," these microalgae and their toxins may adversely impact the marine environment and fisheries and can even harm humans directly or indirectly. Further, such harmful algal blooms (HABs) form food for filter-feeding bivalve mollusks (oysters, mussels, scallops, and clams), gastropod mollusks like *Conus* spp., cephalopods like octopuses, and fishes. Most human illnesses from marine biotoxins are known to occur when people are in direct contact by swimming, breathing in aerosolized toxins in water droplets, or eating toxin-contaminated fish or shellfish. Depending on the nature of poisoning, the marine biotoxins may lead to ciguatera fish poisoning (CFP), paralytic shellfish poisoning (PSP), diarrheic shellfish poisoning (DSP), amnesic shellfish poisoning (ASP), or neurotoxic shellfish poisoning (NSP). According to the producing species and chemical structure, the marine biotoxins are mainly classified into chemical classes, namely, okadaic acid (OA), domoic acid (DA), tetrodotoxin (TTX), saxitoxin (STX), yessotoxin (YTX), ciguatoxin (CTX), pectenotoxin (PTX), brevetoxin (BTX), palytoxin (PlTX), azaspiracid (AZA), cyclic imine (CI), etc.

Recent studies on marine toxins have shown that they may serve as potential therapeutic pharmacological agents for several disorders, including cancer, diabetes, hypertension, dyslipidemia, and obesity. Unfortunately, most of these biotoxins are still at the preclinical and clinical stages in the development of novel marine drugs. However, lead TTX therapeutic candidates—namely, TTX-MC138 and TTX-siPDL1 (for the treatment of rare pancreatic cancer cell lines)—have been granted Orphan Drug Designation (ODD) status by the U.S. Food and Drug Administration (FDA) during 2022–2023.

Though there are a few books presently available on the characteristics and production of marine biotoxins, a comprehensive volume dealing with the therapeutical potential of marine biotoxins has not so far been published. The present title, *Therapeutic Potential of Marine Biotoxins:*

From Poison to Pharmacy, prepared by scientists of both pharmaceutical and marine biology disciplines would be the first of its kind to answer this long-felt need. As blue biotechnology and marine bioprospecting have gained momentum in recent decades for the development of new drugs, we hope that the present book will amply provide the state-of-the-art therapeutic potential of toxins of the different constituents of marine biota, emphasizing the present gaps in the existing knowledge on marine biotoxins.

This new title deals with aspects such as types and chemistry of marine biotoxins; marine phyla-yielding biotoxins; impacts of marine biotoxins on the environment, fisheries, and human health; and marine phycotoxins, invertebrate toxins, and fish toxins and their therapeutic potential. We hope that the present publication will be of great use as a standard text/reference book for teachers, students, and researchers of the disciplines of biomedical sciences, pharmaceutical sciences, marine biology, and fisheries science. It will also serve as a valuable reference for libraries of colleges and universities, and as a guide for the pharmaceutical industries involved in the development of new marine biotoxin-derived drugs.

The authors are highly indebted to Dr. K. Venkataramanujam, former Dean, Fisheries College and Research Institute, Tamil Nadu Dr. J. Jayalalithaa Fisheries University, India, for his valued comments and suggestions on the manuscript. We sincerely thank all of our international friends who were kind enough to share their images for the present publication. The photography and secretarial assistance services rendered by Mrs. Albin Panimalar Ramesh are also gratefully acknowledged.

Introduction

Marine organisms such as fish, shellfish, algae, mollusks, echinoderms, etc. have been a potential source of food for humans. Among these food groups, fish and shellfish form a major category while the other organisms differ in their utilization from country to country. The ever-increasing pollution due to eutrophication and discharge of untreated sewage and waste waters from industries, farming systems, etc. into the sea are the causative factors for the development of biotoxins and their accumulation in edible organisms through the food chain. Indiscriminate catching of food organisms from the sea and their distribution may, therefore, lead to human health problems due to the accumulated biotoxins.

The toxic biological secretions of animals may be classified into three types namely, (i) venoms that are delivered to internal tissues of another animal via the creation of a wound; (ii) poisons that lack a delivery mechanism, that is, they are ingested, inhaled, or absorbed across the body surface; and (iii) toxungens, which are delivered to the body surface without an accompanying wound (Nelson et al., 2014). Marine biotoxicology is the science of poisons from marine organisms and their harmful effects. Marine biotoxins that are the specific types of poison are largely sourced from marine plants such as microalgae and macroalgae (seaweeds); marine invertebrate animals, namely, protozoans, sponges, cnidarians, annelids, crustaceans, mollusks, echinoderms; and chordates, namely, tunicates, elasmobranch fishes, teleost fishes, reptiles, and mammals.

Researchers have been studying marine biotoxins for quite some time, especially their types, chemistry, and source orgasms, and their impacts on the environment, fisheries, and human health. The positive aspect of these marine biotoxins, that is, from poison to pharmacy, is still in an infant stage. The scope of this publication is, therefore, to identify the therapeutic potential of these marine biotoxins for possible clinical applications in the future.

CHAPTER 1

Marine Biotoxins: Origin and Significance

ABSTRACT

This chapter deals with the marine biotoxins and their contributing taxonomic groups; impacts of marine biotoxins on human health and case studies; preventive measures of marine biotoxin- associated seafood poisoning; prospects in the development of new marine biotoxin-based drugs; and challenges in the discovery of new drugs from marine biotoxins.

Marine biotoxins are basically poisons and are produced by certain species of marine phytoplankton (microalgae) and are normally found in small amounts to be harmful to coastal resources, aquatic ecosystems, and human health. However, conditions such as eutrophication (nutrient-rich), sunlight, and warm temperatures can create marine harmful algal blooms (MHABs). However, human activities, climate change (e.g., global warming), the appearance of exogenous species, and habitat modification may also be the causative factors in the spread of MHABs (Morabito et al., 2018). Transportation of encysted algae to new coastal environments is also believed to be responsible for the geographical expansion of these algal blooms. The MHAB phenomenon is worsening in certain geographical localities, and this is largely due to the expansion of aquaculture and associated coastal pollution due to the discharge of large amounts of nutrient-rich wastewaters. The HAB events affecting the marine environments, fisheries, and humans have been collated in the Harmful Algae Event Database (HAEDAT) since 1985 and during the last 33 years, a total of 9503 (as of 10 December, 2019) HAB events have been reported worldwide (http://haedat.iode.org).

North America including the Caribbean recorded 37.6% of MHAB events followed by Australia/Oceania (17.6%), Asia (15%), Europe

Therapeutic Potential of Marine Biotoxins: From Poison to Pharmacy. Santhanam Ramesh, Ramasamy Santhanam, & Subbiah Balasundari (Authors)

DOI: 10.1201/9781003610991-1

(8.2%), Africa (3.4%), and South America (2.6%), in that order (Young et al., 2020). Among the 5000 species of marine phytoplankton described so far, about 200 species, viz., 31 species of diatoms, 105 species of dinoflagellates, 37 species of cyanobacteria (blue-green algae), 8 species of coccolithophorids, 3 species of silicoflagellates, 2 species of pelagophytes, and 6 species of red algae, are known to cause MHABs, the toxins of which harm seafood and human health (Hallegraeff et al., 2021). MHABs are also commonly called "red tide" which is largely due to the rapid reproduction of such phytoplanktonic organisms. The term "red tide" is normally associated with neurotoxic paralytic shellfish toxin poisoning (PSP) which is caused by certain species of dinoflagellates producing the alkaloid, saxitoxin. Further, coastal waters with massive algal blooms may appear clear; and red tide blooms are normally believed to be harmless to humans (https://doh.wa.gov/community-and-environment/shellfish/recreational-shellfish/illnesses/biotoxins). The development of such MHABs and the associated biotoxins are the causative factors for mass mortality of fishes especially in coastal waters; and red tides may directly or indirectly cause illness in humans. The filter-feeding bivalve mollusks such as mussels, clams, and oysters feed largely on the aforementioned phytoplankton, thereby the biotoxins are getting accumulated largely in their systems. Apart from these mollusks, other organisms in the food chain are also likely to get affected by this biotoxin. It is worthy of mention here that accumulated biotoxins do not harm these bivalves which, however, cause illness or death in humans who eat them. Further, it has also been reported that cooking may kill the phytoplankton producing the toxins but the latter is not destroyed by cooking and they remain in the tissue of shellfish. As there is no antidote for biotoxin poisoning, the affected people are normally allowed to wait till the toxin is flushed from the body naturally. In extra-ordinary cases, life support systems include oxygen and respirators to keep the affected stable and alive.

1.1 MARINE BIOTOXINS AND THEIR POISONINGS

The marine biotoxins are either water- or fat-soluble and based on their chemistry, they are grouped into 10 groups, viz., domoic acid (DA), okadaic acid (OA), azaspiracid (AZA), brevetoxin (BTX), saxitoxin (STX), yessotoxin (YTX), pectenotoxin (PTX), ciguatoxin (CTX), cyclic imine (CI), and palytoxin (PlTX) (EFSA, 2009). In recent years, the MHABs have been reported to be on an increase at the biogeographic level, frequency, and severity thereby causing serious health problems in humans due to the consumption of affected seafood including biotoxin-contaminated fish and shellfish. Such affected people may face a wide range of symptoms depending on the nature of involved biotoxins and their poisonings, viz., ciguatera fish poisoning (CFP) and amnesic (ASP), azaspirazid (AZP), paralytic (PSP), diarrhoetic (DSP), and neurotoxic (NSP) shellfish poisonings. Based on the worldwide occurrence of MHABs and the occurrence of seafood toxin syndromes, PST accounted for 35% followed by DST (30%), CFP (9%), cyanotoxins (9%), and AST (7%); and remaining toxins, such as azaspiracid, aerosol, and NST, accounted for 10% (Hallegraeff et al., 2021).

1.2 IMPACTS OF MARINE BIOTOXINS ON HUMAN HEALTH

The impacts of marine biotoxins and the associated human health issues have been recorded in the world countries. Among the different toxins, PST was found to prevail in South America, North America, the Caribbean, North Asia, and South East Asia; DST in the Mediterranean, Europe, and in USA; NST in Florida (USA); CFP in the tropical Pacific, the Caribbean, the Indian Ocean, and Australia; and marine cyanotoxins from in NW Europe, the Baltic Sea, Australia/New Zealand (ANZ), and the Indian Ocean. However, according to HAEDAT, AST, DST, and PST were mostly associated with seaweed/algal related phycotoxins in seafood and they rarely caused human poisonings unlike CFP which was solely responsible for human poisonings (Hallegraeff et al., 2021). Consumption of biotoxin-contaminated seafoods such as shellfish and fish; body contact with water of the MHABs by fishing or recreational activities; and inhalation of aerosolized biotoxins released by toxic microalgae are all directly responsible for affecting the human health. Toxin-producing phytoplankotn species of MHABs can accumulate these biotoxins through the marine food chain and the marine animals in the top of the food chain are therefore likely to be more contaminated and are unfit for consumption if and when the toxic level of such food organisms is beyond permitted level. Among the different biotoxin-based poisonings associated with MHABs, ciguatera fish poisoning (CFP) has been reported to cause a significant disease burden worldwide as on an average 251 cases per 10,000 persons were found affected per year in some Pacific and Caribbean populations. Further, frequent human deaths with outbreaks of CFP, ASP and PSP have also been reported. However, precise estimations on the MHABs-associated human health impacts are inaccessible due to the absence of proper surveillance systems (Young et al., 2020). Depending on the type of MHABS and their toxins, the impact to humans may vary from loss of memory to damage to the intestinal or nervous system. Among the toxic phytoplankton species of MHABS, the diatoms *Skeletonema costatum*, and dinoflagellate, *Noctiluca scintillans*, have been reported to be largely responsible for the death of fish and other marine organisms which is associated with the discoloration of water and depletion of oxygen. On the other hand, the dinoflagellate genera, viz., *Dinophysis, Alexandrium*, and *Gymnodinium*; and diatom genus *Pseudo-nitzschia* are the major causative organisms in the production of marine biotoxins detrimental to humans (Berti and Milandri, 2014; Bruce et al., 2015).

Lim et al. (2023) reported that 15% of global asthma cases are invariably due to the inhalation of aerosolized toxins emanated from MHABs. Further, according to them, about 4000 emergency department visits and 11,000 hospital admissions occurred during the harmful red tide blooms of Florida in 2012. The human health issues associated with the major biotoxin poisonings are given in the following table.

1.3 SELECTED CASE STUDIES RELATING TO MARINE BIOTOXINS

During 1985 and 2018, the species of the dinoflagellates, viz., *Dinophysis* caused mainly diarrhetic shellfish poisoning (DSP) with a 4×-fold increase; the diatom genus *Pseudo-nitzschia* for amnesic shellfish poisoning (ASP) with a 7×-fold increase; and the dinoflagellate genus *Alexandrium* for paralytic shellfish poisoning (PSP) with 6×-fold increase. For PSP, the other dominant species were found to be *Gymnodinium catenatum* and *Pyrodinium bahamense*. It is also worthy of mention here that about DSP-related 11,000 nonfatal have been reported globally including the major countries, viz., Japan, Europe, and South America. This led to the closure of shellfish harvesting localities due to the observed higher DST level above the prescribed regulatory limits to safeguard human health. Further, after the 1987 incident, there were no human deaths in Prince Edward Island, Canada, due to amnesic shellfish poisoning (ASP) though the causative organisms, *viz., Pseudo-nitzschia* spp. had a widespread distribution. However, ASP was reported to cause mortalities of mammal calves in Argentina. As far as human paralytic shellfish poisoning (PSP), it was responsible for 3800 human poisonings during the period, 1985–2018,

TABLE 1.1 Major Biotoxin Poisonings and Human Health Issues.

Poisoning type	Seafood affected	Short-term effects	Long-term effects	Treatment
CFP	Red snapper; barracuda; grouper; amberjack	Stomach pain; vomiting; diarrhea; nausea	Low blood pressure; pain; weakness	Mannitol I.V. (mannitol injection); supportive care
PSP	All edible marine bivalves; crabs; fish	Irregular heartbeat; numbness in lips and mouth; breathing problems; diarrhea; vomiting; nausea	Unknown	Respiratory support; supportive care
DSP	All edible marine bivalves	Fever; headache; stomach pain; vomiting; diarrhea; nausea	Unknown	Supportive care
NSP	Mussels; scallops; oysters	Numbness in throat, lip, and tongue; stomach pain; vomiting; diarrhea; nausea	Unknown	Supportive care
ASP	All edible marine bivalves; fish	Irregular heartbeat; numbness in lips and mouth; breathing problems; diarrhea; vomiting; nausea; memory loss; coma	Loss of memory	Supportive care
AZP	All edible marine bivalves	Stomach pain; vomiting; diarrhea; nausea	Unknown	Supportive care
Palytoxin	Palythoa toxica (Cnidaria-Anthozoa)	Ccutaneous and muscle problems; fatigue; chills; gastrointestinal disorders; transient irritation; nasal congestion; coughs; shortness of breath; local inflammation; numbness	Longer-term disability; speech disturbance and loss of consciousness; impairment of muscles of the cardiac; respiratory and skeletal systems	Supportive treatments, viz., oxygen supplementation and administration of inhaled corticosteroids

Source: https://oehha.ca.gov/fish/general-info/marine-harmful-algal-bloom-hab-related-illness-tracking; https://ncceh.ca/content/blog/palytoxin-potent-poorly-understood-marine-toxin-found-aquarium-coral.

and the Philippines recoded the maximum number PSP poisonings due to its dependance on aquaculture for human food. Further, PSP was also responsible for 2555 poisonings with 165 human deaths which were due to the highly toxic dinoflagellae species, *viz.*, *Pyrodinium bahamense*. It has also been reported that the global distribution of the CFP causing dinoflagellate species, *viz.*, *Gambierdiscus* spp. and *Fukuyoa* spp. increased very considerably. Though about 10,000–50,000 people are affected by CFP globally, human deaths due to this poisoning are rare according to available reports. While, the occurrence of CFP is new to Canary Islands (off the coast of northwestern Africa), French Polynesia, and the Caribbean recorded stable data and a decrease in Hawaii (Hallegraeff et al., 2021).

1.4 IMPACT OF MARINE BIOTOXINS ON AQUACULTURE AND CASE STUDIES

The deaths of fish in aquaculture systems are largely due to the biotoxins associated with the marine species of the microalgal genera, viz., *Heterosigma, Karenia, Karlodinium, Margalefidinium (=Cochlodinium), Chattonella, Pseudochattonella, Chrysochromulina*, and *Prymnesium*. Besides contaminating the seafood, these occurrences have caused much greater economic damage than MHABs. However, the biotoxins associated with fish kills in aquaculture are of no human health value. As per the available reports, due to impacts of biotoxins in aquaculture, there was a loss of USD290M in China in 2012, USD70M in Korea in 1995, USD100M in Norway in 2019 and USD71M in Japan in 1972 (Hallegraeff et al., 2021).

1.5 PREVENTIVE MEASURES OF BIOTOXIN ASSOCIATED SEAFOOD POISONING

Marine toxins, such as ciguatera toxin, scombrotoxin, and tetrodotoxin, have been reported to be heat-stable, and therefore cooking cannot prevent their poisoning properties. However, a few types of biotoxins are moderately water-soluble and they are likely to be leached out by soaking. It is also advisable to separate the flesh of fish of tropical seas as early as possible from their viscera and head and skin which are known to contain toxins in high concentrations. Further, almost all scaleless, naked fishes are more tetrodotoxic than their counterparts. Larger-sized fish species caught from coral reefs are believed to be more ciguatera-toxic as they collect and accumulate food chain-based toxins throughout their lives. As parrot fish and moray eels are more prone to severe ciguatera and scaritoxic fish poisoning, these fishes should not be eaten. It is also known that early freezing of fish could largely avoid scombroid poisoning. Furthermore, the filter-feeding molluskan shellfish collected during red tide periods should not be eaten as the toxic dinoflagellates are invariably associated with such discolored coastal waters (Warrell, 2020).

1.6 MARINE BIOTOXINS: POISON TO PHARMACY

Among the marine biotoxins, the conotoxins extracted from the gastropod mollusks of the genus *Conus* are of paramount importance owing to their profound therapeutic applications in preclinical models of pain, neuromuscular block, cardio-protection, convulsive disorders, and stroke. Further, conotoxin selectivity has been reported to be useful in the validation of several clinical targets, as in the *N*-type Ca_V channel for neuropathic pain (Layer and McIntosh, 2006). The synthetic analog of the omega-conotoxin MVIIA, viz., zicontide (ω-conotoxin MVIIA) (Prialt®, Elan Pharmaceuticals, Inc.), which is isolated from the venom peptide of the gastropod mollusk, *Conus magus*, was approved by the FDA (USA) and EMA in 2004 for the management of severe chronic pain (Bordon et al., 2020; Schroeder and Lewis, 2006). Further, several peptide toxins derived from other species of conus, viz., *Conus victoriae, Conus geographus, Conus catus,* and *Conus marmoreus*; and anthozoan (Phylum, Cnidaria) *Stichodactyla helianthus*; and organic molecule TTX, tectin, from the species of pufferfish were subjected to clinical trials. However, unfortunately, many of them were found discontinued due to factors such as lack of adequate funding support, disputes relating to the intellectual property, and frequent changes in the leadership of the drug development program. The details of such drugs are given in the following table.

1.7 CHALLENGES AND PERSPECTIVES IN THE DEVELOPMENT OF NEW DRUGS FROM MARINE BIOTOXINS (BORDON ET AL., 2020)

i) Although, few drugs derived the marine biotoxins have entered the clinical trials, there is still a large gap between the number of drug

TABLE 1.2 Marin Biotoxin-Based Drugs Approved/in Clinical Trials.

Source species	Taxonomy	Toxin	Chemistry	Type of production	Therapeutic use	Present status	References
Conus magus	Molluska: Gastropoda	Zicontide (Prialt) (ω-conotoxin MVIIA)	Peptide	Synthetic	Severe chronic pain	Approved, FDA (US) and EMA, 2004	1, 2
Conus victoriae	Molluska: Gastropoda	ACV-1	Peptide	Synthetic	Pain	Phase II-discontinued	2
Conus victoriae	Molluska: Gastropoda	ACV-1	Peptide	Synthetic	Diabetic pain	Phase IIa-discontinued	3
Conus geographus	Molluska: Gastropoda	Conantokin-G (CGK-1007)	Peptide	Synthetic	Uncontrolled epilepsy	Phase I-discontinued	2
Conus geographus	Molluska: Gastropoda	CGX-1160	Peptide	Synthetic	Pain	Phase II-discontinued	2
Stichodactyla helianthus	Cnidaria: Anthozoa	Dalazatide, Shk-186	Peptide	Synthetic	Psoriasis, autoimmune disease	Phase II	4
Conus catus	Molluska: Gastropoda	Lecontide (AM336)	Peptide	Synthetic	Pain	Phase I/IIa-discontinued	2
Stichodactyla helianthus	Cnidaria: Anthozoa	Shk-192	Peptide	Synthetic	Autoimmune diseases	Phase I	2
Pufferfish species	Pisces	Tetrodotoxin (TTX) (Tectin)	Organic molecule	Synthetic	Pain caused by chemotherapy	Phase III	5
Conus marmoreus	Molluska: Gastropoda	Xen 2174 (x-CTX)	Peptide	Synthetic	Post-operative pain	Phase II-discontinued	2

Source: 1, Schroeder and Lewis (2006); 2, Bordon et al. (2020); 3, Coulter-Parkhill et al. (2021); 4, Malaise and de Seny (2019); 5, Gonçalves and Costa (2021).

candidates with significant pharmacological effects and those that are approved as commercial drugs for clinical uses. This is largely due to the extensive in vitro and in vivo tests which a drug candidate pass through to prove its safety before entering the clinical phases. Further, significant amount of money and time get involved in the different stages of the drug development, viz., discovery of a potential drug compound, and preclinical and clinical trials which involve significant amount of money and time.

ii) Many drug development programs have been discontinued due to the lack of important data-based publications.
iii) The hardships and challenges met in the different stages are not made available as much of them are governed by the intellectual property law for molecules that are under development as drugs.
iv) Unlike terrestrial animals such as snakes, the quantity of toxin/venom produced by marine animals is very small and this is considered to be the main limitation faced by the basic researchers in the discovery of biotoxin-based drugs. Aquaculture production of biotoxin-yielding marine animals may be the immediate solution to overcome this limitation.
v) Several problems are being faced in clinical trials relating to the development and approval of a new drug. One such issue is the paucity of patients for participation in clinical trials. For example, the tetrodotoxin (TTX) of pufferfish has already been evaluated in phase-III trial which is believed to be underpowered due to the participation of small number of active patients (González-Cano et al., 2021).

KEYWORDS

- **marine phytoplankton**
- **marine harmful algal blooms**
- **red tides**
- **marine biotoxin groups**
- **marine biotoxin poisonings**
- **seafood poisoning**
- **marine biotoxin-derived drugs**

CHAPTER 2

Marine Biotoxins: Types and Chemistry

ABSTRACT

Aspects such as marine biotoxin types and their chemical characteristics; chemistry of domoic acid, okadaic acid and saxitoxin and their analogs; and chemistry of STX and their analogs produced by the phytoplankton and shellfish are dealt with in this chapter.

Marine biotoxins are poisons derived from certain kinds of microscopic, toxic algae (marine phytoplankton) in very small amounts. Depending on their chemical structures, these toxins are divided into six groups, *viz.*, domoic acid (DA), okadaic acid (OA), saxitoxin (STX), brevetoxin (BTX), azaspiracid (AZA), and tetrodotoxin (TTX). All these six groups of bivalve-associated toxins are known to cause four syndromes, *viz.*, (i) diarrhetic shellfish poisoning, that is, DSP (OA and AZA groups); (ii) amnesic shellfish poisoning, that is, ASP (DA group), (iii) paralytic shellfish poisoning, that is, PSP (STX and TTX groups); and (iv) neurotoxic shellfish poisoning, that is, NSP (BTX group). Other biotoxins such as ciguatoxin, palytoxin, cyclic imines, yessotoxin, pectenotoxin, and so on are found associated with other marine biota (FAO and WHO, 2016). The chemical characteristics of major biotoxins are given in the following table.

2.1 CHEMISTRY OF MAJOR MARINE BIOTOXINS

Domoic acid (DA) and analogs: Domoic acid (DA) is a neuroexcitatory amino acid and it first occurred in the Prince Edward Island, Canada, in 1987. It was originally derived from the red seaweed, *Chondria armata* and

Therapeutic Potential of Marine Biotoxins: From Poison to Pharmacy. Santhanam Ramesh, Ramasamy Santhanam, & Subbiah Balasundari (Authors)

DOI: 10.1201/9781003610991-2

TABLE 2.1 Chemical Characteristics of Major Biotoxins.

Toxin	Chemical class	Molecular formula	Molar wt (g/mol)	Syndrome
Domoic acid	Cyclic amino acid, 3 carboxyl groups	$C_{15}H_{21}O_6$	311	ASP
Okadaic acid	Polyether, spiro-keto assembly	$C_{44}H_{68}O_{13}$	804	DSP
Saxitoxin	Tetrahydro-purine alkaloid	$C_{10}H_{17}N_7O_4$	299	PSP
Azaspiracid	Polyether, second amine, 3 spiro-ring assembly	$C_{47}H_{71}NO_{12}$	841	DSP (1)
Brevetoxin	Polyether with fused rings	$C_{49}H_{70}O_{13}$	867	NSP
Palytoxin	Non-proteinaceous molecule	$C_{129}H_{223}N_3O_{54}$	2680	Neurotoxic
Yessotoxin	Sulphated polyether	$C_{55}H_{82}O_{21}S_2$	1143	DSP
Tetrodotoxin	Guanidinum—derivative of penta-hydroxylated 2,4-dioxaadmantane	$C_{11}H_{17}N_3O_8$	319	PSP (2)
Pectenotoxin	Polycyclic ether and a spiroketal	$C_{47}H_{70}O_{15}$	875	DSP
Pinnatoxin A	Macrocyclic imine	$C_{41}H_{61}NO_9$	712	PSP
Spirolide A	Cyclic imine	$C_{42}H_{61}NO_7$	692	PSP
Pteriatoxin A	Cyclic imine	$C_{45}H_{70}N_2O_{10}S$	831	–
Gambierol	Polyether with eight ether rings	$C_{43}H_{64}O_{11}$	757	–
Gambieric Acid A	Polycyclic ether	$C_{59}H_{92}O_{16}$	1057	–
Gymnodimne	Cyclic imine	$C_{32}H_{45}NO_{4.}$	508	NSP
Ciguatoxin-1	Polyether with 13–14 fused rings	$C_{60}H_{86}O_{19}$	1111	CFP

Note: (1) Chemically azaspiracid differs from okadaic acid but produces similar symptom; (2) TTX is not classified as PSP but based on observed symptoms, this toxin is included in this group.

Source: FAO and WHO (2016); http://www.t3db.ca/toxins/T3D2463; Lawrence et al. (2011).

subsequently in the pennate diatom species, viz., *Pseudo-nitzschia multiseries* and *Pseudo-nitzschia australis* as well as in the shellfish. The DA of *Pseudo-nitzschia australis* was known for its high toxicity and it was found to be responsible for large scale death of sea birds (e.g., cormorants and pelicans) and sea lions, which consumed DA-contaminated anchovies (https://www.sciencedirect.com/topics/immunology-and-microbiology/pseudo-nitzschia). The isomers of DA are epi-domoic acid and isodomoic acids,. A total of 11 DA analogs have been identified including isodomoic acids A–F and 5-epi-domoic acid; and domoilactone A and B (Maeno et al., 2018).

Okadaic acid (OA) and analogs: Okadaic acid (OA) is a polyether toxin and its structural analogs are DTX1, DTX2, and DTX3. These toxins are heat-stable and lipophilic. While the DTX@ and OA differ only by the position of one methyl group in the molecule, DTX1 has an additional methyl group and DTX3 represents the derivatives of OA, DTX1, and DTX esterified with unsaturated and saturated fatty acids (EFSA, 2009).

Saxitoxin (STX) and analogs: A total of 57 STX analogs have been derived and at least 18 of them (e.g., NeoSTX, GTX1-6, C1-4, dcSTX, dcNeoSTX, dcGTX2, and dcGTX3) possesses toxicological relevance. This alkaloid group of toxins is separated into carbamoyl analogs, *N*-sulphocarbamoyl analogs, and decarbamoyl analogs (FAO and WHO, 2016). The STX and its analogs reported from marine phytoplankton and shellfish are given below.

TABLE 2.2 STX and Their Major Analogs Reported from Phytoplankton and Shellfish Organisms.

Toxin	Molecular formula	Molar wt (g/mol)
Saxitoxin (STX)	$C_{10}H_{17}N_7O_4$	299
Neosaxitoxin (NEO)	$C_{10}H_{17}N_7O_5$	315
Gonyautoxin-1,4 (GTX1)	$C_9H_{16}N_6O_8S$	411
Gonyautoxin-2,3 (GTX2)	$C_9H_{16}N_6O_7S$	395
Gonyautoxin -5(GTX5)	$C_{10}H_{17}N_7O_7S$	379
Gonyautoxin-6(GTX6)	$C_{10}H_{17}N_7O_8S$	395
Decarbamoyl GTX-1,4 (dcGTX1,4)	$C_9H_{16}N_6O_8S$	368
Decarbamoyl GTX-2,3 (dcGTX 2,3)	$C_9H_{16}N_6O_7S$	352
12-hydroxy-deoxy-GTX-4 (deoxy-GTX4 12ol)	$C_{10}H_{17}N_7O_8S$	395
Deoxycarbomoyl—Gonyautoxin-2 (doGTX2,3)	$C_9H_{16}N_6O_6S$	336
Decarbamoyl-saxitoxin (dcSTX)	$C_9H_{16}N_6O_3$	256
Decarbomyl-neosaxitoxin (dcNEO)	$C_9H_{16}N_6O_4$	272
Deoxydecarbomoyl-saxitoxin (doSTX)	$C_9H_{16}N_6O_2$	240

Source: FAO and WHO (2016).

Brevetoxin (PbTx) group: Brevetoxins are cyclic polyether toxins which are grouped, based on their backbone structure, into A type and B type. While the toxins of A type have ten transfused ether rings, toxins of B type have eleven. The different toxins of brevetoxin group are given below.

Brevetoxin A

Brevetoxin-1 (PbTx-1) R = –CH2C(=CH2)CHO
Brevetoxin-7 (PbTx-7) R = –CH2C(=CH2)CH2OH
Brevetoxin-10 (PbTx-10) R = –CH2CH(-CH3)CH2OH

Brevetoxin B

Brevetoxin-2 (PbTx-2) R = -CH2C(=CH2)CHO
Brevetoxin-3 (PbTx-3) R = –CH2C(=CH2)CH2OH
Brevetoxin-8 (PbTx-8) R = –CH2COCH2Cl
Brevetoxin-9 (PbTx-9) R = –CH2CH(CH3)CH2OH

Other Brevetoxins

Brevetoxin-5 (PbTx-5)
Brevetoxin-6 (PbTx-6).

Source: https://en.wikipedia.org/wiki/Brevetoxin

Azaspiracid (AZA) and analogs: Azaspiracids are polyether toxins with a unique spiral ring assembly, a cyclic amine, or aza group. About 40 AZA analogs have so far been derived from phytoplankton and shellfish. These azaspiracids have been designated by numbering in the chronological order of their postulation or detection. Among the AZAs, AZA1–10, 26, 33–34, and 36–37 have been fully characterized (FAO and WHO, 2016).

Tetrodotoxin and analogs: About 30 structural analogs of TTX have been described based on the position of the hydroxyl group in the TTX molecules. Madejska et al. (2019) reported on the following families of the TTX analogs found in the puffer fish.

1) Analogs chemically equivalent to TTX (4-epiTTX and 4,9-anhydro TTX);
2) Deoxy analogs (5-deoxyTTX, 11-deoxyTTX, 5,11-dideoxyTTX, 6,11-dideoxyTTX and 5,6,11-trideoxyTTX);
3) 11-CH_2OH oxidized analog (11-oxoTTX);
4) C11-lacking analogs (11-norTTX-6(S)-ol and 11-norTTX-6(R)-ol).

Ciguatoxins (CTX): CTXs are composed of contiguous cyclic ether rings and most of the CTX congeners possess a primary hydroxyl group. The structure of CTXs has been reported to vary according to geographic distribution and accordingly; these toxins are classified as Pacific Ocean (P-CTX), Caribbean Sea (CCTX), and Indian Ocean (I-CTX) ciguatoxins. A total of 40 such CTXs have been identified, viz., 22 P-CTX, 12 CCTX, and 6 I-CTX.

Palytoxin (PTX or PLTX): It is a fatty polyalcohol with large, very complex molecules. It is also a most toxic nonprotein compound containing the longest carbon chain. Some of its analogs are ostreocin D, mascarenotoxins A and B, and ovatoxin with molecular weights of 2635 Da, 2500–2535 Da and 2648 Da, respectively (Ramos and Vasconcelos, 2010).

Yessotoxin (YTX): It is a marine polyether toxin with several YTX congeners, the molecular weight of which is ranging between 955 and 1551 mu. Presently, there are 36 derivatives of YTX. Its important analogs include nor-YTX, homo-YTX, noroxo-YTX, 40-epi-41-keto-YTX, and 41-keto-YTXenone (Paz et al., 2008).

Pectenotoxins (PTXs): Pectenotoxins (PTXs): Pectenotoxins and their congeners are macrocyclic polyether-lactone compounds. A total of 21 PTX analogs, viz., PTX 1,2; PTX2b; PTX2c; PTX3-PTX 11; PTX11b; PTX11c; PTX13; PTX25A; 7-epi-PTX25A; 37-O-acyl PTX25A; 33-O-acyl PTX25A; and 11-O-acyl PTX25A have so far been discovered (Anon., 2020).

Pinnatoxins (PnTXs): These are part of the cyclic imine group of toxins. So far, eight different types of pinnatoxins (PnTX A-H) have been described (Wikipedia). Among these pinnatoxins, PnTX A, has been reported to possess high stability which has implications in its oral toxicity and detoxification (https://hero.epa.gov/hero/index.cfm/reference/details/reference_id/1850440).

Spirolides: Spirolids are macrocyclic compounds and are classified as “fast-acting toxins” as they caused rapid death when administered intraperitoneally in experimental mice. The structure of the spirolides is divided into two parts, viz., an upper part with a 6,7 bicyclic spiroimine unit and a lower unit with a spiro191 acetal or bis spiroacetal moiety. The spirolides and their derivatives are Spirolide A-D; 13-desMeC; 13,19-didesMeC; 13-desMeD; 27-OH-13,19 didesMeC E,F,G; and 20-MeG (Otero et al., 2011).

Pteriatoxins (PtTXs): Pteriatoxins are polyether macrocycles and its analogs, viz., PtTX A-C possess 6,7-spiro; 5,6-bicyclo; and 6,5,6-trispiro ketal rings which as similar to pinnatoxins (Davidson et al., 2015).

Gambierol: It is a marine polycyclic ether toxin with tetracyclic and heptacyclic analogs (Perez et al., 2012).

Gambieric acids (GAs): Gambieric acids A–D are marine polycyclic ethers (Majumder et al., 2006).

Marine biotoxin testing: Initially, the testing of marine biotoxins was a crude process, viz., mouse bioassay (MBA), in which the extract samples were injected into the experimental mice via the peritoneum. This testing method, however, raised not only several ethical issues but was also found to have without correlation between the acute toxicity of the extract samples and MBA results. This lead to the researchers to use purified samples of shellfish toxins for use as analytical standards. For example, the newly developed technique, viz., liquid chromatography-mass spectrometry (LC-MS) helped in the accurate determination of individual concentrations of toxin congeners in shellfish extracts (https://www.newfoodmagazine.com/article/112319/marine-biotoxins/). However, this LC-MS method of quantifying the toxicity levels has been reported to be expensive, complex, and time-consuming. On the other hand, the simpler, assay-based methods are commercially available nowadays and they possess the advantage of utilizing them in the field besides their affordability (McLeod et al., 2015).

KEYWORDS

- **marine biotoxin types**
- **chemistry of marine biotoxins**
- **domoic acid**
- **okadaic acid**
- **saxitoxin**
- **brevetoxin**
- **tetrodotoxin**

CHAPTER 3

Biotoxin-Producing Marine Taxonomic Groups

ABSTRACT

This chapter narrates the characteristics of marine life-produced toxins such as diarrhetic toxins, neurotoxins, paralytic shellfish toxins, dermato-toxins, cytotoxins and hemolytic toxins.

Most of the marine biotoxins are produced by the species of microalgae, phytoplankton, which constitute diatoms, dinoflagellates, cyanobacteria (blue-green algae), and so on. These biotoxins get accumulated chiefly in molluskan bivalves (shellfish) as a result of digestion of the above algae on which these bivalves normally feed. Among the bivalves, the mussels and scallops have been reported to be particularly prone to accumulate biotoxins. Further, the marine biotoxins so produced by phytoplankton may also get transferred to other organisms of higher trophic levels in the marine food chain *viz.* sponges, cnidarians, nemertine worms, annelid worms, crustaceans, echinoderms, tunicates and even fishes of the different marine taxonomic groups. The various types of biotoxins *viz.* diarrhetic toxins, neurotoxins, paralytic shellfish toxins, dermatotoxins, cardiotoxins, cytotoxins, hemolytic toxins and other miscellaneous toxins found in the organisms of the different marine taxonomic groups are given in the following tables.

Therapeutic Potential of Marine Biotoxins: From Poison to Pharmacy. Santhanam Ramesh, Ramasamy Santhanam, & Subbiah Balasundari (Authors)

DOI: 10.1201/9781003610991-3

TABLE 3.1 Diarrhetic Toxins of Different Marine Taxonomic Groups.

Taxonomic group	Toxins	References
Dinoflagellates	Okadaic acid, dinophysistoxins, pectenotoxins, azaspiracid	Visciano et al. (2016)
Sponges	Dinophysistoxin-1, okadaic acid	Blanco (2018); (1)
Bivalve mollusks	Domoic acid, okadaic acid, dinophysistoxin, pectenotoxin	Visciano et al. (2016)
Echinoderms—Starfish	Saponin	Rio et al. (1965); (2)
Echinoderms—Sea urchin	Contractin A, echinochrome A, echinometrin, major yolk protein (MYP), centrocins (I, II(, cathepsin B/X, strongylostatins (I,II), vitellogenin, UT841 toxin, spinochrome, and pedoxin	Mohebbi et al. (2016)
Echinoderms—Holothuroids	Holotoxins-Saponins (holothurin B, holothurin B4, HS-1, holothurin A, holothurin A1 holothurins A3 and A4.)	(3)

(1): https://www.sciencedirect.com/topics/pharmacology-toxicology-and-pharmaceutical-science/dinophysistoxin-1;
(2): https://www.sciencedirect.com/topics/pharmacology-toxicology-and-pharmaceutical-science/saponin;
(3): https://www.sciencedirect.com/topics/chemistry/holothurin

TABLE 3.2 Neurotoxins of Different Marine Taxonomic Groups.

Taxonomic group	Toxins	References
Diatoms	Domoic acids	Visciano et al. (2016)
Dinoflagellates	Brevetoxin, saxitoxins tetrodotoxin, yessotoxin, maitotoxin, palytoxin, ostreocin, gambierol gambieric acid, spirolide, gymonodimine	Visciano et al. (2016)
Blue-green algae (Cyanobacteria)	Kalkitoxin	Visciano et al. (2016); Thornburg et al. (2013)
Seaweeds	Caulerpin, caulerpicin, caulerpenyne, cymopol, dopamine, domoic acid, kainic acid	Visciano et al. (2016); Cheney (2016); Cherry et al. (2019)
Cnidarians—Hydrozoa	PpV19.3, P3, P1, metalloproteinases	D'Ambra and Lauritano (2020)
Cnidarians—Scyphozoa	Proteolytic enzymes, aurelin, cationic protein, CcNT, phospholipase A2	D'Ambra and Lauritano (2020)
Cnidarians—Cubozoa	CmNt, CbTX-I,II, CqTX-A	D'Ambra and Lauritano (2020)
Cnidarians—Anthozoa	Ae I; Bg II and III, Hcr1b-2, Hcr1b-3, Hcr1b-4, Sh1, ShK-186 or dalazatide	Lin et al. (1996); D'Ambra and Lauritano (2020)
Flatworms	TTX	Ueda et al. (2018)
Ribbon worms	TTX	Kwon et al. (2017)
Chaetognaths	TTX	Thuesen and Kogure (1989)
Annelid worms	Phyllotoxins, glycerotoxin (GLTx)	Meunier et al. (2002); Rodrigo et al. (2018); Richter et al. (2017); Cabral (2021)
Crustaceans	Domoic acid, TTX	(1); Tsai et al. (2006); Dao et al. (2009); Zhu e al. (2022)
Bivalve mollusks	TTX	
Gastropod mollusks	TTX, ω-conotoxin GVIA, ω-conotoxin MVIIA	Williams et al. (2012); Hort et al. (2020); Antonelli et al. (2022); Gao et al. (2017)

TABLE 3.2 *(Continued)*

Taxonomic group	Toxins	References
Cephalopod mollusks	TTX, α-cephalotoxin, β-cephalotoxin	Gonçalves and Costa (2021)
Echinoderms—Starfish	TTX	Antonelli et al. (2022)
Echinoderms—Sea urchins	TTX; ciguatoxins (CTXs) -P-CTX-3B, 51-OH-P-CTX-3C, P-CTX-3C, P-CTX-4A, and P-CTX-4B	Antonelli et al. (2022); Darius et al. (2018)
Tunicates	GTX 1-3; STX, neo-STX; TTX	Freitast et al. (1996); Antonelli et al. (2022)
Elasmobranchs	Ciguatoxins (CTXs)-I-CTX-1&2 and I-CTX-3&4; P-CTX-1; Gambieric acid C, D	Diogene et al. (2017); (2)
Teleost fishes—Pufferfish	TTX	Ziegman and Alewood (2015)
Other teleosts	Trachynilysin (TLY)	Ziegman and Alewood (2015)

(1) https://oehha.ca.gov/fish/general-info/domoic-acid-marine-biotoxin-fish-and-shellfish; (2) https://www.sciencedirect.com/topics/nursing-and-health-professions/ciguatoxin

TABLE 3.3 Paralytic Shellfish Toxins of Different Marine Taxonomic Groups.

Taxonomic group	Toxins	References
Dinoflagellates (*Alexandrium* spp. and *Gymnodinium* spp.)	Carbomate toxins—STX, NeoSTX, GTX1-4	Barbera-Sanchez et al. (2004)
-Do-	n-Desulfocarbamoyl (sulfmate) toxins—GTX-5, GTX-6, C1(epiGTX-8, C2(GTX-8), C3,C4	-Do-
-Do-	Decarbamoyl toxins—dcSTX, deneoSTX, dcGTX 1-4	-Do-

TABLE 3.4 Dermatotoxins of Different Tmarine Axonomic Groups.

Taxonomic group	Toxins	References
Blue-green algae (cyanobacteria)	Cyanotoxins (lyngbyatoxins, aplysiatoxin, lipopolysaccharides)	Zanchett and Oliveira-Filho (2013); Thornburg et al. (2013)
Seaweeds (Macroalgae)	Aplysiatoxins	Cheney (2016); Cherry et al. (2019)
Cnidarians—Scyphozoans	Tetramine, Cationic protein	D'Ambra and Lauritano (2020)

TABLE 3.5 Cardiotoxins of Different Marine Taxonomic Groups.

Taxonomic group	Toxins	References
Cnidarians—Hydrozoana	PpV19.3	D'Ambra and Lauritano (2020)
Cnidarians—Cubozoas	CfTX-1,2	D'Ambra and Lauritano (2020)
Cnidarians—Anthozoans	ATX-II, Anthopleurin A-C	D'Ambra and Lauritano (2020)
Elasmobranch fishes	TMAO	(1)
Teleost fishes	SNTX, VTX, neoVTX	Wijanarko et al. (2022); Ziegman and Alewood (2015)

(1) https://biocrates.com/metabolite-of-the-month-tmao/

TABLE 3.6 Cytotoxins of Different Marine Taxonomic Groups.

Taxonomic group	Toxins	References
Dinoflagellates	Karlotoxin, goniodomin A, gymnocin, amphidinolide, amphidinol, amphirionin-5, iriomoteolide	Visciano et al. (2016)
Blue-green algae (cyanobacteria)	Apratoxin	Visciano et al. (2016); Thornburg et al. (2013)
Seaweeds	Haloforms, haloacids, haloketones	Cheney (2016); Cherry et al., (2019)
Cnidarians—Hydrozoans	Oshem1, collagenase	D'Ambra and Lauritano (2020)
Cnidarians—Scyphozoans	Cationic protein, DNase, CcTX-1, phospholipase A2, ClGP-1, metalloproteinases, proteinaceous macromolecules	D'Ambra and Lauritano (2020)
Cnidarians—Cubozoans	CfTX-1,2	D'Ambra and Lauritano (2020)
Cnidarians—Anthozoans	Avt120, U1	D'Ambra and Lauritano (2020)
Nemertine worms	A-III	Luo et al. (2018)
Echinoderms—Brittle stars	Curacin A,E	Ueoka et al. (2016)
Tunicates	Pyridoacridine, bistramides, diazonamide A, didemnin B, eudistomin C, euseynstelamide B, lamellarins, lissoclinamides, mandelalides A & B, meridine, patellamides, patellozoles A–C, ritterazines, tamandarins, thiaplidiaquinones, tubulin, rigidin, vitilevuamide, diazonamide	Watters (2018)
Teleost fishes	Karotoxin, trachinine	Ziegman and Alewood (2015)

TABLE 3.7 Hemolytic Toxins of Different Marine Taxonomic Groups.

Taxonomic group	Toxins	References
Cnidarians—Hydrozoans	Milleporin-1, oshem1, collagenase, elastases, PpV9.4, physalitoxin, phospholipase A2	D'Ambra and Lauritano (2020)
Cnidarians—Scyphozoans	Proteolytic enzymes, phospholipase A2, metalloproteinases, proteinaceous macromolecules, rhizoprotease, rhizolysin	D'Ambra and Lauritano (2020)
Cnidarians—Cubozoans	CAH1, CaTX-A, CaTX-B, haemolysin, CrTX-I,II,III, phospholipase A2, CfTX-A,B, CqTX-A	D'Ambra and Lauritano (2020)
Cnidarians—Anthozoans	Equinatoxin-I-IV, tenebrosin A-C, Avt-I, bandaporin, actinoporin RTX-A, actinoporin RTX-S II, magnificalysins I and II; HMgIII, metridin, PsTX-60A; PsTX-60B, sticholysin-I and II, U1, Up-I	D'Ambra and Lauritano (2020)

TABLE 3.8 Miscellaneous Marine Biotoxins of Different Taxonomic Groups.

Taxonomic group	Toxins	References
Seaweeds	Fucoidans, diethyl peroxides, polycavernoside A	Visciano et al. (2016); Cheney (2016)
Nemertine worms	Delta-actitoxin-Amc1a; mu-theraphotoxin-Hhn2a 4	Göransson et al. (2019)
Annelid worms	Hedistin	Cabral (2021)
Teleost fishes	Cardioleputin, nocitoxin, Sp-CTx, plumieribetin, SP-CL 1-5,SA-HT, TmC4-47.2.Nattectin, toxin-PC, Wap65, dracotoxin	Ziegman and Alewood (2015)
-Do-	Epinecidin-1	Chee et al. (2019)
-Do-	Grammistins	Sugiyama et al. (2006)

KEYWORDS

- **diarrhetic toxins**
- **neurotoxins**
- **paralytic shellfish toxins**
- **dermato-toxins**
- **cytotoxins**
- **hemolytic toxins**

CHAPTER 4

Source Organisms of Marine Biotoxin-Associated Poisonings

ABSTRACT

This chapter deals with the biotoxin-associated poisonings and chemistry and source organisms of various marine life-based poisonings such as seafood poisoning, cyanotoxin poisoning, seaweed poisoning, and fish and turtle poisoning.

4.1 SEAFOOD POISONING

"We must know what we are eating." For example, seafood poisonings cause illnesses, which are invariably by the ingestion of a biotoxin present in the contaminated seafood, including edible finfish and shellfish. Toxicity associated with marine fish species may be inherent to that species, such as fugu (pufferfish) poisoning and other Tetraodontiforms. Further, sea fish toxicity can also result from external contamination such as ciguatera fish poisoning or shellfish poisoning. Furthermore, several gastrointestinal issues relating to seafood poisonings are always due to gastrointestinal infections caused by the ingestion of harmful microorganisms such as bacteria, viruses, or parasites; such organisms are not included here. Therefore, aspects relating to source species, their toxin and chemistry, and the type of seafood poisoning along with the symptoms are discussed in the following table.

Therapeutic Potential of Marine Biotoxins: From Poison to Pharmacy. Santhanam Ramesh, Ramasamy Santhanam, & Subbiah Balasundari (Authors)

DOI: 10.1201/9781003610991-4

TABLE 4.1 Biotoxins Associated with Seafood Poisonings: Their Chemistry and Source Organisms.

Source spp.	Taxonomy	Biotoxin	Chemistry	Poisoning type/symptoms	References
Nitzschia spp.*; Pseudo-nitzschia* spp.	Diatoms	Domoic acid	Heterocyclic amino acid	Amnesic shellfish poisoning (ASP)/neurotoxic	Visciano, et al. (2016)
Dinophysis spp; *Prorocentrum lima*	Dinoflagellates	Okadaic acid	Polyether	Diarrheic shellfish poisoning (DSP)/neurotoxic	-Do-
Dinophysis spp.	-Do-	Pectenotoxin	Polyether macrolide	Diarrheic shellfish poisoning	-Do-
Gonyaulax spinifera; Lingulodinium polyedrum; Protoceratium reticulatum	-Do-	Yessotoxin	Polyether	Diarrheic shellfish poisoning (DSP)	-Do-
Azadinium spinosum; Amphidoma languida	-Do-	Azaspiracid	-Do-	Diarrheic shellfish poisoning (DSP)	-Do-
Pyrodinium bahamense; Alexandrium spp.*; Gymnodinium catenatum*	-Do-	Saxitoxin	Alkaloids	Paralytic shellfish poisoning (PSP)/neurotoxic	-Do-
Karenia brevis	-Do-	Brevetoxin	Polyether	Neurotoxic shellfish poisoning (NSP)	-Do-
Gambierdiscus toxicus; Fukuyoa spp.	-Do-	Ciguatoxin; maitotoxin; scaritoxin**	-Do-	Ciguatera fish poisoning (CFP)/neurotoxic	-Do-
Protoperidinium crassipes	-Do-	Azaspiracid	-Do-	Azaspiracid shellfish poisoning (AZP)	-Do-

4.2 CYANOTOXIN POISONING

The marine cyanobacteria, photosynthetic bacteria (also called blue-green algae), produce water-soluble toxins called cyanotoxins. Exposure to cyanotoxins or blue-green algal blooms may cause cyanotoxin poisoning in humans. Cyanotoxins are a diverse group of compounds with various chemical classes such as alkaloids and peptides (cyclodepsipeptide, hepta-peptides, peptapeptides, lipopeptides, etc.) which can affect the nervous system, liver, and skin.

Chief ways of cyanotoxin exposure include ingestion of cyanobacteria-contaminated water during fishing or other recreational activities and direct skin contact with contaminated water or decayed algal scum along beach. Symptoms of acute cyanotoxin poisoning have been reported to develop commonly within 24 h of exposure. Skin rashes may take up to two days to appear. Symptoms associated with cyanotoxin poisoning generally fall into three major syndromes, viz., (i) gastrointestinal disorders including nausea, vomiting, stomach pain, and diarrhea; (ii) respiratory: problems including cough, sore throat, asthma-like symptoms, and so on; and (iii) dermal diseases such as rash and skin blisters (https://www.dhs.wisconsin.gov/publications/p02198.pdf). The common toxins associated with marine cyanobacterial blooms, their chemistry source species, and poisoning symptoms are detailed in the following table.

4.3 SEAWEED POISONING

The marine multicellular plants containing chlorophyll, called seaweeds or macroalgae, are typically divided into three groups, viz., green algae (Chlorophyta), brown algae (Ochrophyta, Phaeophyceae), and red algae (Rhodophyta).

4.3.1 COMMON TYPES OF EDIBLE SEAWEEDS

The common types of edible seaweeds are: Umibudo (*Caulerpa lentillifera* and *Caulerpa racemosa)* (green seaweeds); Kombu (*Saccharina japonica*), Wakame (*Undaria pinnatifida*, and Hijiki (*Sargassum fusiforme*) (brown seaweeds); Nori (*Porphyra yezoensis*), Ogonori, or sea moss (*Chondrus*

TABLE 4.2 Toxins Associated with Marine Cyanobacterial Blooms, Their Chemistry Source Species, and Poisoning Symptoms.

Toxin	Chemistry	Source species	Poisoning symptoms	References
Anatoxin-a	Alkaloid	*Anabaena; Planktothrix*	Loss of muscular fasciculations; coordination; convulsions; and death by respiratory paralysis	(1)
Aplysiatoxins	Alkaloids	*Lyngbya; Schizothrix; Planktothrix*	Dermatotoxic activity with inflammation of the skin; potent tumor promoters	Zanchett and Oliveira-Filho (2013)
Apratoxin A	Cyclodepsipeptide	*Lyngbya majuscula; Moorea producens*	–	Thornburg et al. (2013)
Apratoxin D	Cyclodepsipeptide	*Lyngbya majuscula; Lyngbya sordida*	Cytotoxicity against H-460 human lung cancer cells	Gutie´rrez et al. (2008)
Apratoxin E,F,G	-Do-	*Lyngbya bouillonii*	–	Mathew et al. (2008)
Apratoxin A–C	Cyclodepsipeptides	*Moorea producens*	–	Thornburg et al. (2013)
Apratoxin H	Apratoxin analog	-Do-	–	-Do-
Apratoxin A sulfoxide	-Do-	-Do-	–	-Do-
Lyngbyabellin B, hectochlorin	Lipopeptides	-Do-	–	-Do-
Hectochlorin	Chlorinated lipopeptide	*Lyngbya majuscula*	–	Ramasamy et al. (2007)
Lyngbyabellin B	Cyclic depsipeptide (lipopeptide)	-Do-	Antifungal activity against Candida albicans	Milligan et al. (2000)
Lyngbyatoxin a	Alkaloid	*Lyngbya* sp.	Potent tumor promoter	Zanchett and Oliveira-Filho (2013)

TABLE 4.2 *(Continued)*

Toxin	Chemistry	Source species	Poisoning symptoms	References
Microcystins	Cyclic heptapeptides	*Anabaena; Planktothrix; Nostoc, Anabaenopsis*	Gastrointestinal disorders; kidney and liver damage; neurotoxicity; skin toxicity	Zanchett and Oliveira-Filho (2013)
Nodularins	Cyclic pentapeptídes	*Nodularia*	-Do-	Zanchett and Oliveira-Filho (2013)
Oscillatoxin	Polyketide	*Phormidium nigroviride (=Osillatoria nigroviridis); Schizothrix*	–	Zanchett and Oliveira-Filho (2013)
Saxitoxins	Carbamate alkaloids	*Dolichospermum sigmoideum (= Anabaena circinalis); Microseira wollei (=Lyngbya wollei)*	Neurotoxic; block sodium channels in nerve axons	Zanchett and Oliveira-Filho (2013)
Lipopolysaccharides	–	Blue-green algae (cyanobacteria)	Gastrointestinal irritants; inflammatory agents	Zanchett and Oliveira-Filho (2013)

(1) https://en.wikipedia.org/wiki/Anatoxin-a.

crispus) (Irish sea moss); *Gracilaria* (Jamaican sea moss); and *Eucheuma cottonii* (gusô) and Dulse (*Palmaria palmata*) (red seaweeds).

4.3.2 HEALTH BENEFITS OF SEAWEEDS (https://www.forbes.com/health/body/health-benefits-of-seaweed/)

i) Seaweeds are low in calorie content and rich in many health-promoting bioactive compounds such as antioxidants, phenols, and flavonoids.
ii) Seaweeds contain several vitamins, such as A, B, C, D, and E, and minerals such as iodine, calcium, potassium, magnesium, phosphorus, iron, and zinc. Compared to the mineral content of spinach (20% per dry weight), seaweeds may have minerals up to 40% per dry weight.
iii) Edible seaweeds may also provide a good amount of fiber in a person's diet.
iv) The seaweeds are known to contain important amino acids and their protein content is believed to be similar to egg protein.

4.3.3 METHODS OF EATING SEAWEEDS

i) By adding seaweed flakes to dishes such as rice bowls, soups, or stews.
ii) The most common way of eating seafood is with sushi or salad.
iii) Seaweed may also be eaten as a roasted snack.

The toxins, and their chemistry and poisoning symptoms of green, brown, and red seaweed species are shown in the following tables.

4.4 FISH AND TURTLE POISONING

i) **Shark poisoning:** Shark liver has been implicated in massive human introxication which is believed to be due to two shark toxins, viz., carchatoxin-A and B (Yasumoto, 1998). Further, the gut sample of the bull shark, *Carcharhinus leucas,* has been reported to possess multiple phycotoxins, viz., microcystin,

TABLE 4.3 Green Seaweed Toxins: Their Chemistry and Poisoning Symptoms.

Source	Taxonomy	Toxin(s)	Chemistry	Poisoning symptoms	References
Caulerpa lentillifera	Green	Caulerpicin	Hydroxy amide	Numbness of the tongue and lips and an anesthetizing sensation	Cheney (2016)
Caulerpa prolifera	-Do-	Caulerpenyne	Sesquiterpene	Neurological disorders	Silva et al. (2021)
Caulerpa racemosa	-Do-	Caulerpicin	Hydroxy amide	Numbness of the tongue and lips and an anesthetizing sensation	Cheney (2016)
Caulerpa scalpelliformes	-Do-	Caulerpin and caulerpicin	Hydroxy amides	–	Vidal et al. (1984)
Caulerpa taxifolia	-Do-	Caulerpenyne	Sesquiterpene	Neurological disorders	Cheney (2016)
Caulerpa sp.	-Do-	Caulerpin	Bisindole alkaloid	–	Cheney (2016)
Codium *decoratum*	-Do-	Caulerpin	Bisindole alkaloid	–	Veale and Davies-Coleman (2014)
Cymopolia barbata	-Do-	Cymopol	Monoterpene-bromohydroquinone	–	Silva et al. (2021)
Halimeda incrassata	-Do-	Caulerpin	Bisindole alkaloid	–	Mollo et al. (2000)
Ulvaria obscura	-Do-	Dopamine (excess)	Hydroxytyramine	Parkinson's disease	(1)
Ulva prolifera	-Do-	Arsenic	Heavy metal	Arsenic poisoning; potential health risk	Cherry et al. (2019)
Ulva lactuca	-Do-	Arsenic	Heavy metal	Arsenic poisoning; potential health risk	Cherry et al. (2019)

TABLE 4.3 *(Continued)*

Source	Taxonomy	Toxin(s)	Chemistry	Poisoning symptoms	References
Ulva rigida	-Do-	Polychlorinated biphenyls	–	Gastrointestinal tract tumors; hepatocarcinomas; leukemia; lymphomas; pituitary tumors	Cheney (2016); (2)
Ascophyllum nodosum	-Do-	Arsenic	Heavy metal	Arsenic poisoning; potential health risk	Cherry et al. (2019, 2016)
Alaria esculenta	-Do-	Arsenic	Heavy metal	Arsenic poisoning; potential health risk	Cherry et al. (2019)

(1) https://coastalscience.noaa.gov/news/common-coastal-seaweeds-release-toxins-harmful-marine-organisms/

(2) https://www.atsdr.cdc.gov/csem/polychlorinated-biphenyls/adverse_health.html

TABLE 4.4 Brown Seaweed Toxins: Their Chemistry and Poisoning Symptoms.

Source	Taxonomy	Toxin(s)	Chemistry	Poisoning symptoms	References
Cladosiphon okamuranus	Brown	Fucoidans	Sulfated polysaccharides	Fever; prolonged clotting time	Cheney (2016); Gideon and Rengasamy (2008)
Fucus vesiculosus	-Do-	Arsenic	Heavy metal	Arsenic poisoning; potential health risk	Cherry et al. (2019)
Fucus spiralis	-Do-	Arsenic	Heavy metal	Arsenic poisoning; potential health risk	Cherry et al. (2019)
Nemacystus decipiens	-Do-	Toxic polysaccharides	–	Fever	Cheney (2016)
Saccharina latissima	-Do-	Arsenic	Heavy metal	Arsenic poisoning; potential health risk	Cherry et al. (2019)
Sargassum fusiforme (=Hizikia fusiforme)	-Do-	Arsenic	–	Arsenic poisoning	Cheney (2016)
Sphaerotrichia divaricata	-Do-	Diethyl peroxides	Organic peroxides	Fever	Cheney (2016)
Undaria pinnatifida	-Do-	Excess iodine	–	Excess iodine in breast milk	Cheney (2016)
Laminaria dgitatata	-Do-	Cadmium	Heavy metal	Arsenic poisoning; potential health risk	Cherry et al. (2019)
Laminaria japonica	-Do-	Cadmium	Heavy metal	Arsenic poisoning; potential health risk	Cherry et al. (2019)
Laminaria spp.	-Do-	Arsenic	–	Memory loss; fatigue; gastrointestinal disorders; rashes in legs	Cheney (2016)

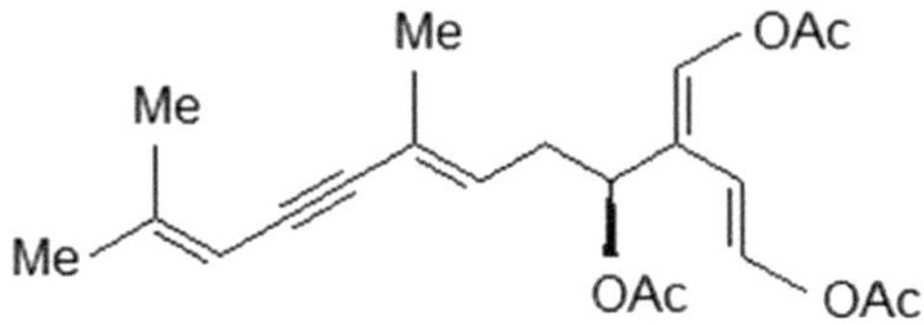

Caulerpenyn.

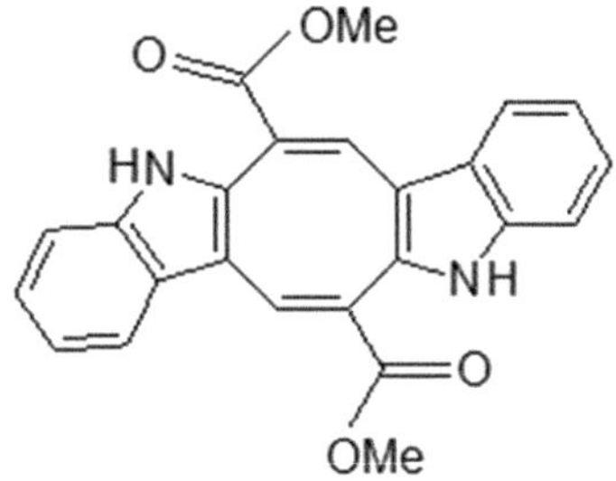

Caulerpin

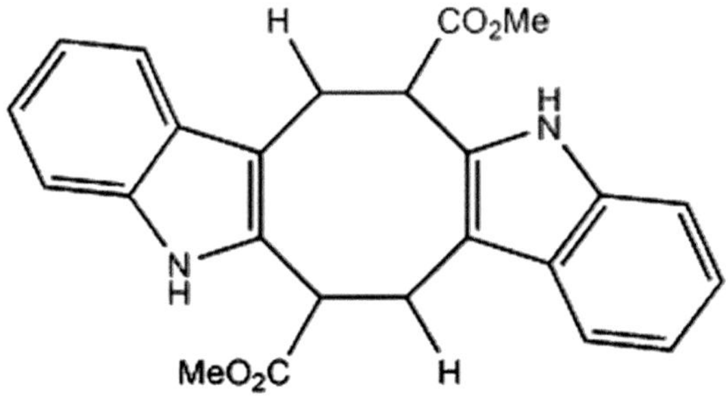

Caulerpicin

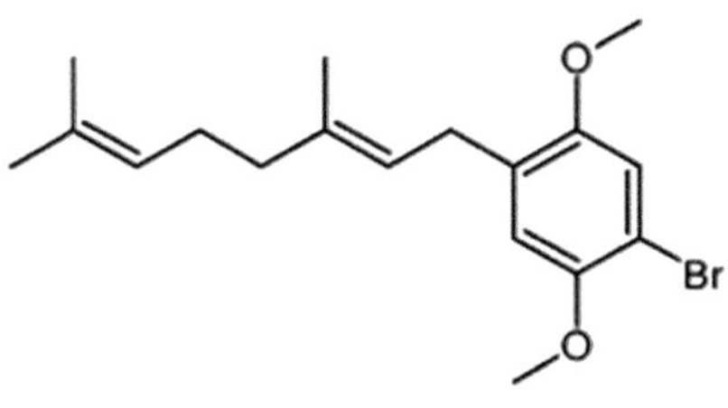

Cymopol.

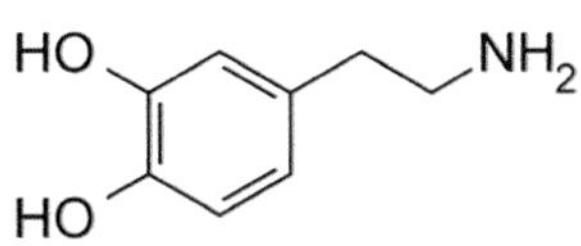

Dopamine

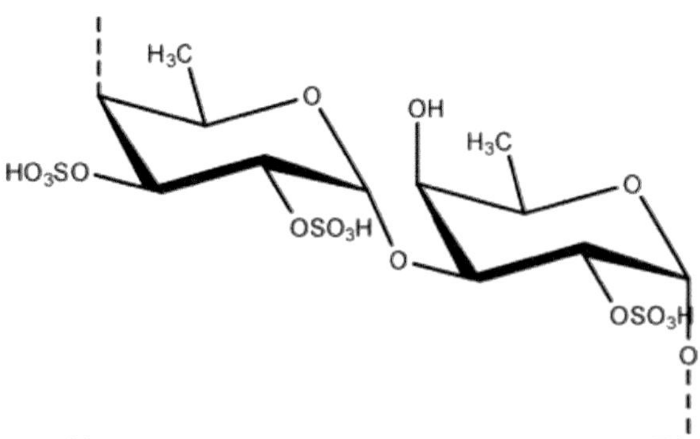

Fucoidan.

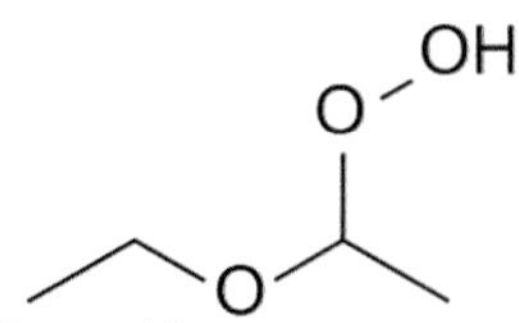

Diethyl peroxide

TABLE 4.5 Red Seaweed Toxins: Their Chemistry and Poisoning Symptoms.

Source	Taxonomy	Toxin(s)	Chemistry	Poisoning symptoms	References
Acanthophora specifera	Red	Polycavernoside A	Glycoside	Sick or death; gastrointestinal and neurological disorders	Cheney (2016)
Alsidium corallinum	-Do-	Domoic acid	Amino acid	Neurotoxic	Bates (2000)
Amansia glomerata	-Do-	-Do-	-Do-	-Do-	-Do-
Amansia glomerata	-Do-	Caulerpin; caulerpicin	Hydroxy amides	Numbness of the tongue and lips and an anesthetizing sensation	Vidal et al. (1984)
Asparagopsis armata	-Do-	Haloforms; haloacids; haloketones	Halogenated compounds	Cytotoxic	Silva et al. (2021)
Caloglossa leprieurii	-Do-	Caulerpin	Bisindole alkaloid	–	Clinton et al. (2014)
Chondria armata	-Do-	Domoic acid	Amino acid	Neurotoxic	Higa and Kuniyoshi (2000)
Chondria armata	-Do-	Palytoxin	Fatty alcohol	Nausea; vomiting; diarrhea; abdominal cramps; paresthesia; bradycardia; cyanosis; respiratory distress; renal failure	Ramos and Vasconcelos (2010)
Chondria armata	-Do-	Caulerpin	Bisindole alkaloid	–	Clinton et al. (2014)
Chondria armata	-Do-	Domoic acid	Amino acid	Neurotoxic	Bates (2000)
Chondria baileyana	-Do-	-Do-	-Do-	-Do-	-Do-
Chondrus crispus (product)	-Do-	Unidentified toxin	–	Gastrointestinal disorders (ulcerations and colon cancer)	Cheney (2016)
Chondrus crispus	-Do-	Arsenic	Inorganic	Arsenic poisoning; potential health risk	Cherry et al. (2019)
Chondrus crispus	-Do-	Palytoxin	Fatty alcohol	Gastrointestinal disorders; respiratory distress; renal failure	Ramos and Vasconcelos (2010)

TABLE 4.5 *(Continued)*

Source	Taxonomy	Toxin(s)	Chemistry	Poisoning symptoms	References
Digenea simplex	-Do-	Kainic acid	Kainoid (amino acid)	Neurotoxic	Higa and Kuniyoshi (2000)
Digenea simplex	-Do-	Domoic acid	Amino acid	-Do-	Bates (2000)
Gracilaria coronopifolia	-Do-	Aplysiatoxins	Macrolides	Dermatoxic	Higa and Kuniyoshi (2000)
Gracilaria coronopifolia	-Do-	Aplysiatoxin; debromoaplysiatoxin	Macrolides	Vomiting and diarrhea; burning sensation of the mouth and throat	Cheney (2016)
Gracilaria tsudae	-Do-	Polycavernosides	Glycosidic macrolides	Fatal human poisoning	Higa and Kuniyoshi (2000); Iwasaki et al. (2017)
Gracilaria vermiculophylla	-Do-	Arsenic	Inorganic	Potential health risk	Cheney (2016)
Gracilaria sp.	-Do-	–	–	Seriously ill or dying	Cheney (2016)
Gracilariopsis chorda *(=Gracilaria chorda)*	-Do-	–	–	Nausea; vomiting; stomach pain; diarrhea	Cheney (2016)
Gracilariopsis longissima *(=Gracilaria verrucosa)*	-Do-	Prostaglandin E2	Eicosanoid (PUFA)	Tension; bleeding; nausea; diarrhea	Higa and Kuniyoshi (2000)
Gracilariopsis longissima *(=Gracilaria verrucosa)*	-Do-	Prostaglandin E2	Eicosanoid (PUFA)	Tension; bleeding; nausea; diarrhea	Cheney (2016)
Hydropuntia edulis (=Gracilaria edulis; Gracilaria lichenoides)	-Do-	Prostaglandin E2	Eicosanoid (PUFA)	Tension; bleeding; nausea; diarrhea	Cheney (2016)

TABLE 4.5 *(Continued)*

Source	Taxonomy	Toxin(s)	Chemistry	Poisoning symptoms	References
Hydropuntia edulis	-Do-	Polycavernoside A and B	Glycosidic macrolides	Vomiting; gastrointestinal and neurological disorders; fatal poisoning	Cheney (2016)
Hypnea concornis	-Do-	Caulerpin	Bisindole alkaloid	–	Clinton et al. (2014)
Laurencia majuscula	-Do-	Caulerpin	Bisindole alkaloid	–	Clinton et al. (2014)
Palmaria palmata	-Do-	Kainic acid	Kainoid (amino acid)	Neurotoxic	FAO and WHO (2022)
Palmaria palmata					
Plocamium lepitophyllum	-Do-	Halogenated monoterpenes	–	Cytotoxic	Silva et al. (2021)
Vidalia obtusiloba	-Do-	Domoic acid	Amino acid	Neurotoxic	Bates (2000)
Eucheuma sp. (product)	-Do-	Unidentified toxin	–	Gastrointestinal disorders (ulcerations and colon cancer)	Cheney (2016)
Porphyra umbilicalus	-Do-	Arsenic	Inorganic	Arsenic poisoning; potential health risk	Cherry et al. (2019)
Porphyra sp.	-Do-	Cadmium	Heavy metal	Cadmium poisoning; potential health risk	Cheney (2016)
Kappaphycus sp. (product)	-Do-	Unidentified toxin	–	Gastrointestinal disorders (ulcerations and colon cancer)	Cheney (2016)

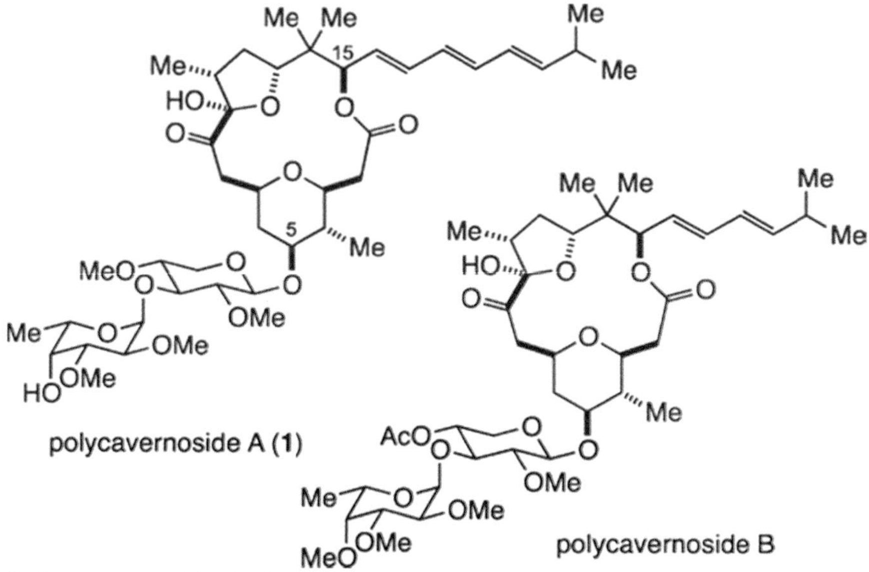

Polycavernoside A, B

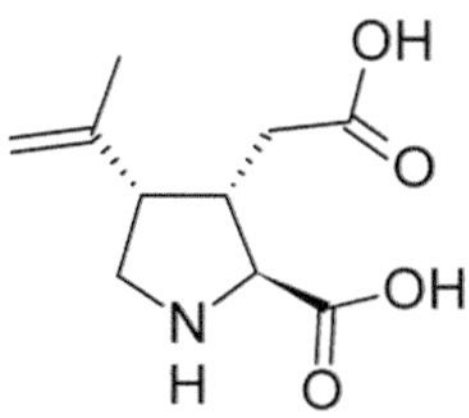

Kainic acid.

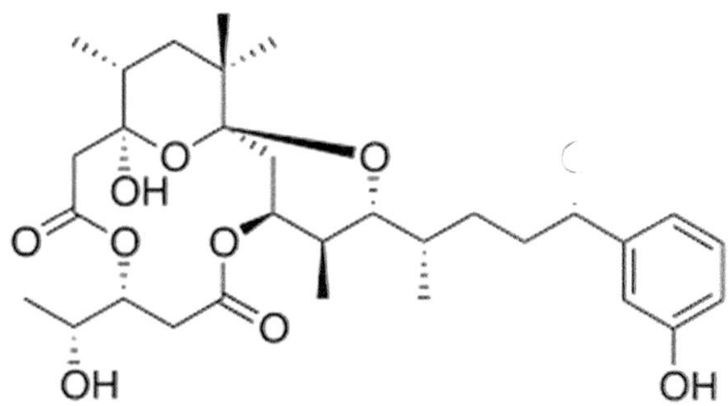

Debromoaplysiatoxin

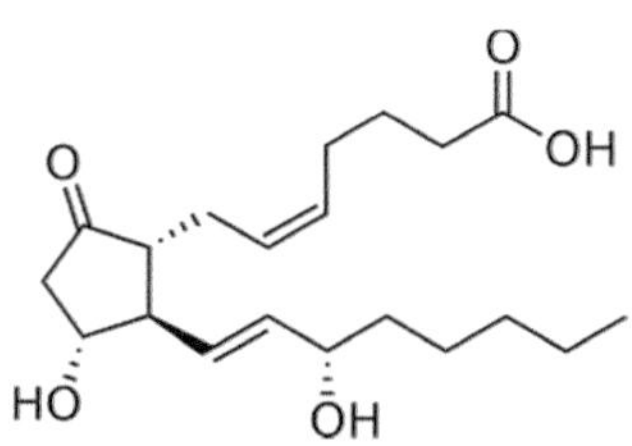

Prostaglandin E2

TABLE 4.6 Miscellaneous Poisonings.

Source spp.	Taxonomy	Biotoxin	Chemistry	Poisoning symptoms	References
Vulcanodinium rugosum	Dinoflagellate	Pinnatoxin	Cyclic imine; polyether	Respiratory distress and paralysis	Kim et al. (2022); Delcourt et al. (2019)
Karenia selliformis (=Gymnodinium selliformis)	-Do-	Gymnodimine	-Do-	Cytotoxicity; public health problem	Kim et al. (2022); Dragunow et al. (2005)
Alexandrium ostenfeldii	-Do-	-Do-	-Do-	-Do-	-Do-
Alexandrium peruvianum	-Do-	-Do-	-Do-	-Do-	-Do-
Alexandrium spp.	-Do-	Spirolide	-Do-	-Do-	-Do-
Prorocentrum spp.	-Do-	-Do-	-Do-	-Do-	-Do-
Vulcanodinium spp.	-Do-	-Do-	-Do-	-Do-	-Do-
Karenia spp.	-Do-	-Do-	-Do-	-Do-	-Do-
Pteria penguin	Molluska; bivalvia	Pteriatoxin	-do-	No report on humans	Kim et al. (2022); Molgó et al. (2015)
Pinna muricata	-Do-	-Do-	-Do-	-Do-	-Do-
Vulcanodinium rugosus	Dinoflagellate	Portimine	-Do-	-Do-	-Do-
Trichodesmium sp.; mullets (pisces: mugilidae)	Cyanobacterium (blue-green alga)	Palytoxin; its analog, 42-hydroxy-palytoxin	Fatty polyalcohol	Clupeotoxism; health hazard to swimmers and fishermen	Kerbrat et al. (2011)
Ostreopsis lenticularis; Ostreopsis siamensis; Ostreopsis ovata; Ostreopsis mascarenensis	Dinoflagellates	Palytoxin; palytoxin-b; 42-hydroxy-palytoxin; homo-palytoxin; bis-homo-palytoxin; neo-palytoxin; deoxy-palytoxin	-Do-	Adverse effects on human health including acute inflammatory reactions	Kim et al. (2022); Crinelli et al. (2012)

TABLE 4.6 *(Continued)*

Source spp.	Taxonomy	Biotoxin	Chemistry	Poisoning symptoms	References
Ostreopsis ovata	Dinoflagellates	Ostreocin-a, b, d, e (a palytoxin analogs)	-Do-	Respiratory illness	Kim et al. (2022); Ramos and Vasconcelos (2010)
Ostreopsis mascarenensis	Dinoflagellates	Mascarenotoxin- a,b,c (a palytoxin analogs)	-Do-		Kim et al. (2022)
Palythoa caribaeorum; Palythoa toxica	Cnidaria; Anthozoa	Palytoxins	-Do-	Nausea; vomiting; diarrhea; abdominal cramps; paresthesia; bradycardia; cyanosis; respiratory distress; renal failure	Ramos and Vasconcelos (2010)
Demania reynaudii	Crustacea; crab	Palytoxins	-Do-	Human fatalities	Ramos and Vasconcelos (2010)
Sardines (*Herklotsichthys quadrimaculatus)*; herrings and anchovies	Clupeid fishes	Palytoxins	-Do-	Clupeotoxism—human fatalities	Ramos and Vasconcelos (2010)
Decapterus macrosoma; *Epinephelus* sp.; *Scarus ovifrons* (smoked)	Fishes	Palytoxins	-Do-	Near fatal	Ramos and Vasconcelos (2010)

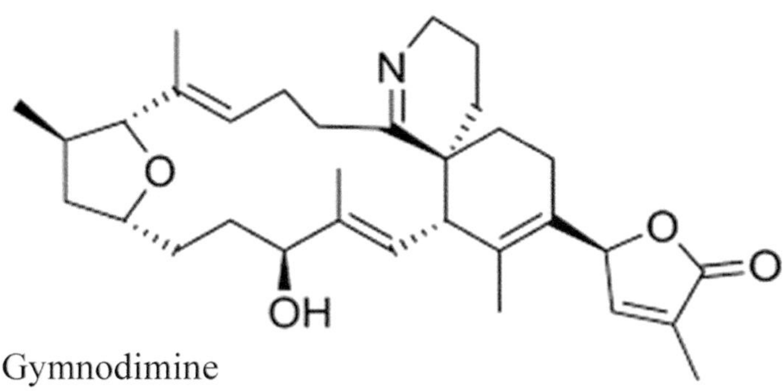

Gymnodimine

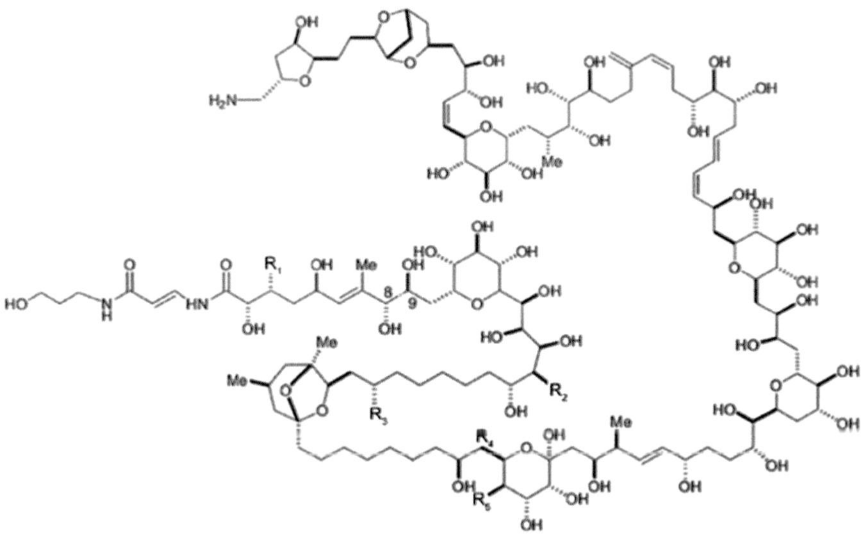

R1=H; R2=H; R3=H; R4=OH; R5=H

Ostreocin-d

nodularin, teleocidin, cylindrospermopsin, domoic acid, okadaic acid, and brevetoxin. Further, except the brevetoxin, all other toxins were present in the plasma sample of the bull shark. It is suggested that the presence of these toxins in these bull sharks may be due to their food fish, *viz.*, mullets which make simultaneous presence with the bull sharks.

ii) Sardine poisoning (clupeotoxism): A new toxin, *viz.*, clupeotoxin has been isolated from the head of the sardine, *Herklotsichtys quadrimaculatus*. This toxin, implicated in human fatality, was identified as palytoxin, based on its chromatographic results (Yasumoto, 1998).

iii) **Turtle poisoning:** An unidentified toxin has been isolated from about 90 g of the left-over turtle meal. From the weak basicity, low polarity, and inflammatory nature of this turtle toxin, a resemblance was pointed out between this turtle toxin and the cyanotoxin, viz., lyngyatoxin-A produced by the marine cyanobacterial species, *Moorea producens*. Further research may therefore be needed on the turtle toxin (Yasumoto, 1998).

KEYWORDS

- **biotoxin-associated poisonings**
- **seafood poisoning**
- **cyanobacterial toxins**
- **seaweed toxins**
- **fish and turtle poisoning**

CHAPTER 5

Marine Biotoxins and Human Health (Case Studies)

ABSTRACT

This chapter deals with the marine biotoxins and human health with case studies; transfer pathways of marine biotoxins through the natural marine food web to humans; and on the occurrence of major seafood poisonings and PSP, DSP, NSP, ASP, CFP, and TTX outbreaks in the world.

Since ancient times, it has been well known that certain species of finfish and shellfish are poisonous/venomous and they can cause death when eaten or contacted. The first Chinese pharmacopoeia, of 2800 BC, reported on the injunctions against ingesting pufferfish (WHO, 1984). Ingestion of contaminated seafood such as finfish or shellfish (crustaceans such as shrimps, crabs, and lobsters; and mollusks including clams, mussels, and cephalopods); skin contact with toxin-contaminated sea water; or the inhalation of aerosolized toxins are the causative factors of human toxicity. The lethal doses of certain marine biotoxins determined through intra-peritoneal (i.p) or oral administration to experimental mice, and the lethality of various PSP toxins, are given in the following tables.

5.1 TRANSFER PATHWAYS OF BIOTOXINS THROUGH THE NATURAL MARINE FOOD WEB TO HUMANS

Biotoxin-producing organisms such as dinoflagellate species are first bioaccumulated by shellfish, which are not, however, affected by toxins such as saxitoxin. Ingestion of the above-contaminated shellfish is the

Therapeutic Potential of Marine Biotoxins: From Poison to Pharmacy. Santhanam Ramesh, Ramasamy Santhanam, & Subbiah Balasundari (Authors)

DOI: 10.1201/9781003610991-5

TABLE 5.1 Lethal Doses (LD50) of Marine Biotoxins For Experimental Mice.

Toxin type	Toxin	Administration	LD50 (ug/kg)
Pinnatoxins	Pinnatoxin A	i.p.	135–180
-Do-	Pinnatoxin B & C	-Do	22
-Do-	Pinnatoxin D	-Do-	400
-Do-	Pinnatoxin E	-Do-	34–75
-Do-	Pinnatoxin F	-Do-	10–16
-Do-	Pinnatoxin G	-Do-	35–68
-Do-	Pinnatoxin H	-Do-	67
-Do-	Pinnatoxin E	Oral	2380–3000
-Do-	Pinnatoxin F	-Do-	19–72
-Do-	Pinnatoxin G	-Do-	105–470
Gymnodimines	Gymnodimine A	-Do-	600– >7500
-Do-	Gymnodimine A	i.p	79–118
-Do-	Gymnodimine B	-Do-	800
Pteriatoxins	Pteriatoxin A	-Do-	100
-Do-	Pteriatoxin B/C	-Do-	8
Spirolides	Spirolide A	-Do-	37
-Do-	Spirolide B	-Do-	99
-Do-	Spirolide C	-Do-	8
-Do-	Spirolide E	-Do	>1000
-Do-	Spirolide F	-Do-	>1000
-Do-	13-Desmethyl spirolide C	-Do	7–28
-Do-	13,19-Didesmethyl spirolide C	-Do-	32
-Do-	27-Hydroxy-13-desmethyl spirolide C	-Do-	>27
-Do-	27-oxo-13,19-didesmethyl spirolide C	-Do-	>35
-Do-	20-Methyl spirolide G	-Do-	63
-Do-	Spirolide mixture	Oral	1000
-Do	Spirolide C	-Do-	176–780
Palytoxins	Palytoxin	i.p	50*
-Do-	Ostreocin-D	-Do-	750*
-Do-	Mascarenotoxin-A	-Do-	900

**ng/kg.*

Source: Kim et al. (2022).

Lethal effects of the various PSP components

Based on single dose intraperitoneal administration to mice models, the lethal effects of the various PSP components such as saxitoxins and gonyautoxins have been determined. The saxitoxin, neosaxitoxin and decarbomoyl saxitoxin showed the values of lethality as 2045, 1617 and 1378 MU/μ mol respectively. The lethality values for Gonyautoxins I, II, III, IV, V, VI and VIII were found to be 1638, 793, 2234, 673, 136, 108 and 277 MU/μ mol respectively (WHO, 1984).

common way of diarrhetic (DSP) or paralytic poisoning (PSP). On the other hand, ciguatera fish poisoning (CFP) occurs indirectly. For example, the toxin-producing dinoflagellates such as *Gambierdiscus* or *Prorocentrum* spp. may symbiotically attach to organisms of higher trophic levels such as seaweeds or corals which are first eaten by fish. Humans are subsequently affected when they eat such a contaminated fish. In certain countries of the temperate latitudes, viz., North America, South America, Australia, South Africa, and Asia, diarrhetic (DSP), neurotoxic (NSP), paralytic (PSP), amnesic (ASP), and azaspiracid (AZP) shellfish poisonings have been reported to be most common (Berdalet et al., 2015). The biotoxin transfer pathways through marine food web are shown in the following figure.

5.2 WORLDWIDE OCCURRENCE OF MAJOR SEAFOOD POISONINGS

Shellfish poisoning and ciguatera fish poisoning are believed to be unknown hazards for travelers, especially, in the countries of tropics and subtropics. Further, climate change, the spread of MHABs, growing seafood consumption, and expansion of international tourism and trade have been reported to increase the risk of global biotoxin-related human poisonings.

Amnesic shellfish poisoning (ASP): Outbreaks of ASP have been reported in Europe (Spain, Belgium, France, Ireland, Scotland, and Portugal), the Americas (Chile and Canada), and the Pacific (New Zealand and Australia). The causative organisms for these outbreaks include mussels, scallops, and razor clams; and edible marine crustaceans such as shrimps and crabs.

Diarrheic shellfish poisoning (DSP): Diarrheic shellfish poisoning outbreaks have been reported in Europe (Spain, Belgium, France, Scandinavia, and Ireland).

Neurotoxic shellfish poisoning (NSP): It was a common phenomenon in the Western Hemisphere (Gulf of Mexico, Caribbean, and SE United States). NSP outbreaks have also been reported in New Zealand.

Paralytic shellfish poisoning (PSP): It is known to occur worldwide but this phenomenon is the most common one in the temperate waters off the Pacific and Atlantic coasts of North America. Further, countries such as New Zealand, Australia (Pacific); China and Philippines (Asia); and Chile (America) also reported PSP cases.

Ciguatera fish poisoning (CFP): About 50,000 cases of CFP have been reported worldwide annually. Although CFP has been a widespread phenomenon in tropical and subtropical waters (between the latitudes of 35°N and 35°S), it has been reported to be a common biotoxin-related poisoning in the Indian and Pacific Oceans and the Caribbean Sea. Madeira and the Canary Islands (Mediterranean) and the western Gulf of Mexico are found to be newly recognized areas of CFP. While the CFP incidence in travelers to highly endemic areas has been found to be 3%; the overall human mortality due to CFP was found to be <0.1% which may; however, vary according to the dose of toxin and availability of medical care to the affected patients (Ansdell, https://wwwnc.cdc.gov/travel/yellow-book/2024/environmental-hazards-risks/food-poisoning-from-marine-toxins). The worldwide occurrence of major seafood poisonings is given in the following image.

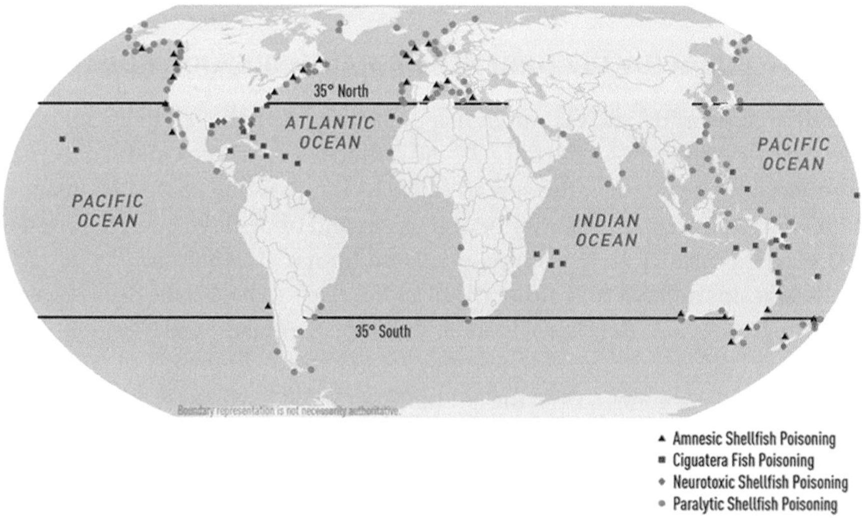

Worldwide occurrence of major seafood poisonings.

Image credit: Public domain

5.3 CASE STUDIES RELATED TO MARINE BIOTOXIN-RELATED HUMAN POISONINGS

5.3.1 *CYANOTOXIN POISONING*

Cyanotoxin poisoning with associated canine deaths is becoming common in freshwater ecosystems all over the world and this is invariably due to the occurrence of blue-green algal blooms containing the biotoxin, microcystin. On the other hand, in marine ecosystems, deaths of sea otters have been recorded in coastal waters where cyanobacterial blooms occur. Further, in the marine and estuarine environments, very rarely humans are found affected due to marine cyanobacterial blooms. Preece et al. (2017) reported on the impacts of microcystins in marine and estuarine environments on human health. According to them, in two cases in China, microcystins were detected in the human blood serum and it was due to the consumption of contaminated shrimp, snails, and fish. Results from one affected case suggested that chronic microcystin exposure may be associated with childhood liver damage.

5.3.2 *SEAWEED POISONING*

i) Seven patients were admitted to the emergency room of the Guam Memorial Hospital after ingesting the red seaweed, *Gracilaria tsudai*. They had complaints of shortness of breath, numbness, and tingling sensation in the face, and general weakness. Further, they also had abdominal disorders such as nausea, vomiting, cramps, and diarrhea. Some had excessive tears and perspiration with skin cold to the touch. Five of them had developed low-grade lowered blood pH, elevated blood CO_2 and muscle fasciculations. Three of these five patients subsequently had tetanic seizures, enhanced respiratory difficulty, hypotension, and cardiac arrhythmias before death (Roos, 1993).

ii) About 11,000 suspected seaweed poisoning cases have been reported from Guadeloupe and Martinique due to the inhalation of toxic gases, namely, hydrogen sulfide (H2S) gas and ammonia from the decomposed brown seaweed species of *Sargassum*.

The affected persons complained of dizziness, vertigo, headache, shortness of breath, heart palpitations, and skin rashes (https://

danboater.org/travel-health-and-safety/sargassum-toxicity-health-hazard-what-you-need-to-know.html).

iii) On September 9, 1994, a 60-year-old woman visited the Maui County Health Department and complained about a burning sensation in her throat and mouth which occurred 15 min after she consumed the seaweed ("ogo") preparation. The species was subsequently identified as a red alga, *Gracilaria coronopifolia* (https://www.cdc.gov/mmwr/preview/mmwrhtml/00039077.htm).

iv) A total of 3404 healthy Korean adults were found to have high urinary arsenic concentrations after the consumption of seaweeds. According to the 2021 review published in the *European Thyroid* journal, more frequent consumption of iodine-rich seaweeds, such as kelp, may cause side effects, particularly in pregnant women and newborns and those people with pre-existing thyroid disorders (https://www.forbes.com/health/body/health-benefits-of-seaweed/).

v) Three fatal seaweed poisoning incidents due to the ingestion of red alga, *Gracilaria edulis*, and *Acanthophora specifera* occurred in the Philippines during 2002–2003 and these poisonings caused eight deaths among 36 patients. The biotoxins involved in these incidents were found to be polycavernosides (PA) and aplysiatoxins. Among these toxins, values of PA estimated in two different methods were found to be 84 and 72 nmol/kg, respectively (Yotsu-Yamashita et al., 2004).

5.3.3 DIARRHETIC SHELLFISH POISONING (DSP)

Although DSP occurs worldwide, outbreaks have been reported in Asia (Japan and China), Europe (Ireland, Scandinavia, Belgium, France, and Spain), and the Americas (United States, Canada, Chile, and Uruguay) (Berdalet et al., 2015). However, the most highly affected areas were found to be Europe and Japan (https://www.whoi.edu/science/B/redtide/notedevents/foreign/DSP_Spain_9-25-02.html).

i) A total of 11 DSP outbreaks involving 45 individuals were reported in France during June 2009. These individuals had ingested contaminated mussels harvested in Vilaine Bay (NW France). On examination, the mussels had okadaic acid (OA) group toxins which were about eight times higher than the European regulatory limit.

Further, based on the consumption data of all 11 cases, the lowest observable adverse effects level (LOAEL) (50 μg OA equivalents per person) was recorded. The LOAEL was established by the European Food Safety Authority in 2006 (Hossen et al., 2009).

ii) Many cases of mussel poisoning were reported in the Netherlands in 1961, 1971, 1976, 1979, and 1981 (Kat, 1983).

iii) Since 1970, more than 1300 DSP cases have been reported from Japan, especially, during April–September. Several outbreaks were also recorded in Europe, South America, and the Far East. In Spain, more than 5000 cases were reported during 1981. In France, over 2000 DSP cases were reported every year in 1984 and 1986. In Scandinavia, 300 cases in 1984; more than 100 cases in Chile during 1991; 400 cases in 1990 through the export of mussels from Denmark to France; and several cases in Uruguay during 1992. In these DSP cases, no neurotoxicity or fatal cases have been reported. However, diarrhea was found to be the chief symptom (92%) followed by nausea (80%) and vomiting (79%). Further, complete clinical recovery was recorded even in severe cases within three days (https://www.whoi.edu/science/B/redtide/notedevents/foreign/DSP_Spain_9-25-02.html).

iv) In the United States, DSP-like illnesses have been recorded on the East Coast since 1980. As per the report of WDOH which was received by the Public Health–Seattle & King County in July 2011, four persons were affected by DSP, and all the ill persons recovered within 96 h without medications and medical care (https://wwwnc.cdc.gov/eid/article/19/8/12-1824_article).

v) Trainer et al. (2013) reported that the ingestion of mussels collected from Sequim Bay State Park (US Pacific Northwest) caused illness for three persons in 2011. Further, between 1976 and 1982, about 1300 DSP cases were reported from Japan; about 5000 cases from Spain in 1981; 3300 cases from France in 1983; and 60 cases from Salt Spring Island, British Columbia (western Canada) during July–August, 2011.

vi) More than 220 probable or suspected cases of DSP were identified in an outbreak of in Zhejiang Province, China, on May 27, 2011. Further, an additional 57 cases were also reported from three collective units in Cangnan County, Zhejiang province (Chen et al., 2013).

vii) Stamman (1987) reported on the DSP outbreaks worldwide and the details are given in the following table.

TABLE 5.2 Worldwide DSP Outbreaks.

Country	Period of occurrence	No. of affected
SW coast of Norway	1968	na
-Do-	1984	100s
US (New York & New Jersey)	1980, 181, 1982, & 1983	31, 210, 1332, & 1951, respectively
Japan	1976–1981	800
Sweden	1983	10
-Do-	1984	10
The Netherlands	1961, 1971, & 1976	NA, 100, & 25, respectively
Spain	1978, 1979, & 1981	5000

Source: Stamman (1987).

5.3.4 AMNESIC SHELLFISH POISONING

ASP outbreaks have been reported in Europe (Scotland, Spain Belgium, France, Ireland, and Portugal), the Americas (Chile and Canada), and the Pacific (New Zealand and Australia,). Shellfish implicated in ASP include mussels, scallops, and razor clams; and other crustaceans. In 1987, an outbreak of ASP in humans involving 107 (47 men and 60 women) illnesses and three deaths were reported from the Prince Edward Island (Canada). Further, an additional 38 probable cases were also reported (Berdalet et al., 2015). Cases of ASP in humans have also been reported (https://www.ncbi.nlm.nih.gov/books/NBK470225/).

5.3.5 AZASPIRACID SHELLFISH POISONING (AZP)

Twiner et al. (2008) reported on the outbreaks of AZP in certain countries during 1995–2007 as shown in the following table.

TABLE 5.3 Reported Cases of AZP During 1995–2007.

Country	Area of AZP outbreak	Period	No. of affected
United Kingdom	Bantry Bay, Ireland	August 2000	12–16
Ireland	Arranmore Island, Ireland	September/October 1997	20–24
France	Bantry Bay, Ireland	September 1998	20–30
Italy	Clew Bay, Ireland	September 1998	10
Netherlands	Killary Harbour	November 1995	8

Source: Twiner et al., 2008.

5.3.6 PALYTOXICOSIS (PALYTOXIN—PTX POISONING)

i) Taniyama et al. (2002) reported on the palytoxin poisoning of 11 persons due to ingestion of a serranid fish *Epinephelus* sp. in Kochi Prefecture, Japan, during October 30 and November 4, 2000. Affected persons had higher levels (700–23,800 IU/L) of serum creatine phosphokinase.

ii) Human fatalities were suspected due to the ingestion of PTX-contaminated crab *Demania reynaudii* in the Philippines and sardine *Herklotsichthys quadrimaculatus* in Madagascar. PTX is also now considered to be the cause of clupeotoxism which is a poorly understood toxic syndrome and is related to the consumption of clupeoid fish, such as herrings, sardines, and anchovies. Further, human fatal cases have also been reported due to the ingestion of PTX-contaminated smoked fish (*Decapterus macrosoma*) in Hawaii; and due to groupers (*Epinephelus* sp.) and blue humphead parrotfish (*Scarus ovifrons*) in Japan (Ramos and Vasconcelos, 2010).

iii) Cases of human poisoning through the consumption of PTX-like toxin contaminated fishes such as *Scarus ovifrons*, *Calotomus japonicus*, *Epinephelus* sp., and *Opsiscarus ovifrons*; PTX-contaminated *Opsiscarus ovifrons;* and Palytoxin and its analog-contaminated *Herklotsichthys quadrimaculatus* have been reported in Japan (Kim et al., 2022).

iv) In August 2019, a family of five people from Shropshire, England, was hospitalized as they were PTX poisoned with corals while cleaning a tropical aquarium. Similarly, two poisonings with corals were reported in Quebec in 2018 and 2019. Further, seven members of a family were affected by PTX while cleaning a salt-water aquarium in Adelaide, Australia, in 2017 (https://ncceh.ca/content/blog/palytoxin-potent-poorly-understood-marine-toxin-found-aquarium-coral).

v) Berdalet et al. (2015) reported that the recurrent *Ostreopsis* blooms which are known for their rich PTX contents contaminated seafood and the people of the nearby Mediterranean coast were at potential risk.

vi) Hoffmann et al. (2008) reported on a case of human poisoning by palytoxin after contact with the zoanthid corals (*Parazoanthus* sp.)

in a saltwater aquarium. His clinical symptoms include numbness, swelling, and paraesthesia around the injury site. After symptomatic treatment, he recovered within three days.

vii) On August 12, 2014, a middle-aged male patient arrived in the emergency department of an Alaska hospital with possible palytoxin exposure to a live zoanthid coral in his home aquarium. He complained of cough, fever, weakness, bitter metallic taste, and muscle pain (Hamada et al., 2015).

viii) When handling *Palythoa* zoanthids in an aquarium, one man died in Texas and his death was due to the palytoxin of the zoanthids (https://reefbuilders.com/2021/02/24/texas-man-dies-from-apparent-exposure-to-palytoxin/).

ix) A marine aquarium hobbyist visited a hospital in Atlanta, Georgia, USA, in October 2006. He had symptoms of lightheadedness, chest pain, and numbness of the left arm which according to him due to the contact of his left hand with a (palytoxin-containing) red and pink zoanthid (Deeds and Schwartz, 2010).

x) In February 2007, a marine aquarium hobbyist in Virginia, USA, was found affected by the palytoxin-containing zoanthids collected from the shore. As the patient experienced difficulty in breathing, coughing, and chest pain within 4 h post exposure, he was admitted to a local hospital (Deeds and Schwartz, 2010).

xi) The clupeotoxin isolated from the head of the Goldspot sardine, *Herklotsichtys quadrimaculatus*, was identified as palytoxin which was implicated in human fatality. A fatal case due to clupeotoxin occurred in Kaua'i in 1978 due to the ingestion of the Marquesan Sardine (*Sardinella marquesensis*) (Randall, 2005).

xii) Deeds and Schwartz (2010) reported on a case of fish poisoning in Japan which was due to the ingestion of blue humphead parrotfish (*Scarus ovifrons*). Further, according to them, 11 people became ill after ingesting the viscera, liver, and boiled muscle of a single large grouper (*Epinephelus* sp.) which was caught in 2000 near Kashiwa Island in southern Japan. Furthermore, a near-fatal case of fish poisoning was reported in 1986 on the island of Kauai, Hawaii, and this according to these authors was due to the ingestion of smoked mackerel (*Decapterus macrosoma*) imported from the Philippines. For all these fish poisoning-associated fatalities, palytoxin is suspected.

5.3.7 PARALYTIC SHELLFISH POISONING (PSP)

PSP is known to occur worldwide but this phenomenon has been reported to be the most common one in the temperate waters off the Atlantic and Pacific coasts of North America, including Alaska. Other countries that reported PSP cases include Americas (Chile); Asia (the Philippines and China); Europe (Scotland and Ireland); and the Pacific (New Zealand and Australia) (Berdalet et al., 2015).

i) In the United States, several cases of PSP have been recorded in New England, Alaska, and the West Coast. Similarly, PSP cases have also occurred in New Zealand, Australia, Europe, Asia (the Philippines, Malaysia, Brunei and Indonesia, and Papua New Guinea), and South Africa (Berdalet et al., 2015; https://www.ncbi.nlm.nih.gov/books/NBK470225/).
ii) In the United Kingdom, the first likely PSP cases were reported in 1827 in Leith and in 1888 in Liverpool. In British Columbia (Canada) and Norway, the first recorded PSP outbreaks were in 1793 and 1901, respectively. In the Philippines, 1995 cases with 117 deaths were linked to PSP toxicity between 1983 and 1999 (Berdalet et al., 2015).
iii) WHO (1984) reported on the occurrence of PSP in various countries along with the number of people affected and deaths, the details of which are shown in the following table.

TABLE 5.4 Occurrence of PSP in Countries.

Country	No. of people affected	No. of deaths
Canada	7	1
Germany	19	0
India	98	1
Malaysia	201	4
Mexico	20	3
Norway	4	0
South Africa	23	2
Switzerland	23	0
Thailand	62	1
United Kingdom	78	0
USA	110	1
Venezuela	180	11
West Europe.	120	0

Source: WHO (1984).

5.4 MAXIMUM PERMITTED LEVEL PRESCRIBED FOR BIOTOXIN-CONTAMINATED SEAFOOD

The maximum permitted level for marine biotoxin-contaminated seafood and its products has been prescribed as a regulatory measure in different countries. However, standard values for each of such biotoxins are the need of the hour. The following are the standard values as per an available report (https://www.seafish.org/trade-and-regulation/legislation-to-control-contaminants-in-seafood/marine-biotoxins).

TABLE 5.5 Maximum Permitted Level for the Marine Biotoxin-Contaminated Seafood.

Biotoxin-contaminated seafood	Maximum permitted level
Paralytic shellfish poisoning (PSP)	800 µg saxitoxin/kg (STX)
Diarrheic shellfish poisoning (DSP): okadaic acid (OA), dinophysistoxins (DTXs), and pectenotoxins (PTXs)	60 µg okadaic acid equivalents/kg
Amnesic shellfish poisoning (ASP)	20 mg domoic acid/kg (DA)
Yessotoxins	3.75 mg yessotoxin equivalent/kg
Azaspiracids	160 µg azaspiracid equivalents/kg

Source: https://www.seafish.org/trade-and-regulation/legislation-to-control-contaminants-in-seafood/marine-biotoxins/

5.5 POISONING AND THEIR PERMITTED LEVELS AS PRESCRIBED BY EUROPEAN UNIIN (EU)

As per the EU (No 853/2004), three groups of toxins are established toward the monitoring of shellfish and complying with the regulations before such shellfish are marketed. These three groups of toxins are (i) ammesic shellfish poisoning (the ASP toxins and domoic acids); (ii) paralytic shellfish poisoning (the PSP toxin; saxitoxin group); and lipophilic marine toxins including diarrheic shellfish poisoning (DSP toxin okadaic acid and dinophysistoxins), azaspiracids (AZA toxins), pectinotoxins (PTX toxins), and yessotoxins (the YTX toxins). Further, Regulation EC No 853/2004 prescribes the maximum permitted level of biotoxins for the different poisonings. They are 800 ug/kg of the algal toxins that cause PSP; 20 ug/kg of domoic acids which cause ASP; 160 ug okadaic acid equiv./kg expressed as a sum of okadaic acid, dinophysis toxins, and pectenotosins; 1 mg yessotoxin equiv./kg; and 160 ug azaspiracid equiv./kg expressed as the sum of azaspiracid-1, 2, and 3 (https://www.marine.ie/site-area/areas-activity/seafood-safety/shellfish-biotoxins).

5.5.1 NEUROTOXIC SHELLFISH POISONING (NSP)

The brevetoxin-associated NSP has been reported to be a major illness in the Western Hemisphere (Gulf of Mexico, the Caribbean, and southeastern coast of the United States (Florida). But this disease has also been reported in New Zealand (Berdalet et al., 2015; https://www.ncbi.nlm.nih.gov/books/NBK470225/).

In the Gulf of Mexico, and on the South east Atlantic coast of the USA, several human poisonings occurred due to the ingestion of brevetoxin-contaminated shellfish, and also respiratory problems in asthmatics which was caused by the inhalation of biotoxin in the form of an aerosol. These poisonings were largely due to the development of frequent MHABs with *Karenia brevis* as the major dinoflagellate species. Fortunately, NSP has not so far been linked to human fatalities. It is worthy of mention here that the occurrence of toxic *Karenia brevis* red tides was recorded as early as 1648 in the Western Gulf of Mexico and 1840s in Florida (USA) (Berdalet et al., 2015). Lawrence et al. (2011) reported that biotoxin-contaminated shellfish and associated PSP outbreaks were responsible for about 20% of seafood-borne diseases in the USA during 1990–1998. The PSP, DSP, NSP, and ASP outbreaks in different countries during this period are shown in the following table.

TABLE 5.6 PSP, DSP, NSP, and ASP Outbreaks in Different Countries.

Country affected	Poisoning type	Causative organisms	No. of cases
Belgium	DSP	*Mytilus edulis* (blue mussel)	>300
Norway, Sweden	DSP	*Mytilus edulis* (blue mussel)	>300
Norway	DSP	*Cancer pagurus* (brown crab)	200
Guatemala (Central America)	PSP	Clams (bivalves)	187
Japan	DSP	Scallops and mussels (bivalve mollusks)	164
U.K.	DSP	*Mytilus edulis* (blue mussel)	159
Canada	ASP	*Mytilus edulis* (blue mussel)	107
North Carolina (USA)	NSP	*Crassostrea virginica* (Eastern oyster)	48
Ireland	AZP	*Mytilus edulis* (blue mussel)	24

Source: Lawrence et al., 2011

5.5.2 CIGUATERA FISH POISONING (CFP)

The ciguatera fish poisoning (CFP) has been reported to be the most common food poisoning worldwide and it is known to occur throughout the tropics, particularly in the island nations of the Pacific and the Caribbean. However, the real extent of illness due to this poisoning has not been adequately documented due to under-reporting and misdiagnosis. Data from the Health and Fisheries Authorities state that during 1973–1983, the mean annual incidence was 104 cases per 100,000 people. Based on this estimate, the recorded CFP incidence in the South Pacific was up to 194 cases per 100,000 people between 1998 and 2008. Further, the CFP illness rates exhibited high variability. For instance, it was up to 497 cases per 10,000 population per year in French Polynesia including Raivavae Island (Austral archipelago) between 2000 and 2008, and up to 440 cases per 10,000 population per year in the Caribbean between 1996 and 2006 (Berdalet et al., 2015). WHO (1984) reported on the occurrence of CFP in various countries along with the number of people affected and deaths, the details of which are shown in the following table.

TABLE 5.7 Occurrence of CFP Outbreaks in Certain Countries.

Country	**No. of people affected**	**No. of deaths**
Bahamas	14	0
Canada	2	0
Cuba	100	0
Fiji	925	1
France	2	0
French Polynesia	3009	3
Jamaica	250	0
USA (Florida)	129	0
USA (Maryland)	12	0
USA (Shipboard)	24	0
US Virgin Islands	84	0

Source: WHO (1984).

5.5.3 SHARK POISONING

Food poisoning due to tiger sharks (*Galeocerdo cuvier*) and bull sharks (*Carcharhinus leucas)* is a new phenomenon. Cases of shark poisoning due to the ingestion of sharks (flesh and/or liver) are on the increase from the 19th century to the present day. Human fatal cases have been reported occasionally in the Gilbert Islands, French Polynesia, New Caledonia, the Cook Islands, Madagascar (in November 2013 and February 2014), and Reunion Island (in 1993). The tiger shark was implicated in shark food poisoning in the Gilbert Islands. In Madagascar, flesh from a tiger shark was found associated with at least one case of food poisoning; and flesh from a bull shark with two cases. Genetic analysis of the shark involved in the food poisonings which occurred in November 2013 concluded that it was a bull shark. The symptoms recorded in these shark food poisonings were more related to that of ciguatoxins, produced by a dinoflagellate genus, *Gambierdiscus*. However, some authors suggested that other toxins with similar properties, known as "carchatoxins" might be responsible, but their structure is yet to be characterized. According to these authors, these carchatoxins could be new analogs of ciguatoxins, for example, highly oxidized forms (https://www.anses.fr/en/system/files/ERCA2013sa0198EN.pdf).

5.5.4 PUFFERFISH POISONING

An outbreak of pufferfish poisoning occurred in Nosy Be, an island located in the north-western part of Madagascar in July 1998. During this poisoning, four people were affected and one of them died. Further mouse bioassay analysis revealed that this poisoning was due to the tetrodotoxin content (16 MU (mouse unit) per gram meal) of the pufferfish (Ravaonindrina et al., 2001).

5.5.5 TTX POISONING

TTX poisonings occur due to marine fishes such as pufferfish and gobiids; gastropod mollusks; and horse-shoe crabs. About 30–50 cases of TTX intoxications are known to occur every year. While a regulatory limit of 2 mg equivalent of (eq) TTX/kg has been fixed for TTX in Japan, this compound is not yet regulated in Europe as it is considered as an emerging toxin. However, poisonous fishes including the pufferfish are not allowed to be placed on the European markets.

TABLE 5.8 Human Cases Due to TTX Intoxication During the Last Three Decades.

Period	Locality/country	No. of cases	Causative species	Commonname/taxonomy
Dec. 1979	Shizuoka, Japan	1	*Charonia sauliae*	Triton snail; Gastropod mollusk
1986	Hawaii, USA	1	*Diodon hystrix*	Spot-fin porcupinefish; Diodontidae
Apr.1996	California, USA	3	Pufferfish	–
Oct. 1996	Nagasaki, Japan	1	*Takifugu poecilonotus*	Fine-patterned puffer; Tetraodontidae
Jul. 1998	Nosy Be, Madagascar	4	–	–
Jan. 2000	Chungua, Taiwan	5	*Takifugu niphoble*	Grass puffer; Tetraodontidae
Apr. 2001	Taiwan Strait	6	–	–
Apr. 2004	Tungsa Island, Taiwan	6	*Nassarius glans*	Acorn dog whelk; Gastropod mollusk
Sept. 2007	China	202	8 strains of *Shewanella* spp.	Gram-negative bacterium
1998–2008	Bangladesh	156	13 species of pufferfish	–
Oct. 2008	Nagasaki, Japan	–	*Lagocephalus inermis*	Smooth blaasop; Tetraodontidae
Dec. 2010	Taipei, Taiwan	2	*Hapalochlaena fasciata*	Blue-lined octopus; Cephalopod mollusk
Mar. 2012	Guangdong, China	26	*Amoyacaninus*; *Yongeichthys nebulosus*	Tiger-toothed spine fish; hairfin goby; Gobiidae
Sept. 2013	Coast of Reunion Island (Southwest Indian Ocean)	12	*Lagocephalus sceleratus*	Smooth blaasop; Tetraodontidae
June 2014	Minnesota, USA	2	*Lagocephalus lunaris*	Lunartail puffer; Tetraodontidae
–	Chon Buri, Thailand	71	*Carcinoscorpius rotundicauda*	Mangrove horseshoe crab; Limulidae

Source: Lago et al. (2015).

Lago et al. (2015) reported on TTX intoxication during the last three decades and the affected human cases in the different countries and the details are given in the following table.

KEYWORDS

- **amnesic shellfish poisoning**
- **diarrheic shellfish poisoning**
- **neurotoxic shellfish poisoning**
- **paralytic shellfish poisoning**
- **Ciguatera fish poisoning**
- **TTX-intoxicated human cases**

CHAPTER 6

Therapeutic Potential of Phycotoxins of Marine Algae

ABSTRACT

Aspects such as therapeutic potentials of phycotoxins relating to marine diatoms, dinoflagellates, algae etc; and medical and therapeutic applications of TTX, YTX, maitotoxin, palytoxin, karlotoxin, aplysiatoxin, and lipopolysaccharides (LPS) are presented in this chapter with illustrations.

6.1 PHYCOTOXINS OF MARINE MICROALGAE (PHYTOPLANKTON)

6.1.1 DOMOIC ACIDS OF MARINE PENNATE DIATOMS

Domoic acid (DA), a neuroexcitatory amino acid, is produced by certain species of marine Pennate diatoms (Bacillariophyceae) and red algae and this toxin is biomagnified through the food web. It is a potent neurotoxin that causes amnesic shellfish poisoning. In humans and nonhuman primates, oral exposure to a few mg/kg DomA elicits gastrointestinal effects, whereas slightly higher doses cause neurological symptoms, seizures, memory impairment, and limbic system degeneration. Significant levels of DA have been observed in the viscera of Dungeness crabs and razor clam meat. Consuming DA-contaminated Pacific razor clams (*Siliqua patula*) presented the greatest risk. Further, the razor clams have been reported to retain DA for up to 1 year in the natural environment, or several years after being processed, canned, or frozen (https://www.sciencedirect.com/topics/agricultural-and-biological-sciences/domoic-acid). The highest DA toxicity in mussels was found to be 71 μg/g (http://www.gov.pe.ca/photos/original/af_domoic_acid.pdf).

Therapeutic Potential of Marine Biotoxins: From Poison to Pharmacy. Santhanam Ramesh, Ramasamy Santhanam, & Subbiah Balasundari (Authors)

DOI: 10.1201/9781003610991-6

The species producing domoic acid are given below:

Phylum: Bacillariophyta; Class: Bacillariophyceae;

Order: Bacillariales;

Family: Bacillariaceae

Nitzschia bizertensis.

Nitzschia navis-varingica

Pseudo-nitzschia australis (= Nitzschia pseudoseriata)

Pseudo-nitzschia delicatissima (= Nitzschia actydrophila)

Pseudo-nitzschia fraudulenta (= Nitzschia fraudulenta).

Pseudo-nitzschia multiseries (= Nitzschia pungens forma *multiseries)*

Pseudo-nitzschia multistriata (= Nitzschia multistriata)

Pseudo-nitzschia pseudodelicatissima

(= Nitzschia pseudodelicatissima)

Pseudo-nitzschia pungens (= Nitzschia pungens)

Pseudo-nitzschia seriata (= Nitzschia seriata).

Pseudo-nitzschia subcurvata.

Pseudo-nitzschia turgidula (= Nitzschia turgidula).

Order: Thalassiophysales; Family: Catenulaceae

Halamphora coffeaeformis

(= Amphora coffeaeformis; Frustulia coffeaeformis).

Distribution: These species are found distributed worldwide including the Antarctic waters (*Pseudo-nitzschia subcurvata).*

Habitat: While the species of *Pseudo-nitzschia* are mostly planktonic, *Nitzschia* spp. and *Halamphora coffeaeformis* are invariably benthic. Most of these species occur in coastal waters and few of them are euryhaline thriving in estuarine, backwater, and mangrove habitats. During bloom formations, the highest observed cell concentration of these species was found to be 403,000 cells per liter (http://www.gov.pe.ca/photos/original/af_domoic_acid.pdf).

Toxins and Their Therapeutic Uses: Domoic acid is present in all the species. *Nitzschia navis-varingica* and *Pseudo-nitzschia australis* have also been reported to produce the analogs of DA, viz., isodomoic acid A and B and isodomoic acid C, respectively (Pradhan and Ki, 2022). Tan et al. (2016) reported that the DA contents of the different species ranged between 0.37 and 11.06 pg/cell.

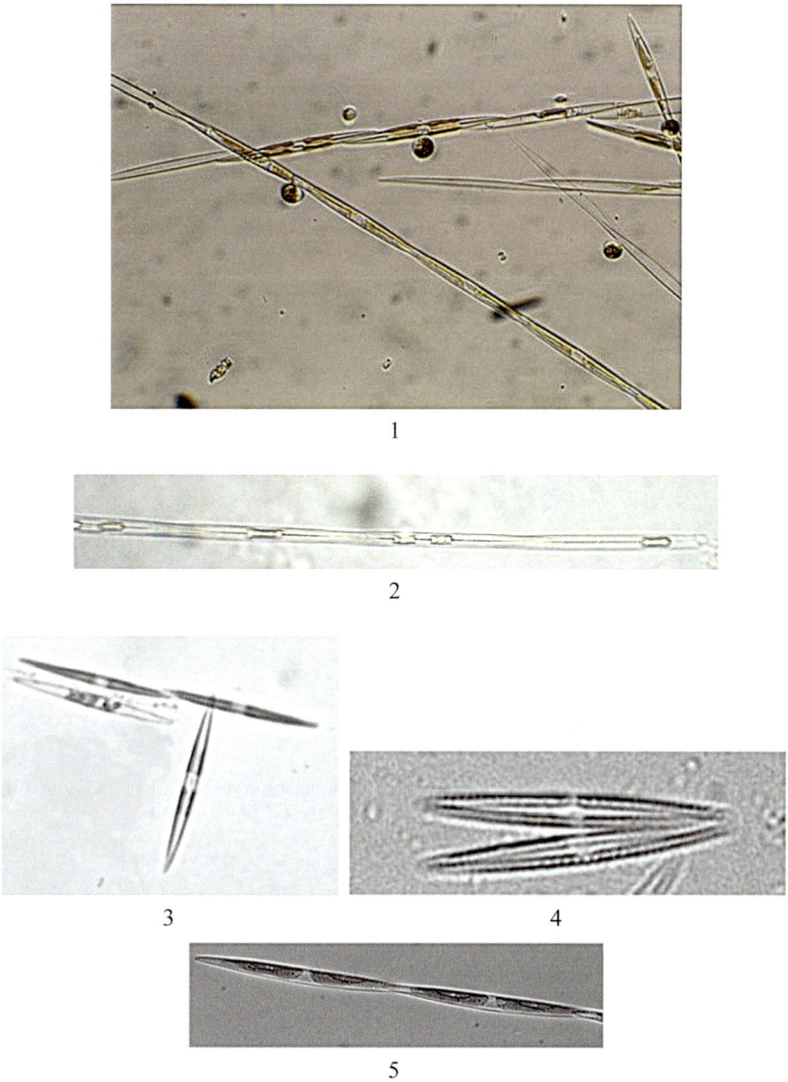

Figure 6.1 Some species producing domoic acid

1. Pseudo-nitzschia australis; 2. Pseudo-nitzschia delicatissima; 3. Pseudo-nitzschia multiseries; 4. Pseudo-nitzschia pseudodelicatissima; 5. Pseudo-nitzschia seriata

1. Image credit: Oregon State University . https://creativecommons.org/licenses/by-sa/2.0/
2. Image credit: Minami Himemiya. https://creativecommons.org/licenses/by-sa/3.0/
3. Source: Reproduced with permission from Dr. Stephen S. Bates
4. Source: Reproduced with permission from Dr. Stephen S. Bates
5. Source: Reproduced with permission from Dr. Stephen S. Bates

Although DA is a potent neurotoxin, it has been reported to possess pharmacological and therapeutic properties. Pinto-Silva et al. (2008) reported that DA has shown promising cytotoxicity and genotoxicity in a human colorectal adenocarcinoma cell line (Caco-2). DA reduced cell viability (IC(50) about 70 ng/mL); apoptosis in Caco-2 cells at 100 ng/mL; and induced direct DNA damage from 15 ng/mL. Pradhan and Ki, (2022) reported that DA displayed anti-proliferative effects on cancer cell lines such as K562 and EA.hy 927 in vitro. DA has also shown antihelmintic properties and curative effectiveness in killing parasitic worms (Pradhan and Ki, 2022; https://www.sciencedirect.com/topics/agricultural-and-biological-sciences/domoic-acid).

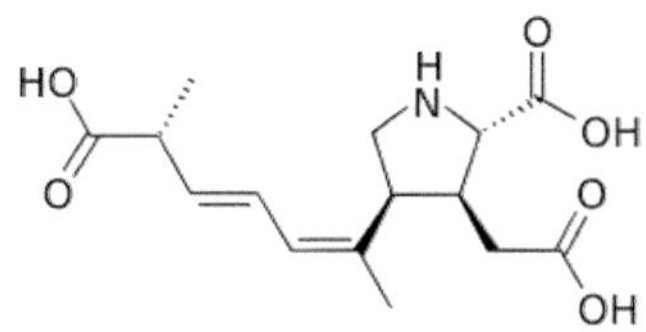

Domoic acid

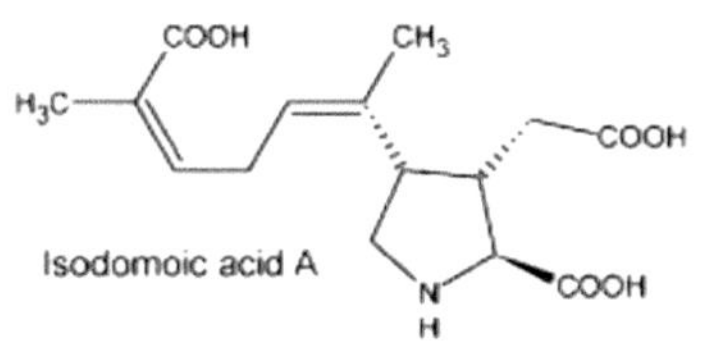

Isodomoic acid A

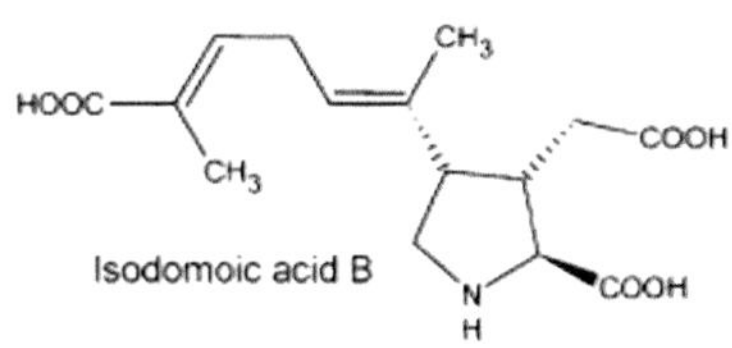

Isodomoic acid B

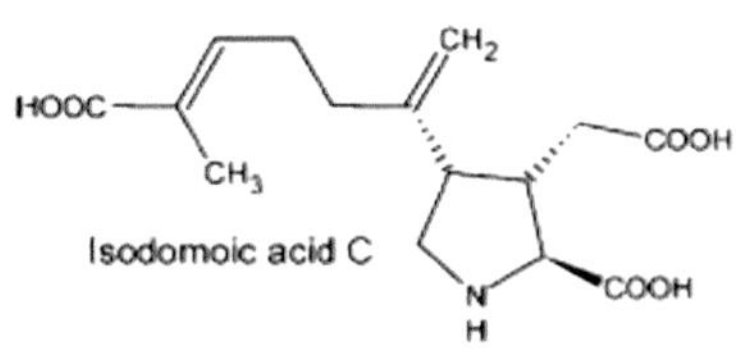

Isodomoic acid C

6.1.2 DINOFLAGELLATE TOXINS

Four major groups of dinoflagellate toxins, viz., ladder-shaped polyether compounds, saxitoxins, long-chain polyketides, and macrolides have been described. The different toxins and their causative organisms are given below.

6.1.2.1 DIARRHETIC TOXINS

6.2.1.1.1 Okadaic Acid (OA) and Its Derivatives, Dinophysistoxins (DTXs)

Okadaic acid and its derivatives, viz., dinophysistoxins 1, 2, and 3 are polyether compounds derived from the benthic marine dinoflagellates such as *Prorocentrum arenarium, Prorocentrum lima, Prorocentrum concavum, Prorocentru belizeanum, Prorocentrum maculosum;* and *Dinophysis acuta, Dinophysis caudata, Dinophysis acuminata, Dinophysis infundibulum, Dinophysis miles, Dinophysis norvegica, Dinophysis ovum, Dinophysis sacculus,* and *Dinophysis fortii* (Assumcao et al., 2017; Botelho et al., 2022).

Therapeutic Uses

Anticancer effects: OA and DTXs have been reported to be highly selective inhibitors of protein phosphatase types 1 (PP1) and 2A (PP2A). Inhibition of protein phosphatase-2A (PP2A), a serine/threonine phosphatase, is known to enhance anticancer immunity. Further, OA and its above analogs are considered to be extremely useful research tools in the investigation of cellular regulation processes, especially those concerned with reversible phosphorylation of proteins—viz., memory, signal transduction, and cell division. Due to the inhibition of protein phosphatase-2A, OA is being used as a tool for further research on Alzheimer's and other neurodegenerative disorders associated with memory impairment. Furthermore, OA has also been reported to possess immunoregulatory properties owing to its inducement in the down-regulation of T-cell receptor expression (Assumcao et al., 2017).

The polyketide toxin, viz., prorocentin-1 derived from *Prorocentrum* spp. showed cytotoxic activity against the human malignant melanoma cell line, RPMI-7951, and colorectal adenocarcinoma cell line (DLD-1) with IC50 values of 16.7 and 83.6 μg/mL, respectively. Further, another polyketide toxin of these species, viz., Limaol, was found to be cytotoxic against human HepG2, HCT-116: and Neuro2a cell lines with IC50 values of 3.7, 7.3, and 9.6 μM, respectively (Cousseau et al., 2020).

Carvalho et al. (2006) reported that at a concentration of 100 ng/mL, OA displayed cytotoxicity and induced apoptotic cell death in Caco-2 cells.

The human colonic (Caco-2) cells incubated with OA (3.75–60 ng/mL, i.e., $4.6 \times 10(-3)-7.5 \times 10(-2)$ μM) induced cytotoxicity and DNA lesions; and caused a significant reduction in cell viability. According to their findings, the cytotoxic activity of okadaic acid and its ability to damage DNA result in cell death, mainly by apoptosis (Carvalho et al., 2006).

In the treatment of Alzheimer's disease (AD): OA may be of great use in the treatment of AD as it is known to inhibit the serine/threonine protein phosphatases 1 and 2A (Botelho, et al., 2022).

Adenosine triphosphatase (ATPase) activity: At a concentration of 1 μM, the polyether toxin, acuminolide-A derived from *Dinophysis* sp. has been reported to yield adenosine triphosphatase (ATPase) activity in skeletal muscles. On hydrolysis, the adenosine triphosphate (ATP) provides energy to facilitate a variety of cellular processes that are essential for life (Cousseau et al., 2022).

Antifungal activity: OA from the species of *Prorocentrum* has been reported to have antifungal activity by inhibiting the growth of *Candida albicans*.

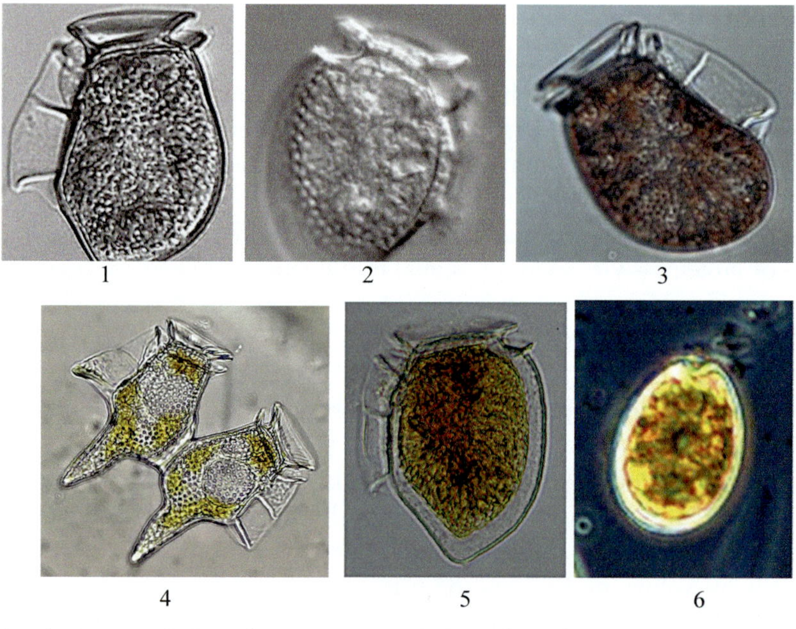

1. Dinophysis acuta; 2. Dinophysis acuminata; 3. Dinophysis fortii; 4. Dinophysis caudata 5. Dinophysis norvegica; 6. Prorocentrum lima

1. Image credit: Alexandra Kraberg. https://creativecommons.org/licenses/by-sa/3.0/
2. Image credit: fjouenne. https://creativecommons.org/licenses/by-sa/3.0/
3. Image credit: Reproduced with permission from Dr. Tara Ivanochko
4. Image credit: Phycoplankton. https://creativecommons.org/licenses/by-sa/4.0/
5. Image credit: Fisheries and Oceans Canada, Dalhousie University, Rajashree Gouda . https://creativecommons.org/licenses/by-sa/4.0/
6. Image credit: Cybergerac. https://creativecommons.org/licenses/by-sa/4.0/deed.en

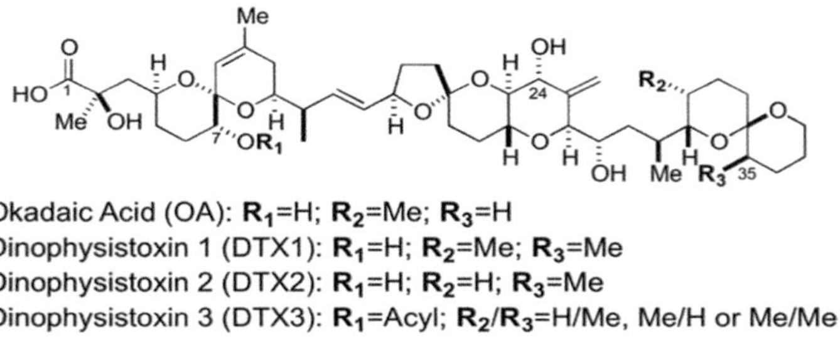

Okadaic acid, Dinophysis toxin 1, 2, 3

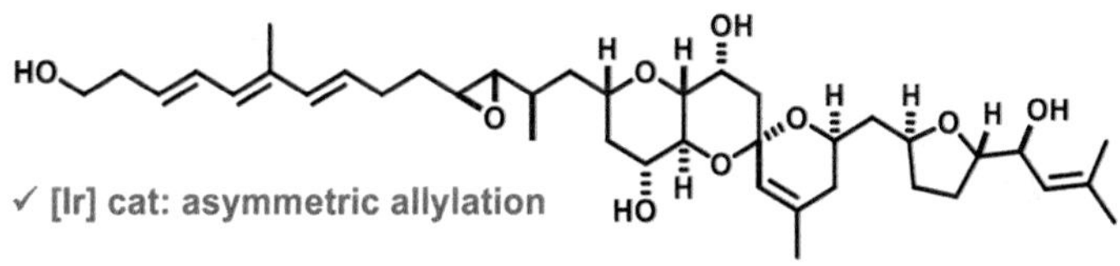

Prorocentin

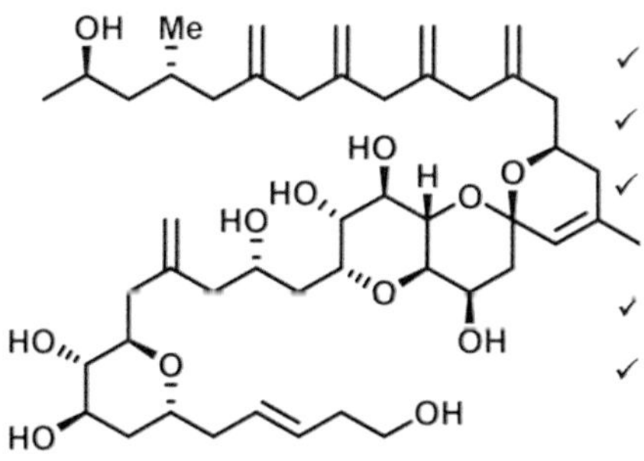

Limaol

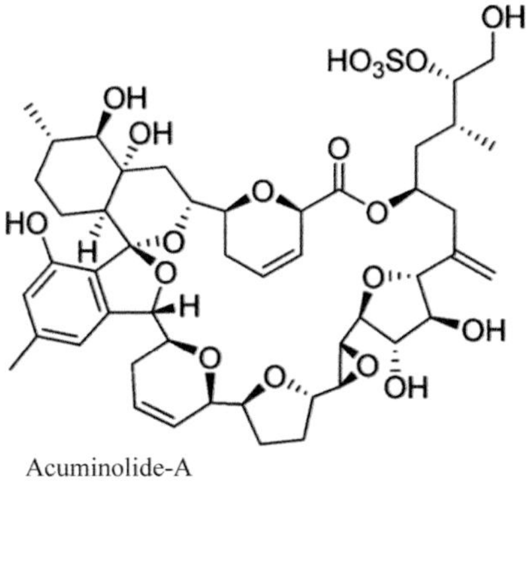

Acuminolide-A

6.1.2.1.2 Pectenotoxins (PTX)

Anticancer activity: The pectenotoxin (PTX), and its 20 analogs are polyether macrolide compounds produced only by the species of *Dinophysis.* viz.. *Dinophysis acuta, Dinophysis tripus, Dinophysis acuminata, Dinophysis caudata, Dinophysis fortii, Dinophysis rotundata,* and *Dinophysis norvegica.* PTXs have been reported to be effective against cancer cell lines with cell cycle arrest and apoptosis. For instance, PTX-2 has displayed antitumor effects against breast cancer and human lung, and colon cell lines, and it could also serve as a potent chemotherapeutic agent against p53-deficient tumors (Assumcao et al., 2017).

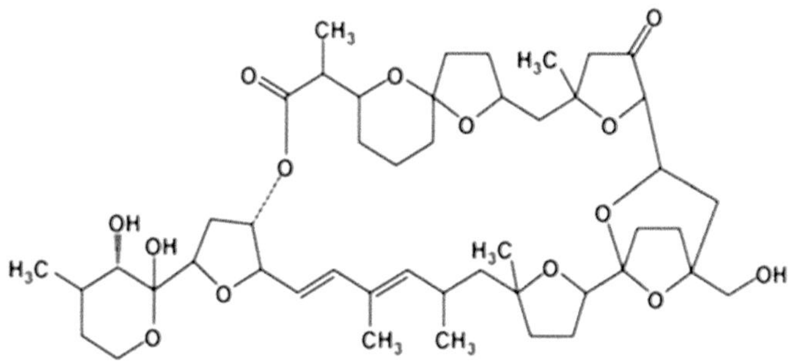

Pectenotoxin

6.1.2.1.3 Azaspiracids (AZA)

The AZAs are nitrogen-containing polyether compounds associated with human incidents of shellfish poisoning. They are produced by the dinoflagellates, viz., *Azadinium spinosum, Azadinium poporum,* and *Azadinium dexteroporum.* Among these species, *Azadinium spinosum* has been reported to be the primary producer of azaspiracid toxins. This species was also shown to produce AZA-1, AZA-2, and an isomer of AZA-2. However, the dinoflagellate *Protoperidinium crassipes* is believed to be the primary source of AZA (Tilmann et al., 2009).

Therapeutic Uses

i) Cytotoxic activity: Azaspiracids-33, 34, 36, and 37 have been reported to possess cytotoxic activity against Human Jurkat T cell lymphoma cell line, ATCC, TIB-152 with EC50 values of 5.2, 0.2, 1.7, and 0.85 nM, respectively (Cousseau et al., 2022).

ii) **Others:** Azaspiracid was found to be a potent activator of c-Jun-N-terminal kinase (JNK) and caspases which are believed to be implicated in stress-signaling pathways such as cell damage, apoptosis, and so on; and as effective modulator of intracellular cAMP(Cyclic adenosine monophosphate) and calcium levels. Further, this toxin inhibits cell cholesterol levels (particularly in T-lymphocyte cells) and affects potassium ion channels which control several cell functions including hormone secretion, cognition, and muscle contraction (Assumcao et al., 2017).

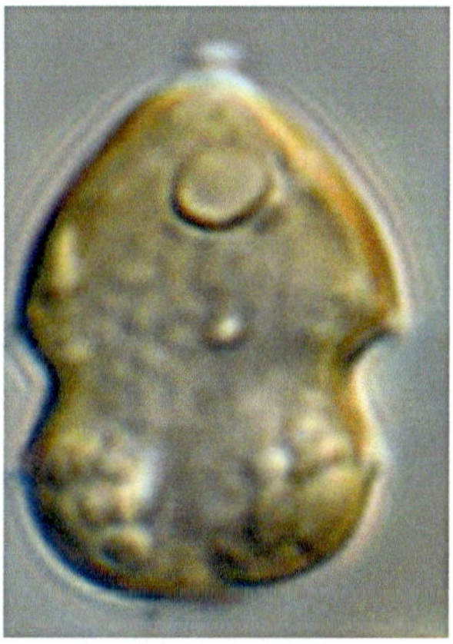

Azadinium spinosum

Image credit: Stephan Wietkamp, Bernd Krock, Dave Clarke, Daniela Voß, Rafael Salas, Jane Kilcoyne, Urban Tillmann. https://creativecommons.org/licenses/by/4.0/

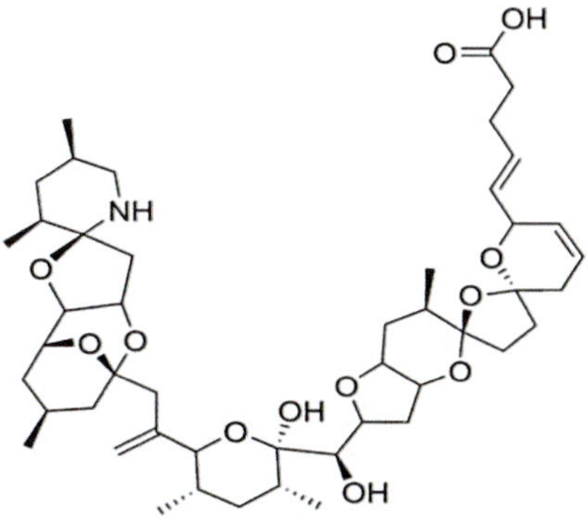

Azaspiracid

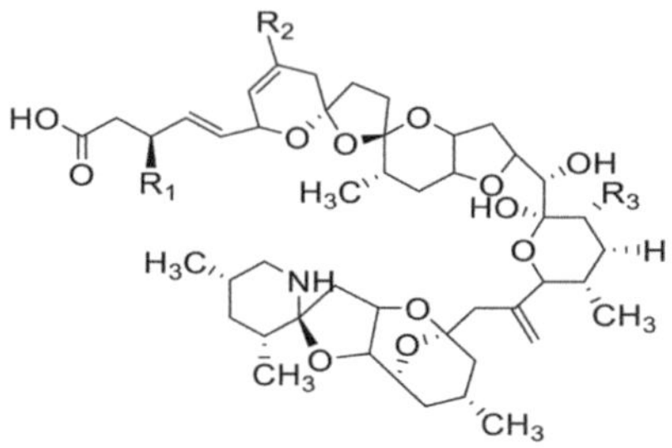

Toxin	R1	R2	R3
AZA1	H	H	CH_3
AZA2	H	CH_3	CH_3
AZA3	H	H	H

Azaspiracid 1,2,3

6.1.2.2 NEUROTOXINS

6.1.2.2.1 Brevetoxin (PbTx)

The brevetoxins consist of two types of lipid-soluble toxins, viz., hemolytic and neurotoxins, and cause human neurotoxic shellfish poisoning (NSP). Brevetoxin A (PbTx-1) and its analogues, PbTx-2, PbTx-3, PbTx-4, PbTx-5, PbTx-6, PbTx-7, PbTx-8, and PbTx-9; brevetoxin B, hemibrevetoxin B, brevisamide, and brevisin are all produced by the dinoflagellate *Karenia brevis (= Gymnodinium breve; Ptychodiscus brevis)*. Among these brevetoxins, the major one produced is PbTx-2, followed by PbTx-1 and PbTx-3, which are produced in lesser amounts (Chand, 2009). In human cases of NSP, the contaminated clams have been reported to possess brevetoxin concentrations in the range of 78–120 μg/mg (Murray, 2021).

Therapeutic Uses

Brevetoxin treatment has been reported to increase both the mean open time and open probability of N-methyl-D-aspartate (NMDA) receptor which is a receptor of glutamate, the primary excitatory neurotransmitter in the human brain. Further, brevetoxin could enhance cerebrocortical neuron neurite outgrowth, dendritic filopodia density, dendritic arbor complexity, and synaptic density. Furthermore, epicortical brevetoxin treatment was found to enhance dendritic arborization and synapse density of cortical layer V pyramidal neurons in the peri-infarct cortex. Brevetoxins were also found to produce a very strong improvement in motor recovery. These findings suggest that brevetoxins may serve as a novel pharmacologic approach to enhance activity-dependent recovery from a stroke (Murray, 2021).

Assumcao et al. (2017) reported that BTX-2 behaved as a neuronal stimulator and can increase neuronal plasticity. This function of brevetoxins could help in pharmacological treatment to recover brain function after stroke or other traumatic brain injury. Further, a therapeutic formulation based on brevetoxin derivatives has now been designed to regulate diseases such as cystic fibrosis and mucociliary dysfunction related to increased mucus transport.

Brevetoxin derivatives could also be used for treating cystic fibrosis, mucociliary dysfunction, and pulmonary diseases (https://patents.google.com/patent/EP1664057B1/en).

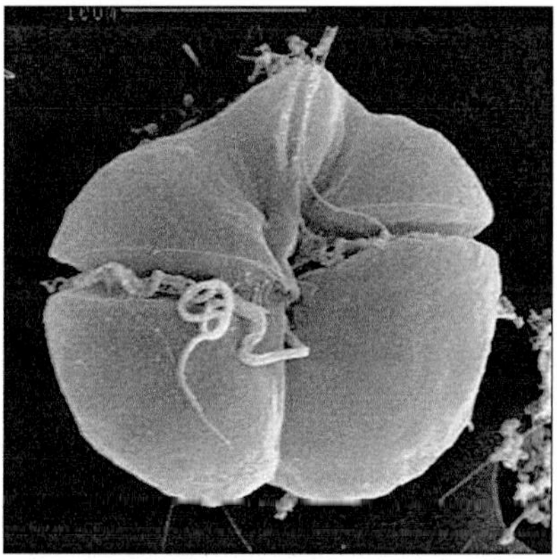

Karenia brevis

Image credit: Florida Fish and Wildlife Conservation Commission; Public domain.

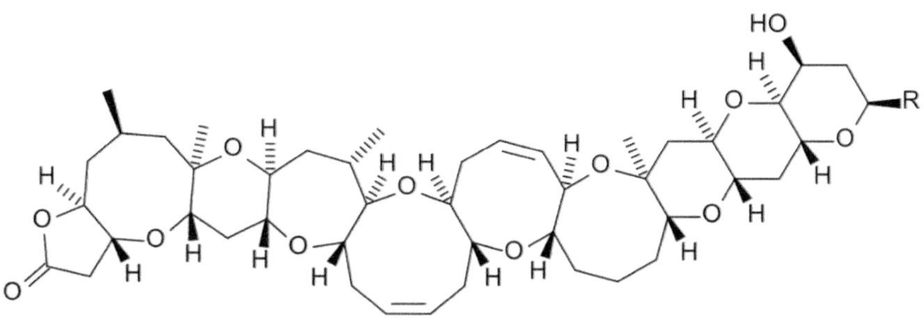

Brevetoxin

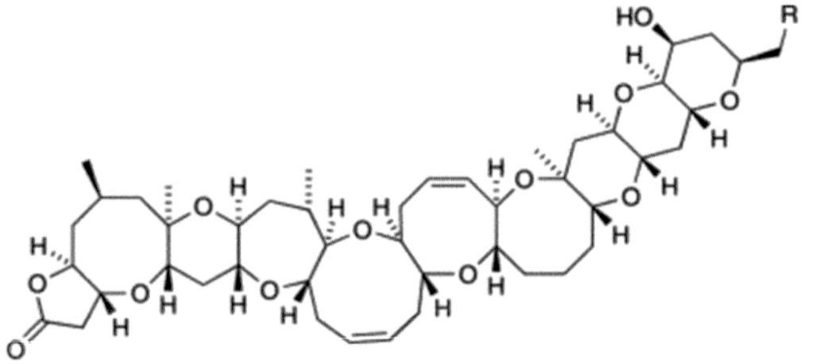

154: R = 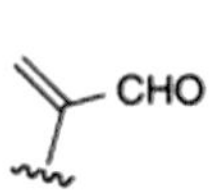

161: R = 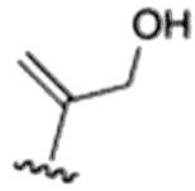

162: R = 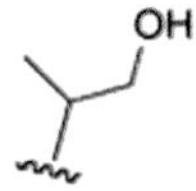

1 2 3

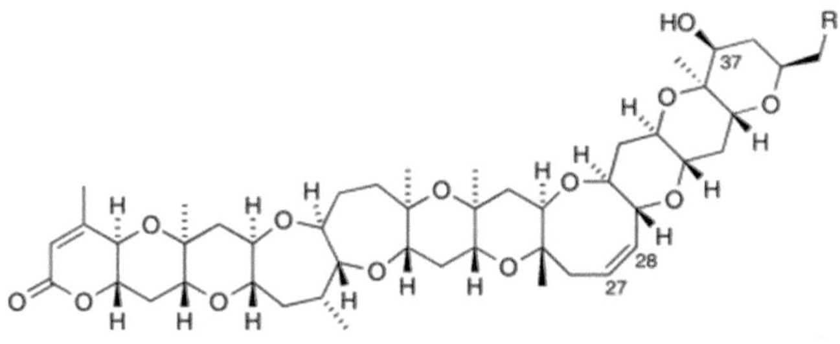

155: R = 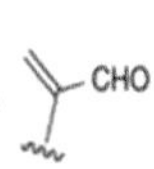

156: R = 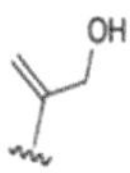

157: R = 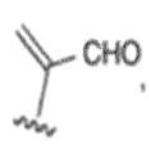

, 37-acetoxy

4 5 6

: R = 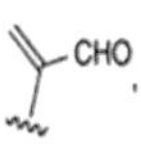

, 27,28-epoxy
159: R = 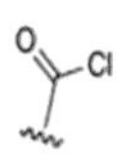

160: R = 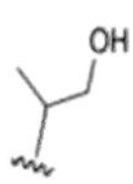

7 8 9

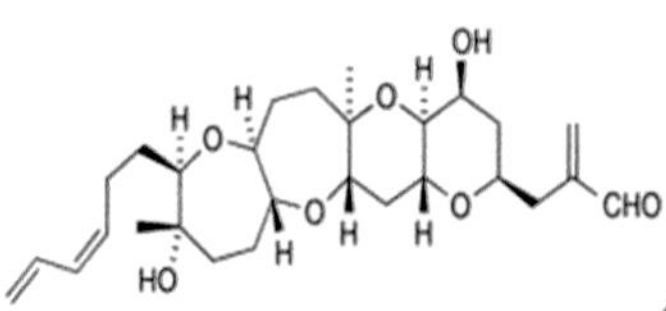

163

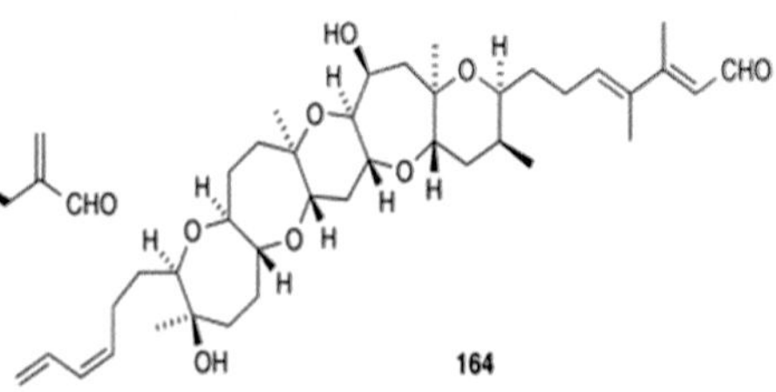

164

10 11

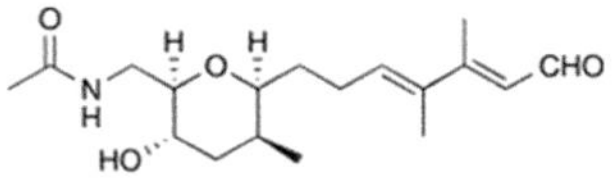

165

12

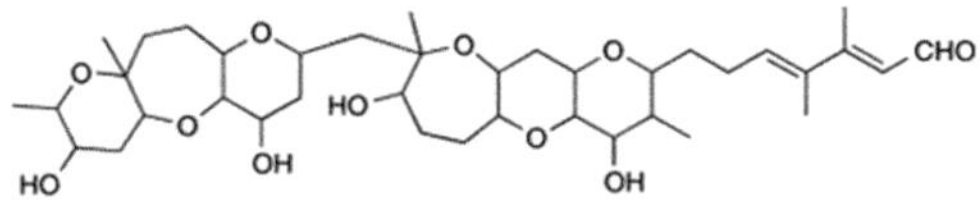

166

13

1,4: Brevetoxin A,B; 2,3,5,6,7,8,9: Brevetoxin analogs; 10: Hemibrevetoxin B; 11: Brevenal; 12: brevisamide; 13: Brevisin

6.1.2.2.2 Ciguatoxins (CTXs)

Ciguatoxins (CTXs), a group of marine polycyclic ether biotoxins, are crystalline, lipophilic, colorless, and heat-stable solids produced by the dinoflagellate species of *Gambierdiscus toxicus.* These toxins get accumulated throughout the marine food chain in several fishes including sharks, moray eels, sturgeon, sea bass, barracuda, parrotfish, jacks, and snappers which are generally larger predatory fish that accumulate these ciguatoxins by feeding on smaller tropical reef fish that feed on the CTX-producing dinoflagellates.

Pharmacologic action of ciguatoxin: The main pharmacologic function of ciguatoxin is an increase in cell permeability to sodium via voltage-gated sodium channels in cell membranes, thereby causing sustained depolarization and current leakage. The depolarization of nerve cells thus leads to neurological symptomology. The mechanism of action in humans, however, is dependent more on the anticholinesterase activity and cholinomimetic action. The species of *Gambierdiscus* and the CTX analogs produced by them are given in the following table.

TABLE 6.1 *Gambierdiscus* spp. and Their CTX Analogs.

Species	CTX analog (s)
Gambierdiscus australes	CTX1B, P-CTX-3C
Gambierdiscus pacificus	51-hydroxyCTX-3C, 2,3-dihydroxyCTX-3C
Gambierdiscus polynesiensis	P-CTX-4A, P-CTX-4B, P-CTX-3C, M-seco-CTX-3C, 49-epiCTX-3C
Gambierdiscus toxicus	P-CTX-3C, 2,3-dihydroxy P-CTX-3C, P-CTX-4A/B

Source: Wang et al. (2022).

Therapeutic Use: Research reveals no clinical applications of ciguatera toxins. However, the ciguatoxin CTX3C has now been chemically synthesized for further study of its therapeutic action (https://www.drugs.com/npp/ciguatera.html).

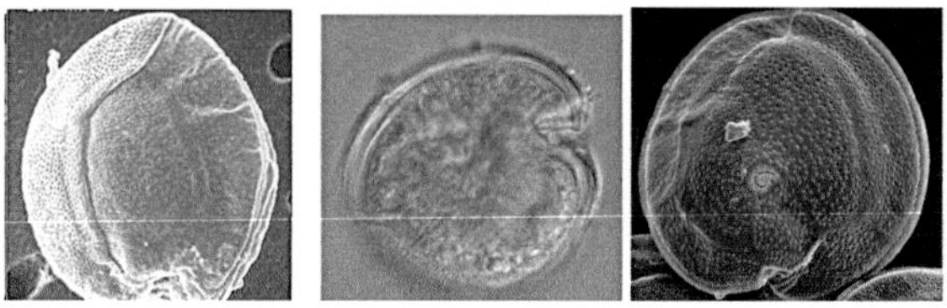

1:Gambierdiscus australes; 2: Gambierdiscus toxicus; 3: Gambierdiscus polynesiensis

1. *Image credit:* Chinain, M & Smithsonian National Museum of Natural History. https://creativecommons.org/publicdomain/zero/1.0/
2. *Image credit*: Reprodued with permission from the United States National Herbarium (US)
3. *Image credit*: Reprodued with permission from the United States National Herbarium (US)

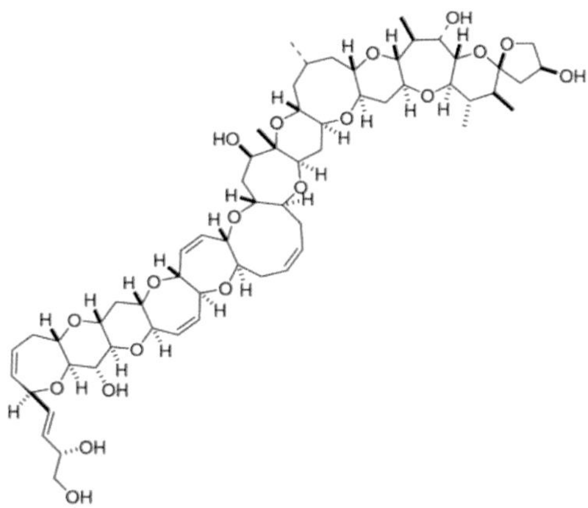

Ciguatoxin

6.1.2.2.3 Saxitoxin

Saxitoxin (STX) and its 57 analogs are neurotoxic alkaloids that are the causative agents of paralytic shellfish poisoning (PSP). These STX and its analogs are structurally classified into a number of classes such as carbamonyl saxitoxins (e.g., STX, neosaxitoxin, gonyautoxin 1 and 4 GTX1 and GTX4); dicarbamoyl saxitoxins (e.g., GTX2 and GTX3); and N-sulfocarbamoyl (i.e., GTX5 and GTX6) saxitoxins. Most of these compounds are produced by the marine dinoflagellates belonging to the genera, *Alexandrium, Gymnodinium*, and *Pyrodinium* (Assumcao et al., 2017; Weise et al., 2010).

Therapeutic potential of saxitoxins

Anticancer activity: The extracts of *Alexandrium andersoni* containing unidentified toxins have shown anticancer activity by inducing high cytotoxicity toward A549 lung cancer and HT29 colorectal cancer cell lines without affecting normal cell viability (Sansone et al., 2018).

As pain killer and anesthetic agent: GTXs have proven as a therapeutic method against acute or chronic anal fissures. These compounds have been reported to help in sphincter relaxation, thus functioning as a pain killer. Further, GTX2 and GTX3 have been used to treat chronic tension-type headaches (Assumcao et al., 2017). Wiese et al. (2010) reported that STX can induce anesthesia through interaction with site 1 of the voltage-gated Na+ channel. Bhattacharya et al. (2009) stated that drugs such as STX which block voltage-gated sodium channels are more suitable for the management of neuropathic pain. The neosaxitoxin (carbamonyl saxitoxin) combination formulation may also be of great use for prolonged local anesthesia (https://patents.google.com/patent/US8975281B2/en).

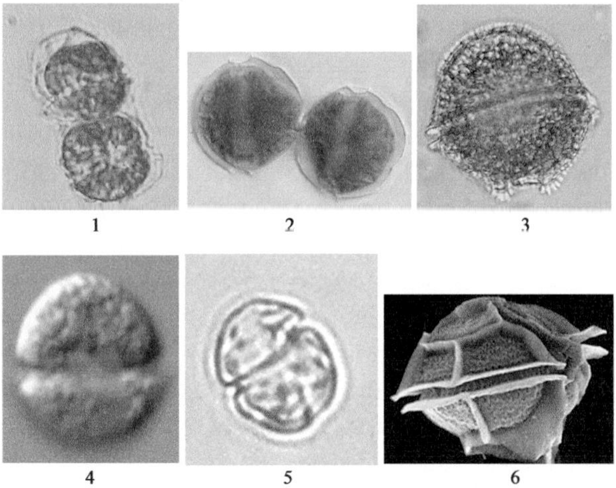

1: Alexandrium catenella; 2: Alexandrium fundyense; 3: Alexandrium minutum; 4: Alexandrium ostenfeldii; 5: Gymnodinium catenatum; 6: Pyrodinium bahamense.

1. *Image credit:* Me.garneau . https://creativecommons.org/licenses/by-sa/3.0/
2. *Image credit*: National Research Council of Canada, Nancy Lewis . https://creativecommons.org/licenses/by-nc-sa/4.0/
3. *Image credit*: Philippe Garcelon. https://creativecommons.org/licenses/by/2.0/
4. *Image credit*: Furche Körpermitte I. https://creativecommons.org/licenses/by/2.0/
5. *Image credit*: Image credit: Minami Himemiya . https://creativecommons.org/licenses/by-sa/3.0/
6. *Image credit*: CSIRO. https://creativecommons.org/licenses/by/3.0/

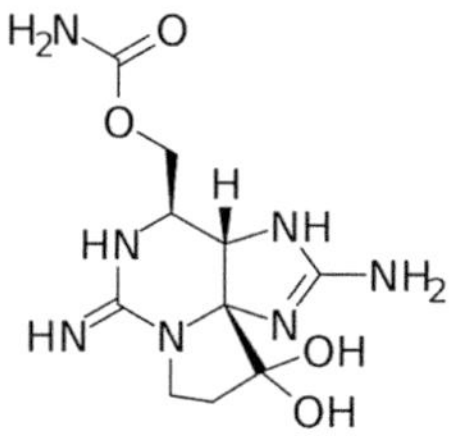

STX

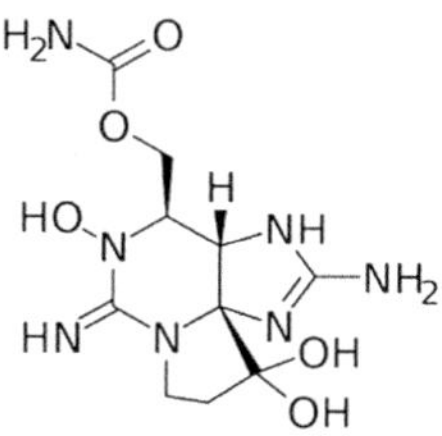

Neosaxitoxin

6.1.2.2.4 Tetrodotoxin (TTX)

This compound is considered the most lethal toxin found in the marine environment. At least 30 structural analogs, with different degrees of toxicity depending on their structure [16], have been described to date (Rossignoli et al., 2023).

Dr. Yoshizumi Tahara was the first to isolate this toxin from pufferfish in 1894 and identified it as TTX in 1909. TTX is considered the most lethal toxin found in the marine environment and it has an increasing concern worldwide. About 30 structural analogs of TTX have been isolated to date (Rossignoli et al., 2023). While this toxin was a threat only in Asian countries mainly from pufferfish (Tetroodontidae), the Pacific and Mediterranean are also facing similar threats due to seafood. Although the marine microbiota such as bacteria and microalgae are believed to be the main producers of TTX, evidence relating to the production of TTX by microalgae is very much limited and so far one study has only suggested the dinoflagellate *Alexandrium tamarense* produced this toxin (Biessy et al., 2019). TTX, for which there is no known antidote till today, has been reported to inhibit sodium channel thereby producing heart attack in many cases and subsequently death. In Japan, a regulatory limit of 2 mg eq TTX/kg in seafood including the delicious "fugu" derived from pufferfish (Lago et al., 2015).

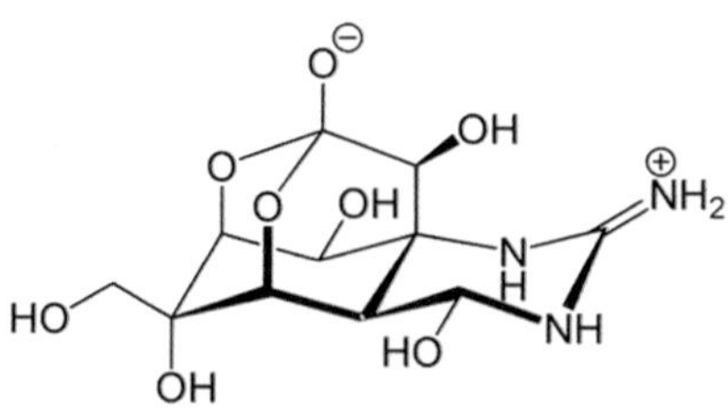

Tetrodotoxin

Medical and Therapeutic Applications of TTX

Like brevetoxins, TTX is also a powerful neurotoxin-based drug with analgesic and anesthetic effects which are due to its sodium channel-blocking properties. For example, a powerful drug from this potent toxin, viz., Tectin®, by Wex Pharmaceuticals in Canada, is currently undergoing phase III clinical trials as a pain controller in cancer patients. Further, this drug is also in Phase II clinical trials for assessing its efficacy against neuropathic pain. Further, another formulation of TTX, viz., TocudinTM, is under investigation for anesthesia, and preclinical testing is to be started soon (Asssumcao et al., 2017).

Anticancer activity: At concentrations of 0.32 μg and 319 μg/L, the TTX inhibited the invasiveness of metastatic prostate cancer with an IC50 value of 5.75 μg/L. Further, at 319 μg/L, the migration of the MAT-LyLu cell line was significantly reduced (Katikou et al., 2022).

TTX as a tumor suppressor: In human non-small cell lung carcinoma cell line (H460), TTX has been reported to reduce cell invasion by 50% without affecting non-invasive wild-type cells (Bucciarelli et al., 2021).

Neuroprotective effects: At concentrations of 6–31.9 μg/L, TTX has been reported to protect the neurons from veratridine-induced toxicity. It is therefore suggested that TTX could be used in the treatment of ischemic neuronal injury by preventing excessive neuronal depolarizations (Katikou et al., 2022).

Anticonvulsant effects: At low concentrations of TTX (1.6, 3.19, or 6.38 μg/L), anticonvulsant effects were noticed in rat hippocampal slices. TTX blocked EGSs (Katikou et al., 2022).

TTX as a pain reliever (Bucciarelli et al., 2021).

- **i) Cancer-related and neuropathic pain:** TTX injections for treating mild-to-severe cancer pain have shown promising results. Further, TTX may also be used as a neuropathic pain treatment in which there are no side effects such as sedation, inflammation, and respiratory depression.
- **ii) Allodynia and hyperalgesia:** TTX yielded a significant antinociceptive effect in the treatment of mechanical allodynia induced by sciatic nerve ligation. Application of TTX post-injury was found to prevent permanently the development of both mechanical allodynia and thermal hyperalgesia in two rat models of neuropathic pain, viz., spared nerve injury and chronic constriction injury.

iii) **Inflammatory pain:** Therapeutic use of TTX has been reported to relieve inflammatory pain.
iv) **Chronic pain:** TTX treatment has been reported to control the hyperexcitable state of the neurons thereby this toxin may be highly effective in the treatment of chronic pain.
v) **Visceral pain:** The subcutaneous injection of TTX at 1–6 μg/kg was found to prevent pain-associated behaviors in the mice models tested.

Use of TTX in brain trauma and spinal cord injury: TTX treatment after a traumatic brain injury prevented the development of post-traumatic epilepsy (Bucciarelli et al., 2021).

TTX as an anesthetic: TTX was found to be a potential anesthetic, especially in union with epinephrine or bupivacaine (Bucciarelli et al., 2021).

TTX prevents heroin and cocaine addiction: A single intramuscular injection of 10 μg TTX has been found to reduce significantly craving in abstinent recovering and cue-induced anxiety heroin addicts without significant effects on heart rate or blood pressure (Bucciarelli et al., 2021).

Antiviral activity: TTX has been reported to be an active compound against SARS-CoV-2 according to ligand-based approach. However, further studies are needed to confirm this possible antiviral activity (Katikou et al., 2022).

6.1.2.2.5 Yessotoxin (YTX)

In the 1960s, the first YTX analog, yessotoxin, was found in the bivalve scallop species, viz., *Patinopecten yessoensis*. Subsequently, numerous yessotoxin analogs were isolated from the marine microalgae, *viz.*, dinoflagellates and molluskan shellfish (https://en.wikipedia.org/wiki/Yessotoxin). About 90 yessotoxin analogs have so far been isolated, although structures for most of these analogs are yet to be determined (Miles et al., 2005). When the environmental conditions are favorable, there will be abundant growth of YTX-producing dinoflagellates, the toxin of which will find entry into the members of the food chain such as bivalve mollusks such as clams, mussels, and scallops. Among the dinoflagellates, *Protoceratium reticulatum, Lingulodinium polyedrum*, and *Gonyaulax spinifera* are known to produce YTX. It is also worth mentioning here that about 90 analogs of YTX have been recovered from the cultures of *Protoceratium reticulatum* (Assumcao et al., 2017).

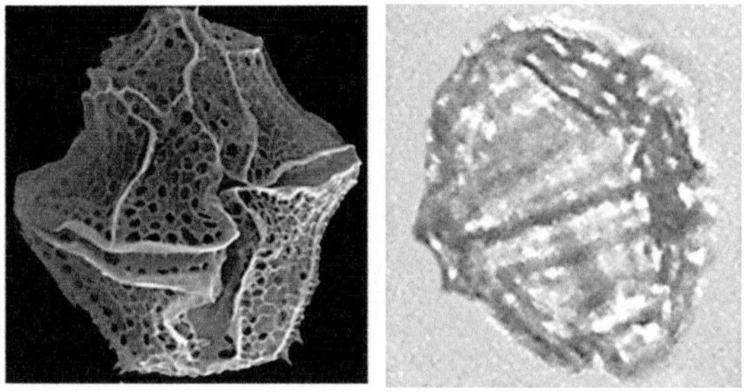

1 2

1: Lingulodinium polyedrum; 2: Gonyaulax spinifera

1. *Image credit:* Morseds. https://creativecommons.org/licenses/by-sa/4.0/
2. *Image credit:* Minami Himemiya. https://creativecommons.org/licenses/by-sa/3.0/

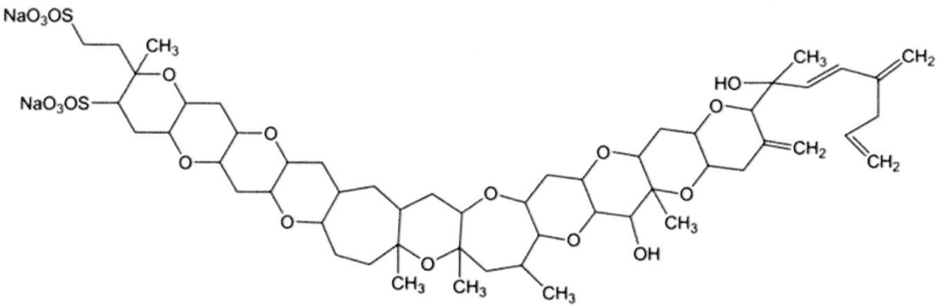

Yessotoxin

Therapeutic Uses of YTX

YTX has been reported to induce non-apoptotic cell death of the cells of BC3H1 muscle cell line; it serves as a potent phospodiesterase (PDE) activator. Phosphodiesterase inhibitors are used as medications in the treatment and management of chronic obstructive pulmonary disease; induce genetic alterations and mitotic catastrophe which control tumor progression; play a minor role as an asthmatic and anti-allergenic drug; and regulate the immune-effect on T-lymphocyte EL-4 cells. YTX and its analogs may be a potential therapy for Alzheimer's disease (AD), a progressive neurodegenerative disorder; help in the prevention/treatment of lipid and glucose metabolism-associated diseases; and possess anti-fungal activity against fungi and yeasts (Assumcao et al., 2017). Botelho et al. (2022) reported that yessotoxin may be of great use in the treatment of AD as it reduces the accumulation of amyloid (Aβ) plaque and hyper-phosphorylated tau tangles. Tobio et al. (2016) reported that the small molecular compound YTX caused a significant reduction in the ability of mast cells to degranulate. Further, according to them, YTX demonstrated selective cytotoxicity against RBL-2H3 mast cell line thereby inducing its apoptosis. Furthermore, YTX was also found to show potential cytotoxicity against human B-chronic lymphocytic leukemia cell line MEC1 and murine melanoma cell line B16F10 (Tobio et al., 2016).

6.1.2.2.6 Maitotoxin

MTXs are a class of large cyclic polyethers that function as calcium channel antagonists. Among the marine dinoflagellates, the only genus Gambierdiscus with its species is known to produce MTX and its analogs, viz., MTX-1 to MTX-7 as given in the following table.

TABLE 6.2 Gambierdiscus spp. and Their MTX Analogs.

Species	MTX analog (s)
Gambierdiscus pacificus	MTX-1, MTX-2, MTX-3
Gambierdiscus australes	MTX, MTX-1, MTX-2, MTX-3, MTX-5
Gambierdiscus belizeanus	MTX-3
Gambierdiscus toxicus	MTX-1, MTX-2, MTX-3
Gambierdiscus cheloniae	MTX-3, MTX-6
Gambierdiscus excentricus	MTX-4
Gambierdiscus honu	MTX-3, MTX-7
Gambierdiscus polynesianus	MTX-1, MTX-3

Source: Wang et al. (2022); Estevez et al. (2021); Murray et al., (2022).

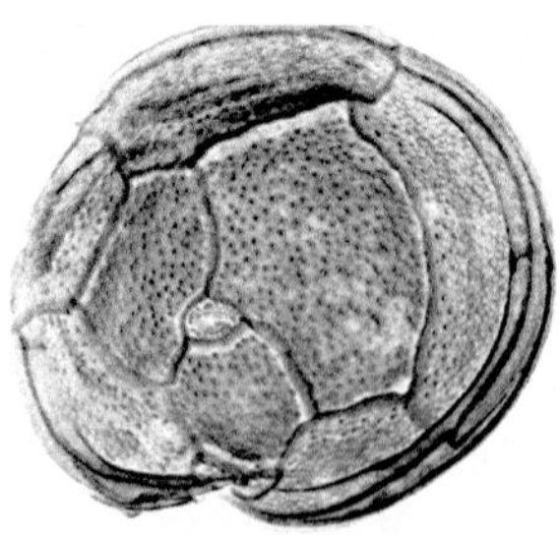

Gambierdiscus excentricus

Image credit: Fraga, Santiago. https://creativecommons.org/licenses/by-nc-sa/4.0/

Therapeutic Uses of Maitotoxins

i) MTX is believed to trigger insulin secretion, and program cell death of certain cancer cell lines besides playing a key role in innate immune responses (Assumcao et al., 2017).
ii) Reyes et al. (2014) also reported that MTX has shown insulinotropic activity in insulinoma cells by serving as a novel blood glucose-lowering agent.
iii) MTX is also believed to serve as an interleukin-1β secretagogue. Interleukin-1 beta (IL-1 beta) has been reported to reduce the severity of gastroduodenal damage and to inhibit acid secretion in the pylorus-ligated rat (Reyes et al., 2014).
iv) MTX is also known to serve as an oncotic death inducer (Reyes et al., 2014).

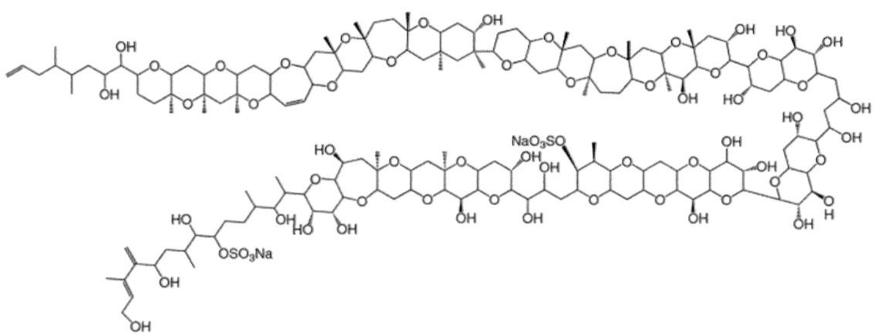

Maitotoxin

6.1.2.2.7 Palytoxin (PLTX) and Ostreocin

PLTX was first isolated from the cnidarian zoanthid species, *Palythoa tubercurosa*, in 1971, and this toxin and its analogs were subsequently reported from the dinoflagellates (genus Ostreopsis), red algae (*Chondria armata*), sea anemones (*Radianthus macrodactyylus*), marine worms (*Hermodice carunculata*) and from mackerel, crabs, or seaweed harvested from affected areas. PLTX is considered to be one of the most potent marine toxins capable of causing cardiac failure and death in human beings within minutes of its exposure (Watters, 2009). Further, PTX poisoning may also cause rhabdomyolysis, a syndrome affecting and subsequent muscle break-down. PLTX and its analogs, viz., homoPLTX, bishomoPLTX, neoPLTX, deoxyPLTX, and 42-hydroxyPLTX, were isolated from Palythoa species.

Moreover, PLTX and its analogs have also been derived from marine dinoflagellates such as ostreocin-d (42-hydroxy-3,26-didemethyl-19,44-dideoxy palytoxin) from *Ostreopsis siamensis*; mascarenotoxin-a, -b, and -c from *Ostreopsis mascarenensis*; Ostreotoxin 1 and 3 from *Ostreopsis. lenticularis*; and ovatoxin-a–f from both *Ostreopsis ovata* and *Ostreopsis fattorussoi* (Ramos and Vasconcelos, 2010; Accoroni, et al., 2016).

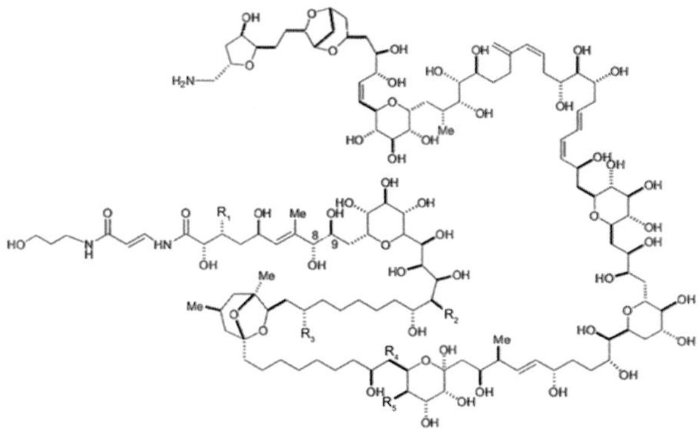

	R_1	R_2	R_3	R_4	R_5
Palytoxin	Me	OH	Me	H	OH
Ostreocin-D	H	H	H	OH	H

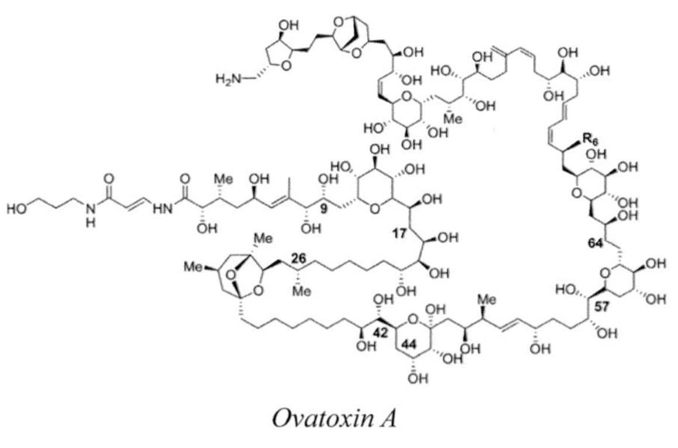

Ovatoxin A

Therapeutic Uses of Palytoxins

Anti-tumor/cytotoxic activity: The polyketide compound, Ostreol-B produced by the species of *Ostreopsis* has been reported to possess cytotoxic activity against HepG2, HCT-116, and Neuro2a cancer cell lines with IC50 values of 4.8, 0.1, and 0.9 μM, respectively. Polytoxins are therefore considered to be promising forms of anti-tumor therapeutics (Cousseau et al., 2022).

Antileukemic activity: The pharmaceutical formulation with PLTX derived from the endosymbiotic dinoflagellate's species of the cnidarian *Palythoa* sp., viz., *Symbiodinium* has been reported to be more suitable for therapeutic uses against myelogenous or lymphoblastic leukemia (Assunção et al., 2017).

6.1.2.2.8 Gambierol, Gambieric Acid, Gambierone, and Gambieroxide

The polyether compounds such as gambierol, gambieric acids A–D and gambieroxide. are produced by the dinoflagellate *Gambierdiscus toxicus*; gambierone and its new analogs, viz., 12,13-dihydro-44-methylgambierone, 38-dehydroxy-12,13-dihydro-44-methylgambierone, 38-dehydroxy-44-methylgambierone, and desulfo-hydroxyl gambierone by *Gambierdiscus balechii* and *Gambierdiscus cheloniae*; 44-Methylgambierone, a gambierone analogue by *Gambierdiscus australes, Gambierdiscus silvae, Gambierdiscus caribaeus* and *Gambierdiscus holmesii*; sulfo-gambierone and dihydro-sulfo-gambierone, analogs of gambierone by *Gambierdiscus excentricus*; and P-Gambierone analogue, putative gambieroxide by *Gambierdiscus australes* (Nagai et al., 1992; Wang et al., 2022; Yon et al., 2021; Murray et al., 2019; Liu et al., 2023).

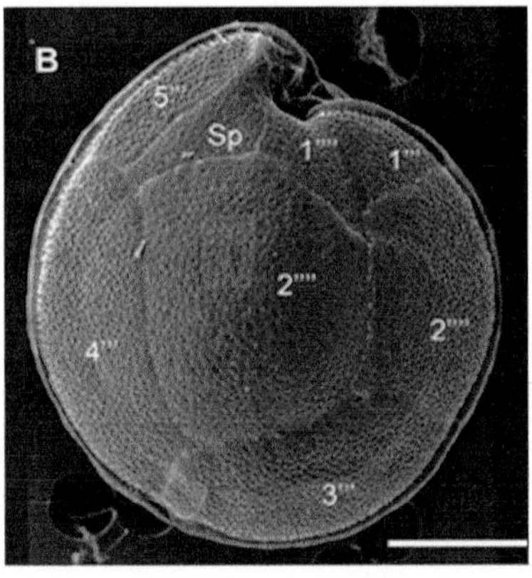

Gambierdiscus silvae

Image credit: Reproduced with permission from Dr. Santiago Fraga

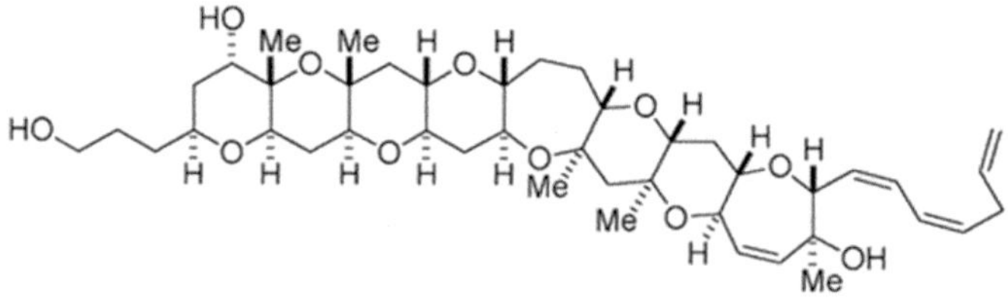

Gambierol

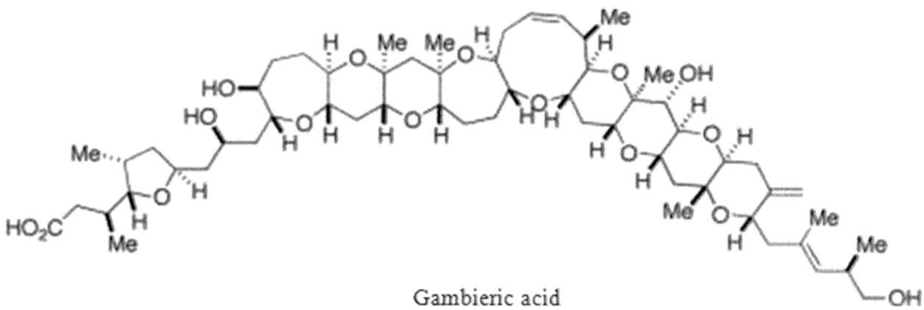

Gambieric acid

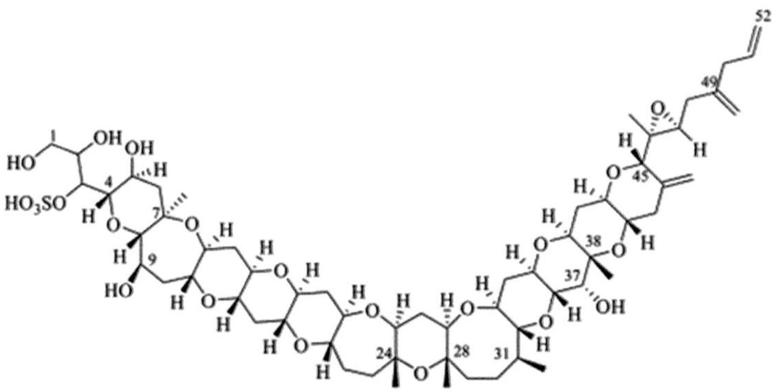

Gambieroxide

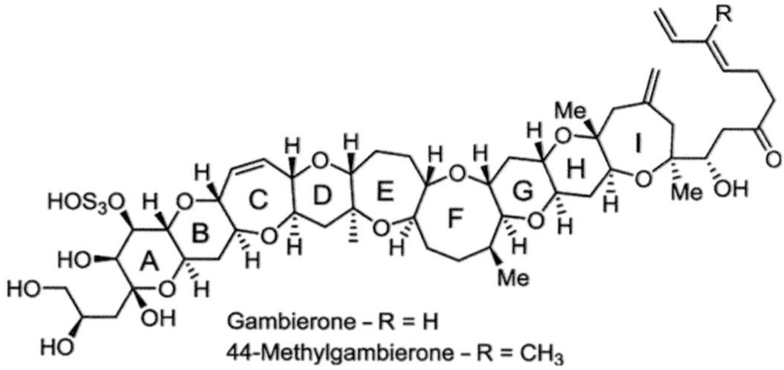

Gambierone (R=H) and 44-Methylgambierone (R=CH3).

Therapeutic Uses of Gambierol

Although the compound gambierol is toxic, its less toxic and synthetic analogs have been reported to be of great use as new drugs for immunotherapy. Further, due to its blocking ability of Kv3.1 channels, gambierol is suggested to be a very useful compound for application as an immunosuppressor in dysfunctional immune system-related diseases, such as diabetes mellitus type 1, multiple sclerosis, and rheumatoid arthritis. Gambierol and its analogs (tetra and heptacyclic forms) are also believed to be promising molecules for modulating Alzheimer's disease (AD) hallmarks in primary cortical neurons. Gambierol is also known to induce the outgrowth of neurites; thereby; it may be a promising drug for victims of neural injury (Assumcao et al., 2017). Botelho et al. (2022) reported that gambierol may be of great use in the treatment of AD as it inhibits voltage-dependent Na and K channels besides reducing the tau hyperphosphorylation and overexpression of B-amyloid.

Therapeutic Uses of Gambierone

Gambierone has been report to activate Voltage-gated sodium channels (VGSCs) which possess great potential as pharmacological targets for treating several disorders in the neuronal and cardiac systems (Luiz and Wood, 2016). Simmons (2007) reported that voltage-gated sodium channel agents are of great use as class I antiarrhythmics and local anesthetics.

Therapeutic Uses of 44-Methylgambierone

Both Gambierone and 44-Methylgambierone have been reported to induce a small rise in the cytosolic (fluid portion of the cytoplasm without organelles and membranes) calcium concentration in the human cortical neurons (Botana et al., 2020).

Therapeutic Uses of Gambieric Acids

Assumcao et al. (2017) reported that gambieric acids A, B, C, and D displayed potent antifungal activity against Aspergillus niger, and this activity is believed to be more than 2000-fold that of amphotericin B, which is a common antifungal drug (Assumcao et al., 2017; Nagai et al., 1992).

6.1.2.2.9 Spirolide and Gymonodimine

The cyclic imine toxin, spirolide (SPX), and its 14 SPX analogs are synthesized by the dinoflagellates *Alexandrium peruvianum, Alexandrium ostenfeldii,* and *Karenia selliformes*. Among the SPX analogs, 13-desmethyl SPX C is the most common one. The lipophilic toxins, Gymnodimine (GYM), and its three analogs, viz., GYM-A, B, and C are derived from the dinoflagellate, *Karenia selliformis (= Gymnodinium selliforme*); and the fourth anaolog, viz., GYM—12-methylgymnodimine has been derived from *Alexandrium ostenfeldii.* A new analog GYM-D has also been found more recently (Assumcao et al., 2017).

Therapeutic Uses of Spirolides (SPX)

Botelho et al. (2022) reported that the 13-desmethyl spirolide C derived from *Alexandrium ostenfelii* and *Alexandrium peruvianum* possesses neuroprotective activity and it may be of great use in treating Alzheimer's disease by reducing the tau hyperphosphorylation

Therapeutic Uses of Gymonodimines (GYM)

A mixture of gym and okadaic acid (OA) has been reported to increase the anti-cancer effects of chemotherapeutic agents. Further, GYM is known to sensitize cells to apoptotic stimuli in Neuro2a (neuroblastoma) cell line. Furthermore, GYM derived from *Karenia selliformis* reduces β-amyloid levels and tau phosphorylation, which may contribute substantially to the treatment of degenerative diseases (Botelho et al., 2022; Assumcao et al., 2017).

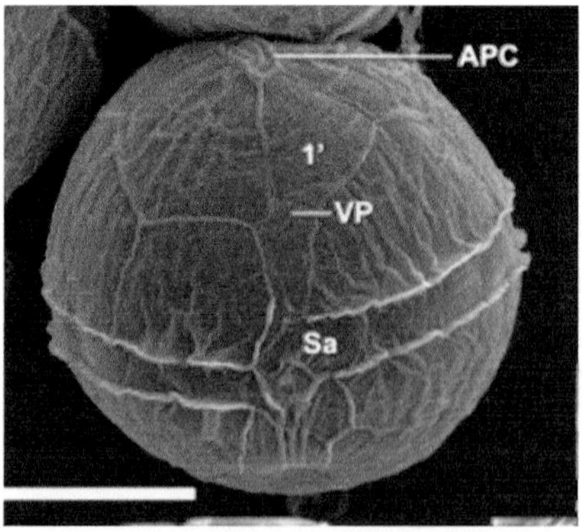

Alexandrium peruvianum

Image credit: Reproduced with permission from Dr. Margarida Costa

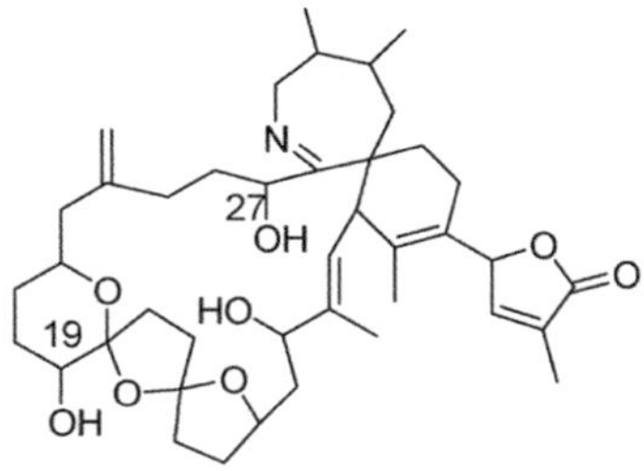

Spirolide.

A

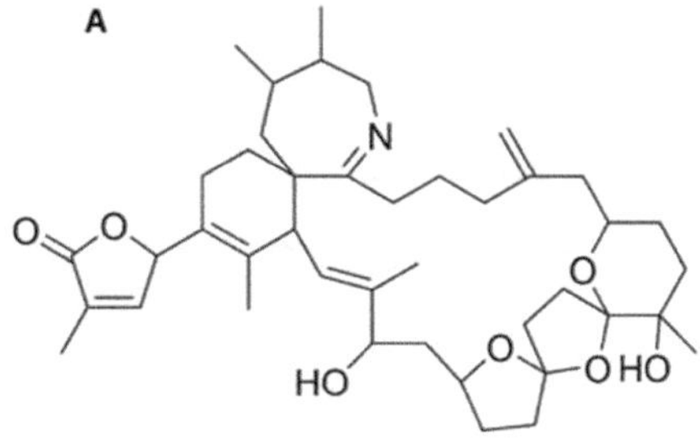

13-desmethyl spirolide C

A

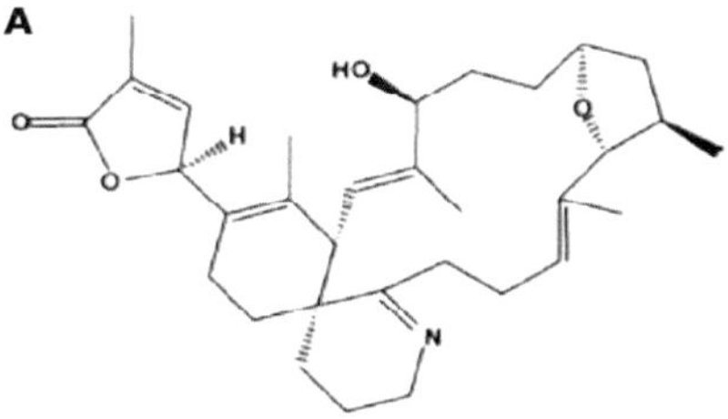

Gymnodimine A.

B

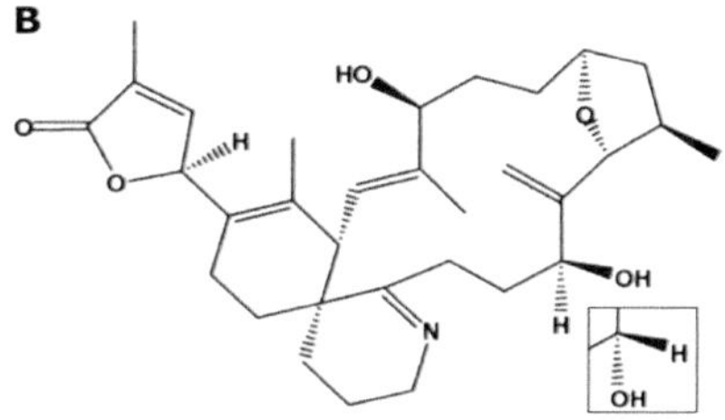

Gymnodimine B,C (inset).

6.1.2.3 CYTOTOXINS

6.1.2.3.1 Karlotoxin (KmTx)

The linear polyketide toxin, karlotoxin is produced by the dinoflagellate, *Karlodinium veneficum.*

Therapeutic Uses of Karlotoxin

KmTx has been reported to possess several bioactivities including cytotoxic and antifungal. As this toxin triggers the formation of pores in cholesterol-containing cell membranes, it has an amazing potential to treat a variety of human diseases including coronary heart disease and cancer. It is also worth mentioning here that in solid tumor lines such as breast and prostate cancer cells, much more lipid rafts are present and they are more sensitive to cholesterol depletion-induced cell death. Keeping these in view, KmTx could be formulated as a novel chemotherapeutic agent for cancer control (Assumcao et al., 2017).

Place et al. (2012) reported that KmTx has cytotoxic properties. According to them, this toxin can generate pores in membranes with desmethyl sterols and enhance the ionic permeability leading to membrane depolarization, osmotic cell swelling, and lysis.

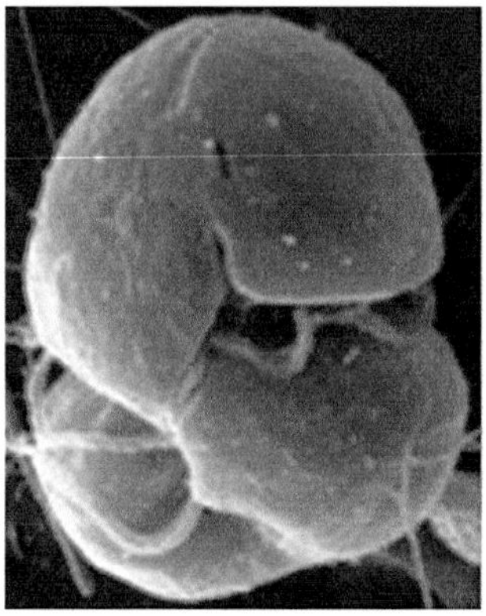

Karlodinium veneficum

Image credit: Reproduced with permission form Dr. Allen Place

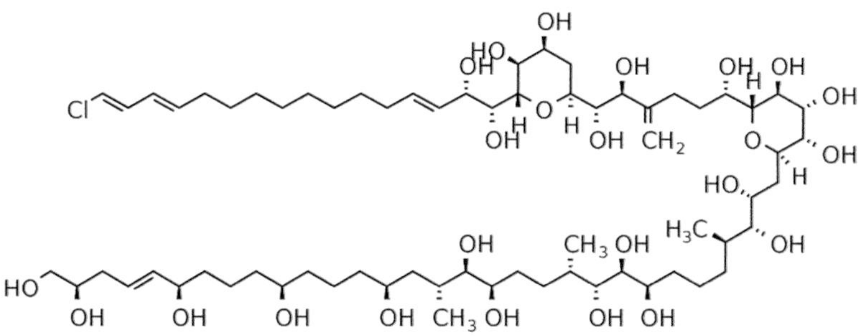

Karlotoxin

6.1.2.3.2 Goniodomins

The polyether macrolide toxins, viz., goniodomin A (GDA) and its analog goniodomin B (GDB) are produced by the dinoflagelllates, *viz., Alexandrium hiranoi, Alexandrium monilatum,* and *Alexandrium pseudogonyaulax.*

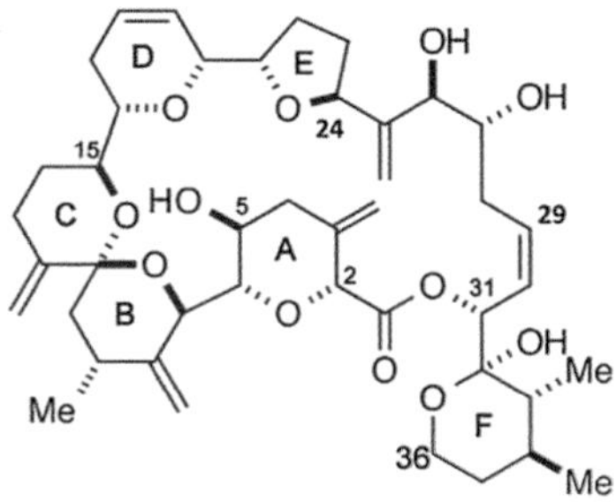

Goniodomin A.

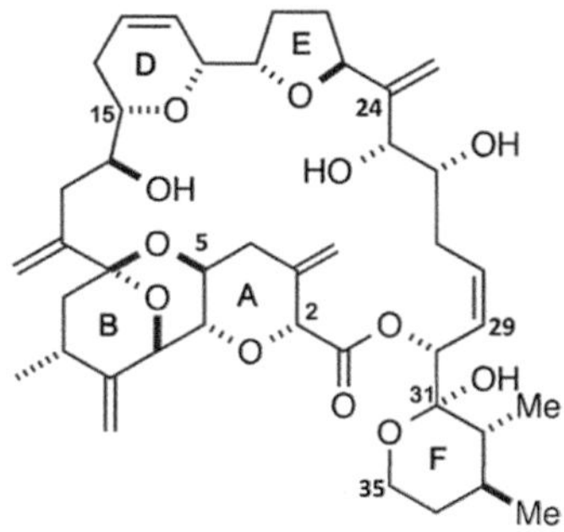

Goniodomin B

Therapeutic Uses of Goniodomins

GDA and GDB have been reported to possess antiangiogenic activities by inhibiting actin reorganization in endothelial cells. Further, these compounds possess cytotoxicity against BE(2)-M17 human neuroblastoma cells and antifungal activity against *Candida albicans* and *Mortierella ramannianus* (Assumcao et al., 2017).

6.1.2.3.3 Gymnocins (GYMs)

Gymnocin A (GYMA) is a complex and rare polyether toxin produced by *Karenia mikimotoi (= Gymnodinium mikimotoi)* (Satake et al., 2002). Gymnocin-B is an analog of GYMA

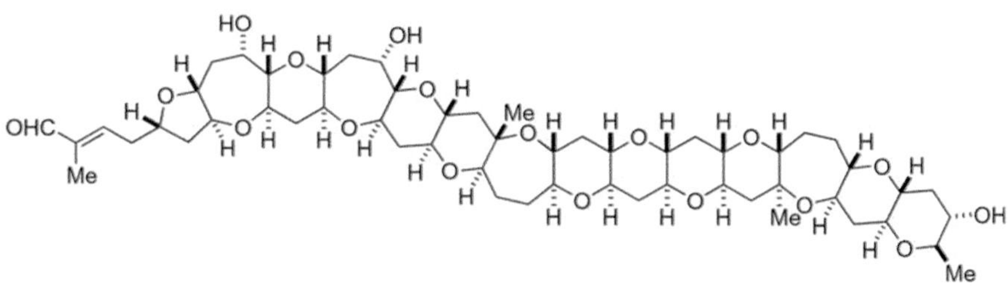

Gymnocin-A

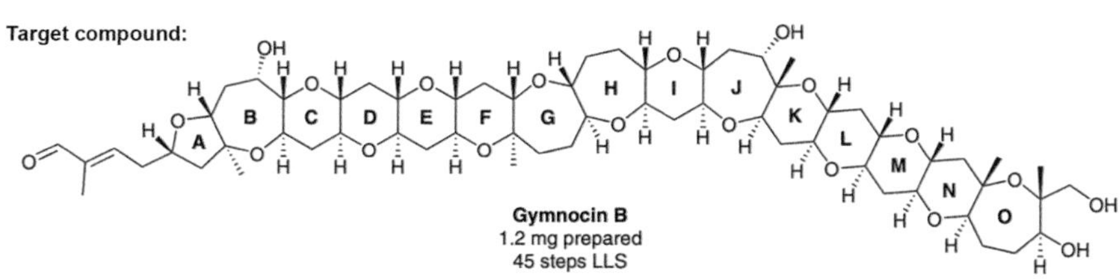

Gymnocin-B

Therapeutic Uses of Gymnosins

Both GYM-A and GYM-B have been reported to possess cytotoxicity against P388 mouse leukemia cells (Assumcao et al., 2017).

6.1.2.3.4 Amphidinolides (AMPs)

The macrolide compounds, Amphidinolides (AMPs), which are believed to have a mixed polyketide origin are produced by the dinoflagellates of *Amphidinium* spp. The various analogs of amphidinolide are amphidinolide A–H, N–P (Furstner et al., 2009; Valot, et al., 2015; Assumcao et al., 2017; Nuzzo et al., 2019).

Therapeutic Uses of Amphidinolides

Antineoplastic activity: The *Amphidinium* sp. derived from the Okinawan flatworm, *Ampiscolops* sp., has been reported to produce amphidinolide C (AMP-C) a 25-membered macrocyclic lactone. This AMP-C showed significant antineoplastic activity against L1210 murine leukemia cells with an IC50 value of 5.8 ng/mL (Kobayashi, et al., 1988).

Cytotoxicity: Furstner et al. (2009) reported that the amphidinolides of the B, D, G, H, and N types exhibited amazing cytotoxicity against various cancer cell lines.

Cytotoxcity and anti-tumor capacity: The AMP-G and AMP-H showed significant cytotoxicity against L1210 murine lymphoma, and KB human epidermoid carcinoma cells with IC50 values of 0.0054 and 0.00048 μg/mL; and 0.0059 and 0.00052 μg/mL, respectively (Martinez et al., 2019; Assumcao et al., 2017). These compounds are also expected to lead in the development of new anticancer drugs Kobayashi et al. (1991) stated that the AMP-G and AMP-H exhibited potent cytotoxicity against several cancer cell lines. Baur et al. (1995) reported that the macrolide caribenolide I isolated from the Caribbean dinoflagellate, *Amphidinium* sp. S1-36-5 demonstrated potent cytotoxicity against the human colon tumor cell line (HCT 116) with an IC50 value of 0.35 ug/mL.

6.1.2.3.5 Amphidinols (AMs)

The polyketide compounds, viz., Amphidinol (AM) and its analogs such as AM02, −03, −04, −05, −06, −07, −09, −11, −14, −15, −17, −18, and −22 are produced by the benthic dinoflagellates species of *Amphidinium* such as

Amphidinium carterae, Amphidinium klebsii, Amphidinium operculatum, Amphidinium eilatiensis, and *Amphidinium massartii* (Durán-Riveroll et al., 2023; Assumcao et al., 2017).

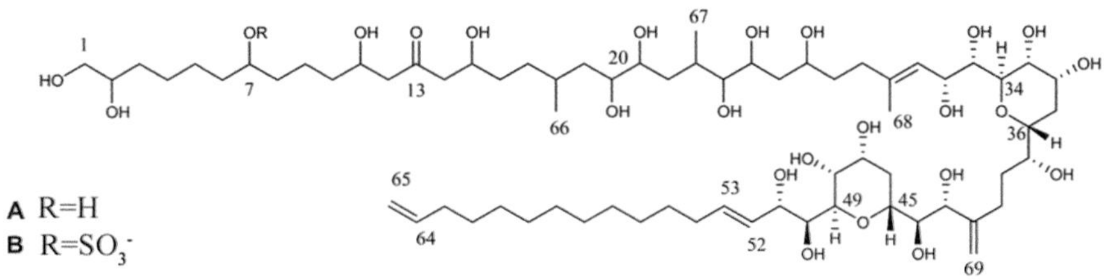

Amphidinolide A,B

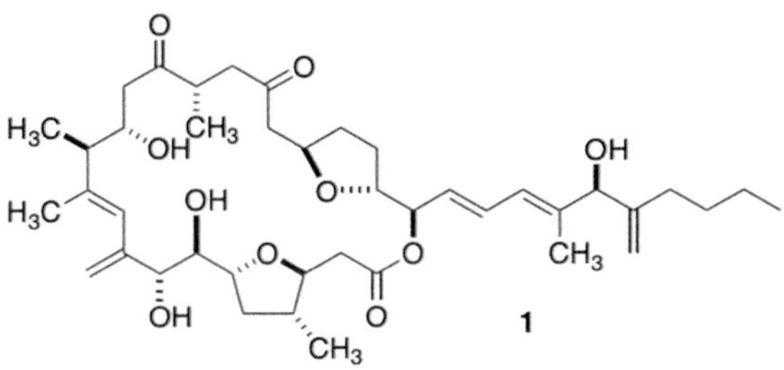

Amphidinolide C

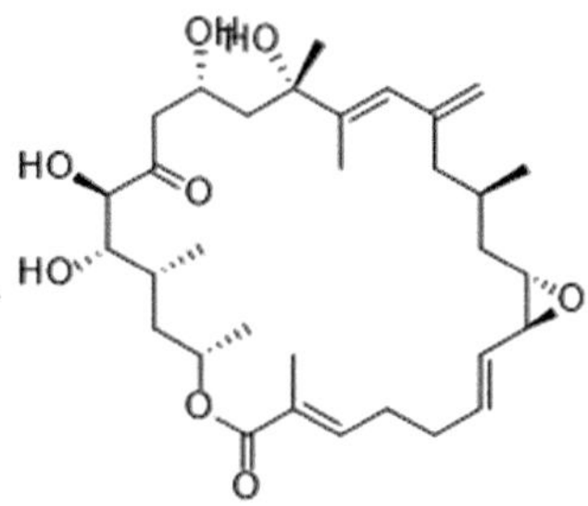

Amphidinolide D

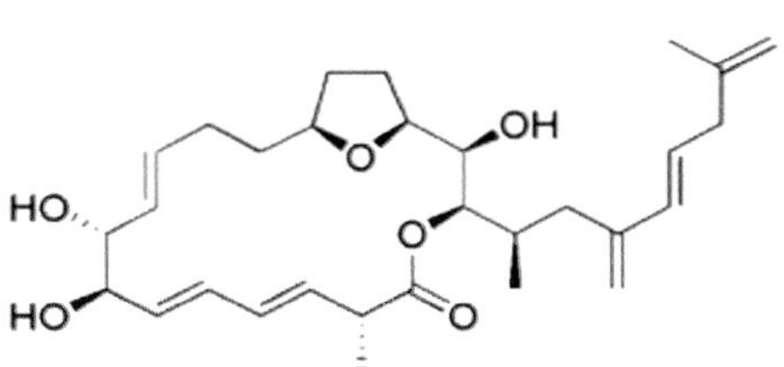

Amphidinolide E

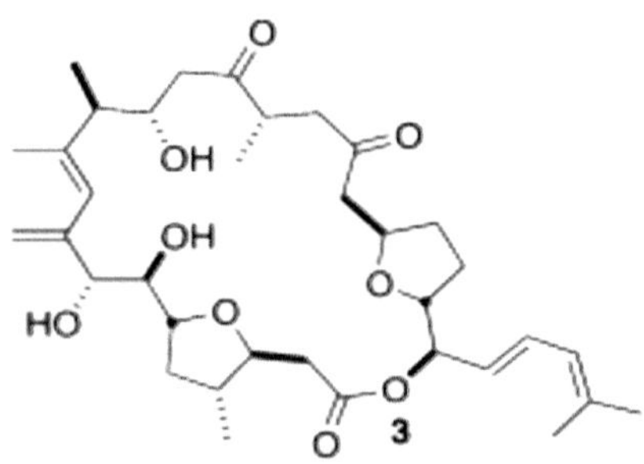

Amphidinolide F

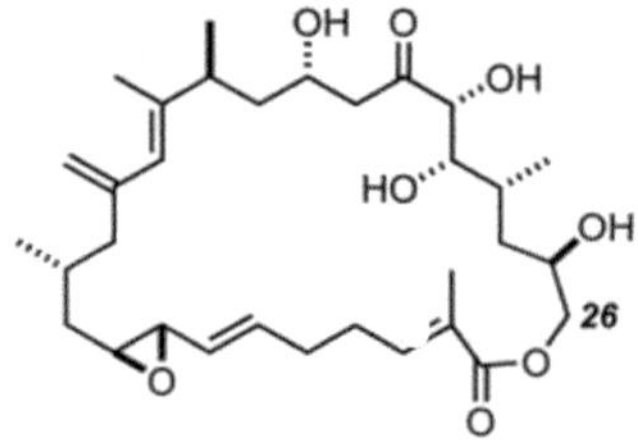

Amphidinolide G

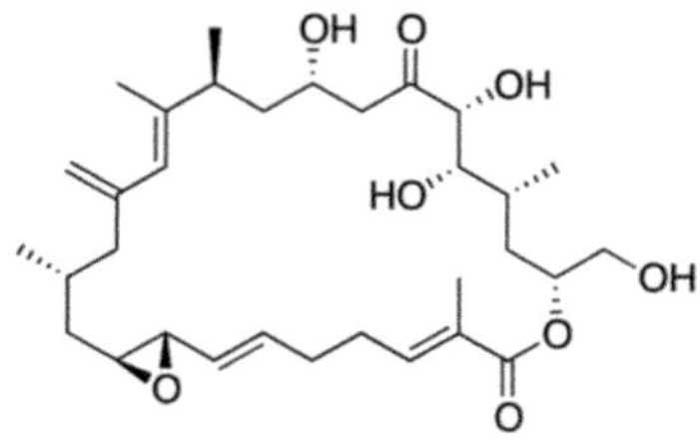

Amphidinolide H

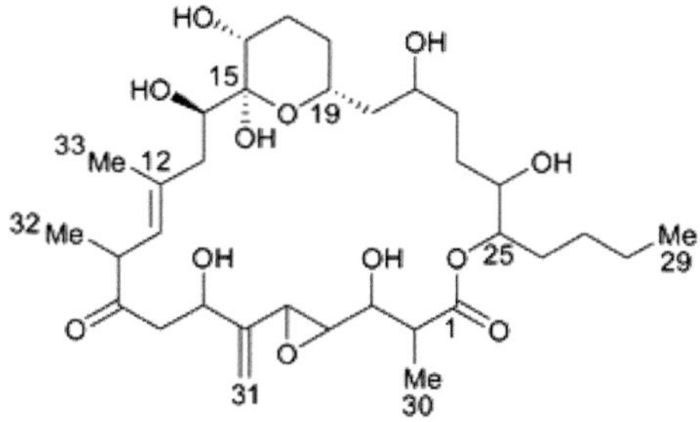

Amphidinolide N

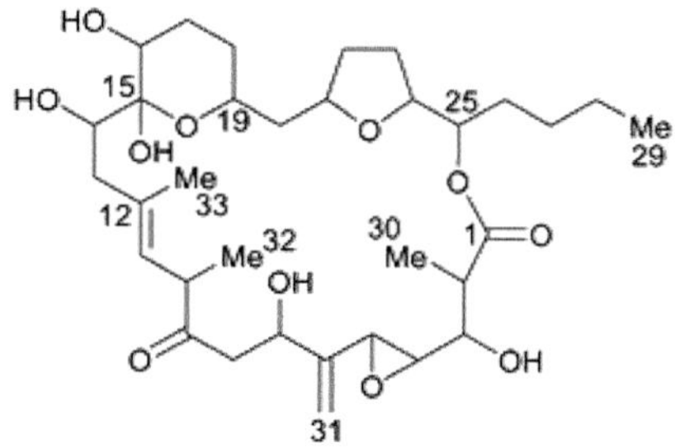

Caribenolide I

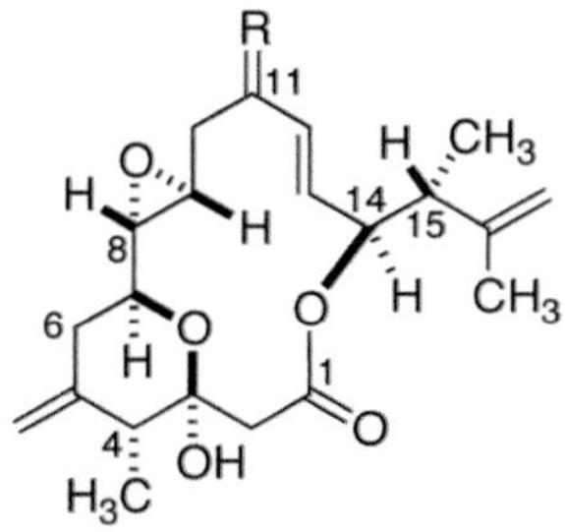

Amphidinolide O (R=O)

Amphidinolide P (R=CH2)

Therapeutic Uses of Amphidinols

Cytotoxicity and anticancer effects: The Amphidinol 22 produced by *Amphidinium carterae* has been reported to display cytotoxicity against A549, A2058, HepG2, MCF7, and MiaPaca-2 cancer cell lines with its IC50 values of 8, 16.4, 6.8, 16.8, and 8.6 μM, respectively (Martinez et al., 2019). Espiritu et al. (2017) reported that amphidinol 2 (AM2) derived from *Amphidinium klebsii* was found to show cytotoxicity against HCT-116, HT-29, and MCF-7 cancer cells with IC50 vlaues of 8.5, 1.2, and 4.1 μg/mL, respectively. Mejía-Camacho et al. (2021) reported that the methanolic crude extracts with unidentified toxins of *Amphidinium operculatum* were found to cause strong growth inhibition in HL-60 cancer cell line at a concentration of 50 μg/mL.

Antifungal activity: As Amphidinol 3 (AM3) derived from *Amphidinium klebsii* shows higher affinity to ergosterol membrane, this compound has provided insights for a new antifungal drug (Assumcao et al., 2017; Hieda et al., 2021). Morsy et al. (2005) stated that AM7 of *Amphidinium klebsii* has shown antifungal activity. Martinez et al. (2019) reported that AM22 displayed antifungal activity against *Candida albicans* and *Aspergillus fumigatus* strains with a Mic value of 64 μg/mL. According to Nuzzo et al. (2014), AM18 of *Amphidinium carterae* was significantly effective against Candida albicans with an MIC value of 9 μg/mL. Amphidinol C of *Amphidinium carterae* has been reported to display moderate fungicidal activity against filamentous fungi and yeast at 8–16 μg mL (Murphy et al., 2022).

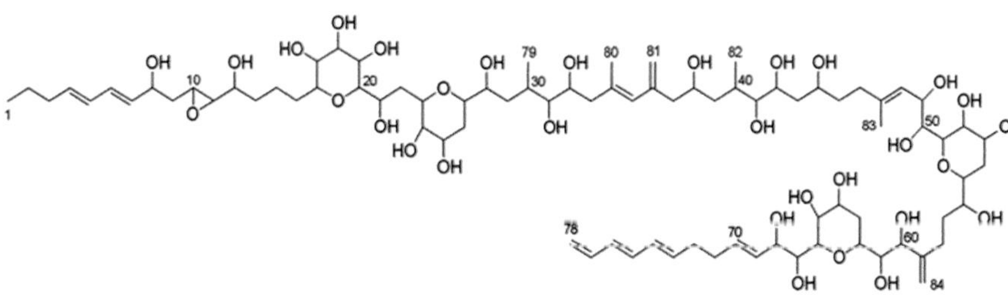

Amphidinol 22

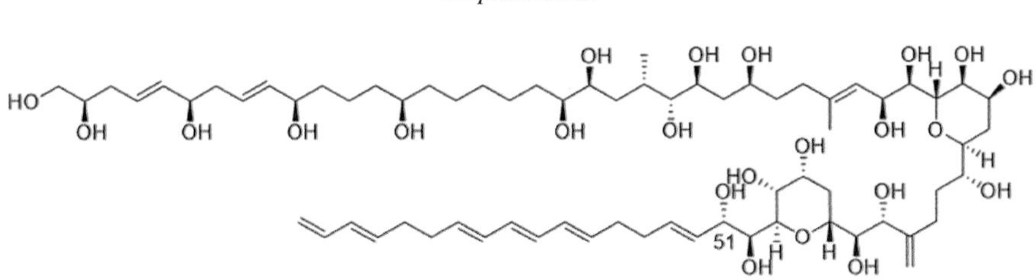

Amphidinol 3

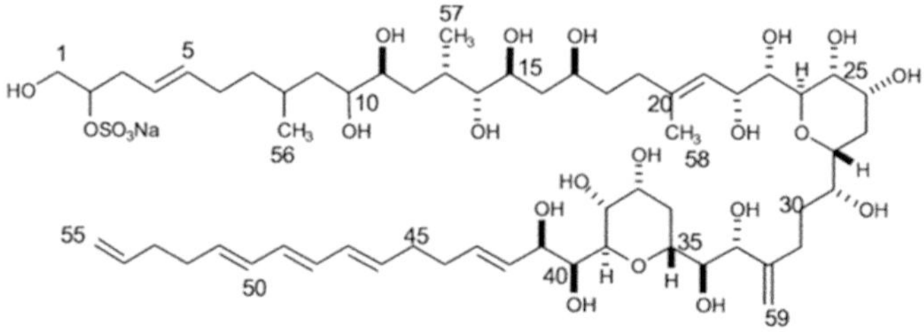

Amphidinol 7

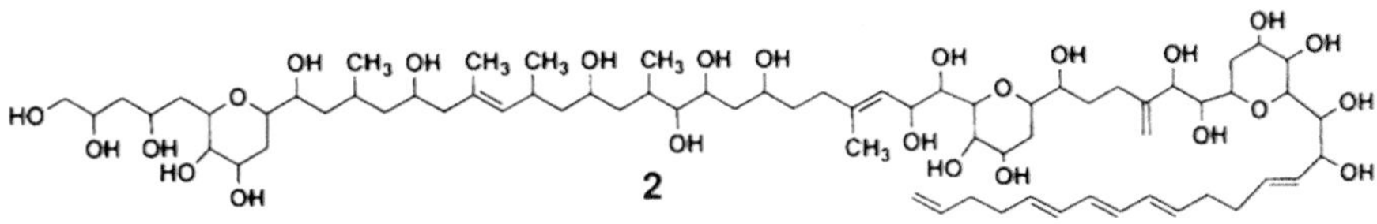

Amphidinol 2

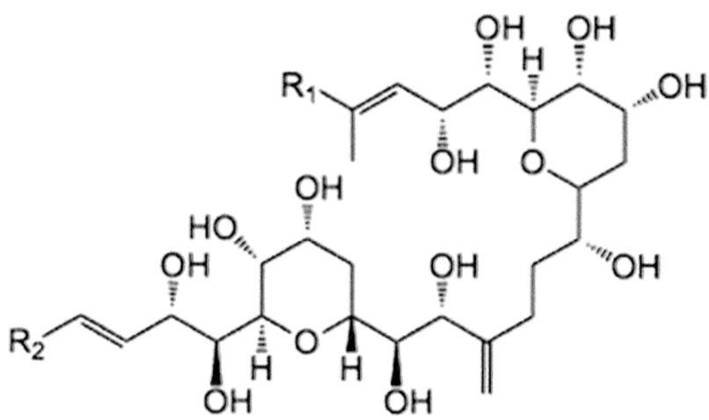

Amphidinol 18

6.1.2.3.6 *Amphirionins and Amphidinins*

Amphirionins (4 and 5) and Amphidinins (A–G) are amphidinol-related linear polyketides produced by the species of *Amphidinium.*

Thereapeutic Uses of Amphirionins and Amphidinins

Proliferating activity of murine bone marrow stromal ST-2 cells: Amphirionins, the linear polyketides, produced by the species of *Amphidinium* have been reported to be of great help in the regeneration of bone and cartilage through their potentiate proliferation of murine bone marrow stromal ST-2 cells and murine osteblastic MC3T3-E1 cells.

Amphirionin-4 displayed extremely significant (950%) proliferation-promoting activity on murine bone marrow stromal ST-2 cells at a concentration of 0.1 ng/mL (Minamida et al., 2014). Similarly, Amphirionin-5 showed 282% increase in the proliferation of murine bone-marrow-derived stromal ST-2 cells and 320% increase in the proliferation of murine osteoblastic MC3T3-E1 cells (Akakabe et al., 2014).

Cytotoxicity: Cousseau et al. (2022) reported that at a concentration of 0.5 mg/kg, Amphirionin-2 displayed cytotoxicity against Caco-2 and A549 cell lines with IC50 values of 0.1 and 0.6 μg/mL respectively.

Antimicrobial activity: Amphidinins A, C, D, E, F, and G have been reported to possess antibacterial and antifungal activity as given in the following table.

TABLE 6.3 Antimicrobial Activity.

Toxin	Antibacterial/antifungal activity	IC50/MIC (μg/mL)
Amphidinin A	Against *Bacillus subtilis* and *Aspergillus niger*	MIC of 16 and 4 μg/mL, respectively
Amphidinin C	Against *Staphylococcus aureus, Bacillus subtilis, Aspergillus niger,* and *Trichophyton mentagrophytes*	MIC of 32 μg/mL, MIC of 32 μg/mL, IC50 of 12 μg/mL, and IC50 of 16 μg/mL, respectively
Amphidinin D	Against *Trichophyton mentagrophytes*	IC50 of 16 μg/mL
Amphidinin E	Against *Staphylococcus aureus, Bacillus subtilis, Aspergillus niger,* and *Trichophyton mentagrophytes*	MIC of 32 μg/mL, MIC of 32 μg/mL, IC50 of 16 μg/mL, and IC50 of 16 μg/mL, respectively
Amphidinin F	Against *Trichophyton mentagrophytes*	IC50 of 16 μg/mL
Amphidinin G	Against *Trichophyton mentagrophytes*	IC50 of 8 μg/mL

Source: Cousseau et al. (2022).

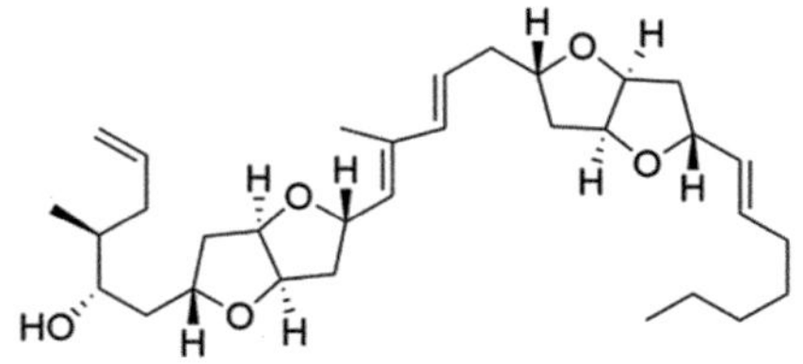

Amphirionin-2

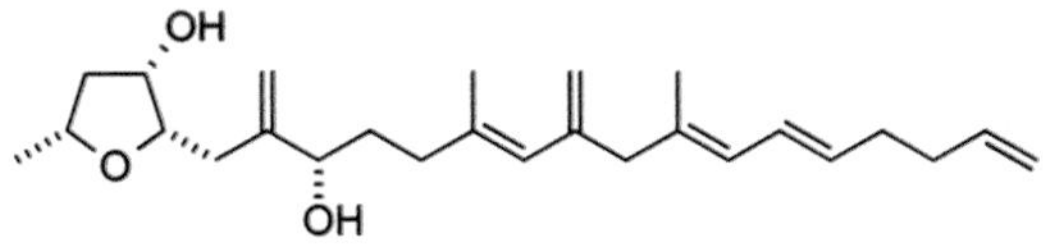

Amphirionin-4

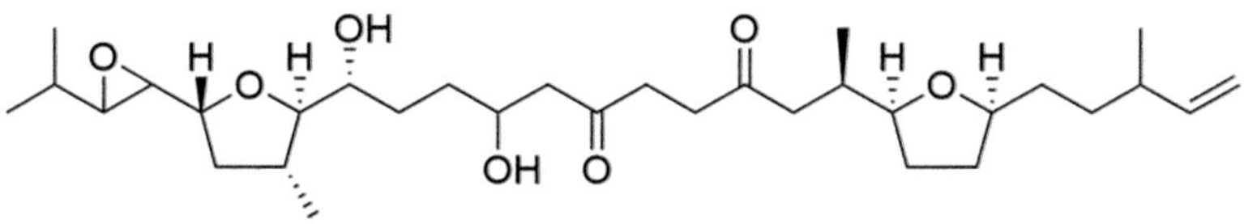

Amphirionin-5

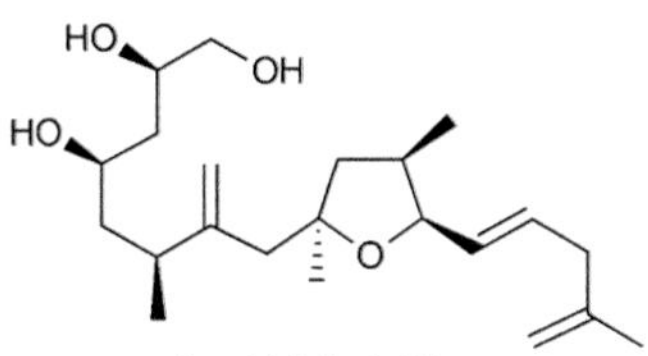

Amphidinin A (**1**)

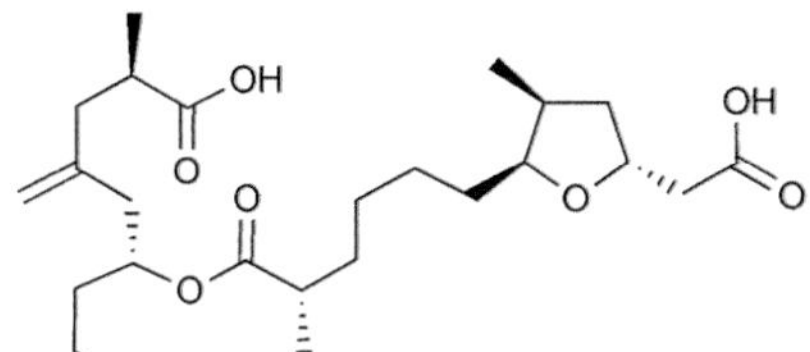

Amphidinin B (**2**)

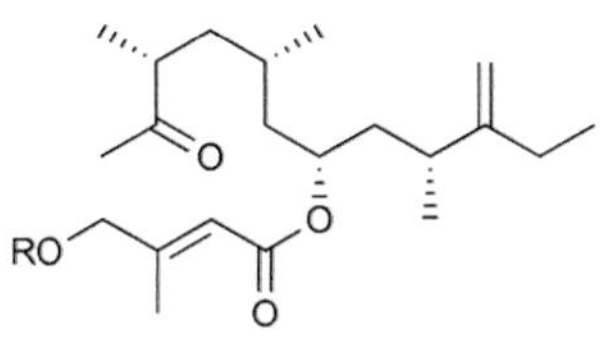

Amphidinin C (**3**) R = H

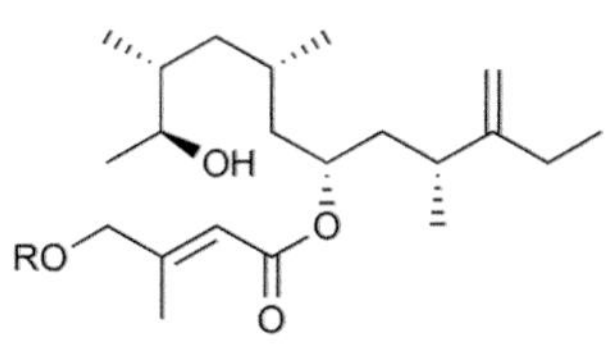

Amphidinin E (**5**) R = H

Amphidinin D (**4**) R =

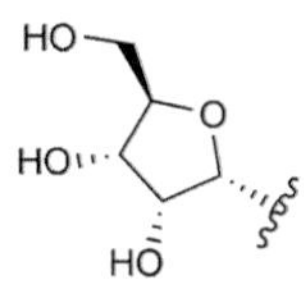

Amphidinin F (**6**) R =

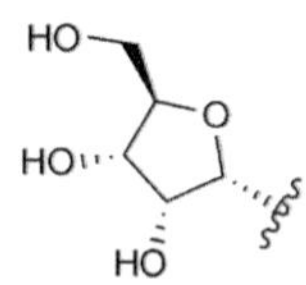

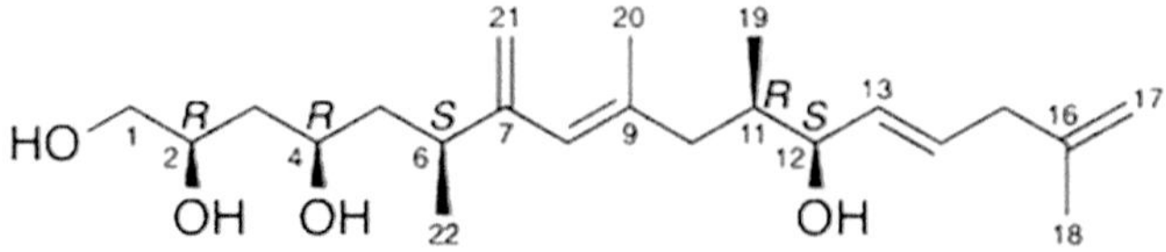

Amphidinin G

Patenting of Dinoflagellate Toxins

Certain toxins produced by marine dinoflagellates have been patented and the details are given in the following table.

TABLE 6.4 Patents Relating to Dinoflagellate Toxins and Their Potential Therapeutic Values.

Toxin patented	Therapeutic value	Year of patent
Tetrodotoxin, saxitoxin (analogs and derivatives)	Sodium channel blockers for treatment of neuropathic pain	2017
10′,11′-modified saxitoxins	For treatment of pain	2017
Gonyautoxin	Sodium channel blocker for treatment of loss of superficial sensitivity	2016
Brevetoxin, maitotoxin, yessotoxin, gambierol	Ladder-frame polyether conjugates	2016
Saxitoxin	For treatment of loss of sense of touch	2014
Neosaxitoxin	Formulations for prolonged local anesthesia	2014
Palytoxin	For its isolation for medical use	2014
Yessotoxin (analogs and derivatives)	For treating and/or preventing neurodegenerative diseases	2013
Yessotoxin and derivatives	For treatment of gliomas	2013
Gambierol	For treating and/or preventing neurodegenerative diseases	2012
Gymnodimine (analogs and derivatives)	For treating and/or preventing neurodegenerative diseases	2012
Yessotoxin and derivatives	For treatment and/or prevention of metabolic diseases	2012
Saxitoxin, gonyautoxin, and other analogues	For studying, imaging, and treating pain	2011
Yessotoxin	Human tumor cell growth inhibitor	2011
Tetrodotoxin, saxitoxin	Sodium ion channel blocker in treating biology drug resistance of antibiotic	2009
Brevetoxins and derivatives	For treating cystic fibrosis, mucociliary dysfunction, and pulmonary diseases	2008
Tetrodotoxin, saxitoxin (analogs and derivatives)	Sodium channel blockers for treatment of preterm labor	2007
Brevetoxin and derivatives	For treating neurotoxic shellfish poisoning and ciguatera fish poisoning	2005
Karlotoxin	Methods of isolation and uses	2005
Gonyautoxin	For treating wounds	2005
Yessotoxin	For treating allergic and asthmatic processes	2003

Source: Assunção et al. (2017).

Potential Therapeutic Applications of Marine Phytoplankton

The potential therapeutic applications of phycotoxins of marine phytoplankton such as diatoms and dinoflagellates have been reported by Pradhan and Ki (2022) and are given in the following table.

TABLE 6.5 Anticancer Activity of Marine Phytotoxins.

Phycotoxin	Applications
Okadaic acid (OA)	Cancer cell lines relating to breast, lungs, hepatic, intestinal, blood, brain, and leukemia
Azaspiracid (AZA)	T-lymphocyte cell line
Caribenolide I	Human colon tumor cell lines (HCT 116 and HCT 116/VM 46)
Gymnocin-A (GYMA)	P388 murine leukemia
Gymnocin (GYM)	Neuroblastoma cell line
GYM and OA	Many cancer cell lines
Amphidinolides and colopsinols	human epidermoid carcinoma KB and Murine lymphoma L1210
Palytoxin (PLTX)	Myelogenous leukemia or lymphoblastic cell lines
Palytoxin (PLTX) and Ostreocin-D	Neuroblastoma and intestinal cell lines
Protoceratin I-IV	Human colon cancer cell lines
Pectenotoxin (PTX)	Breast, lung, and colon cancer cells
Yessotoxins (YTXs)	Epithelial cancer cells
YTX and its analogues	Melanoma tumor cells
Brevetoxin (BTX)	Jurkat E6-1 cell lines
Ciguatoxin (CTX)	Gastrointestinal cell lines
Karlotoxin (KmTx)	Breast and prostate cancer cell lines

Source: Pradhan and Ki (2022).

6.1.2.3.7 Iriomoteolides

The iriomoteolides are amphidinol -related macrolide toxins isolated from the benthic dinoflagellate species of Amphidinium.

Therapeutic Uses of Iriomoteolides

Cytotoxic activity: The iriomoteolides have been reported to possess cytotoxic activity against HeLa human cervix adenocarcinoma cells and DG-75 human lymphoblastoid cell line (Assumcao et al., 2017). The IC50 values recorded for the different iriomoteolides are given in the following table.

Other Bioactivities of Marine Phytotoxins.

Phycotoxin	Disease	Mode of action/applications
Okadaic acid (OA) Gambierol	Diabetes	–
Okadaic acid (OA)	AIDS	–
Okadaic acid (OA); YTX and its analogues	Alzheimer	Inhibits t- and β-amyloid
Gambierol, GYM, spirolides, Gonyautoxins (GTX), GTX2, GTX3 and TTX	Pain	–
Okadaic acid (OA)	Schizophrenia	–
YTX and its analogues	Allergy and Asthma	–
BTX-2	Brain disorder	–
Symbioimine	Osteoporosis	Postmenopausal women
Symbioimine	Inflammation	Treatment of cyclooxygenase-2-related disorders
Gambierol	Brain injury, autoimmune disorders,	–
Karlotoxin (KmTx)	Coronary heart disease	–
Saxitoxin (STXs)	Bacterial, fungal, and protozoal diseases	–
Okadaic acid (OA) Karlotoxin (KmTx)	Fungal disease	Inhibit growth of Candida albicans

Source: Pradhan and Ki (2022).

TABLE 6.6 Cytotoxicity of Iriomoteolides on Human Cancer Cell Lines.

Toxin	Cell line	IC50 value
Iriomoteolide-1a	DG-75	0.004 μg/mL
Iriomoteolide-1b	DG-75	0.9 μg/mL
Iriomoteolide-1c	DG-75	0.004 μg/mL
Iriomoteolide-2a	DG-75 & HeLa	0.006 μg/mL and.03 μM, respectively
Iriomoteolide-3a	DG-75	0.08 μg/mL
Iriomoteolide-3b	DG-75	0.02 μg/mL
Iriomoteolide-4a	DG-75	0.8 μg/mL
Iriomoteolide-5a	DG-75	1 μg/mL
Iriomoteolide-9a	HeLa	15 μM
Iriomoteolide-10a	DG-75, HeLa & MH134	1.5, 1.5, and 3.3 μM, respectively
Iriomoteolide-11a	DG-75 & HeLa	1.5 and 2 μM, respectively
Iriomoteolide-12a	DG-75	50μM

Source: Cousseau et al. (2022).

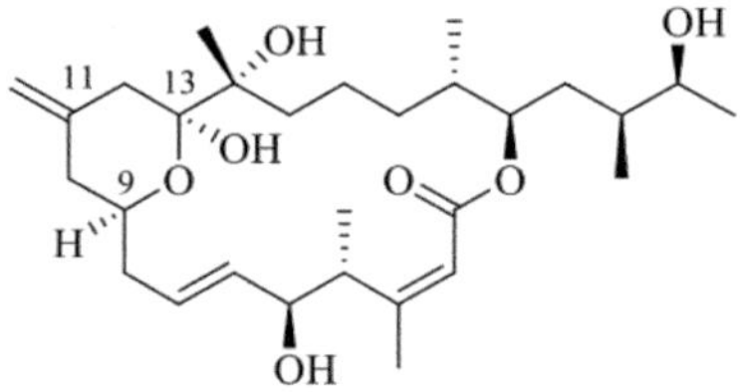

Iriomoteolide -1a.

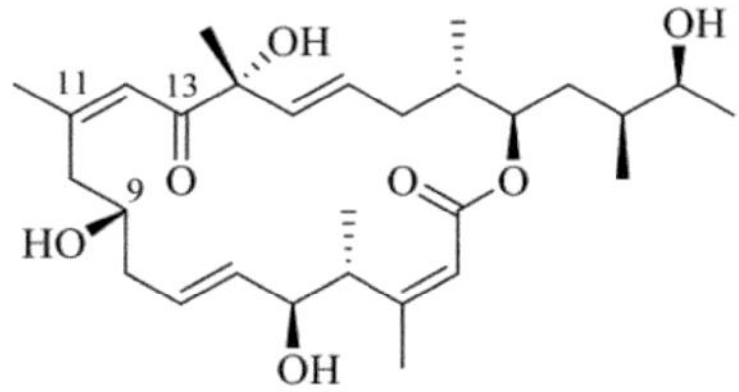

Iriomoteolide-1b

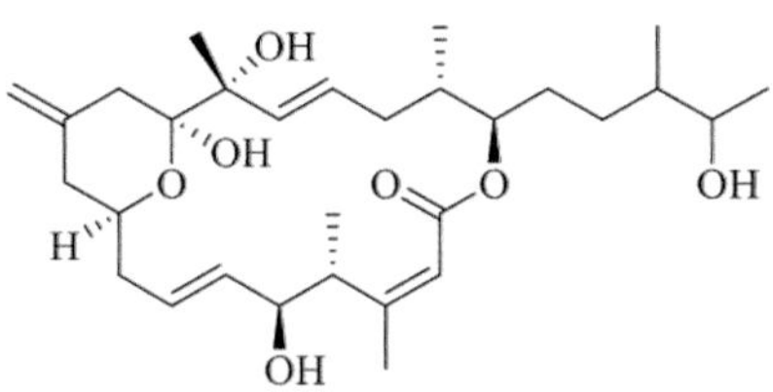

Iriomoteolide.-1c

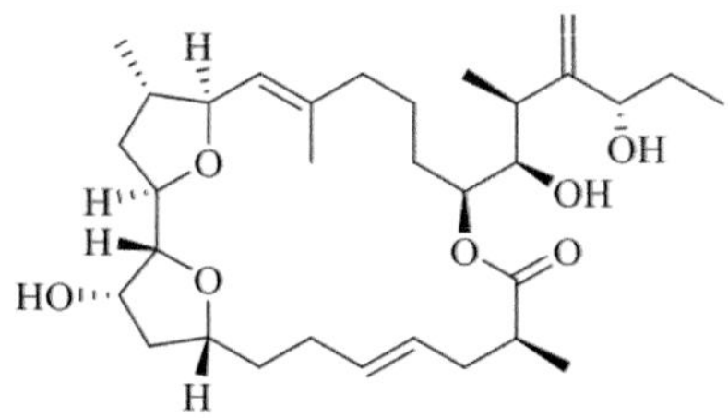

Iriomoteolide-2a

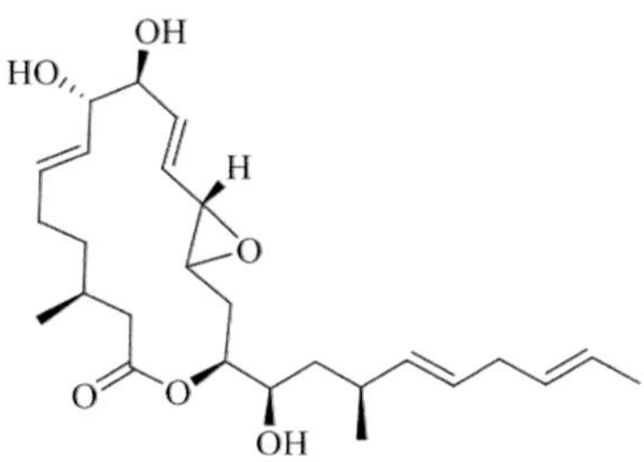

Iriomoteolide-3a.

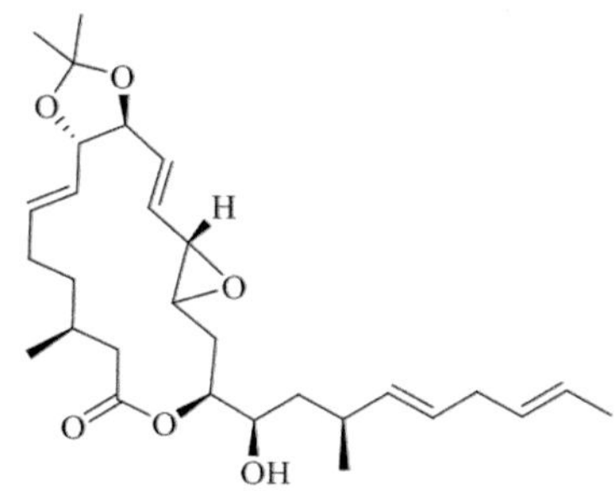

Iriomoteolide-3b

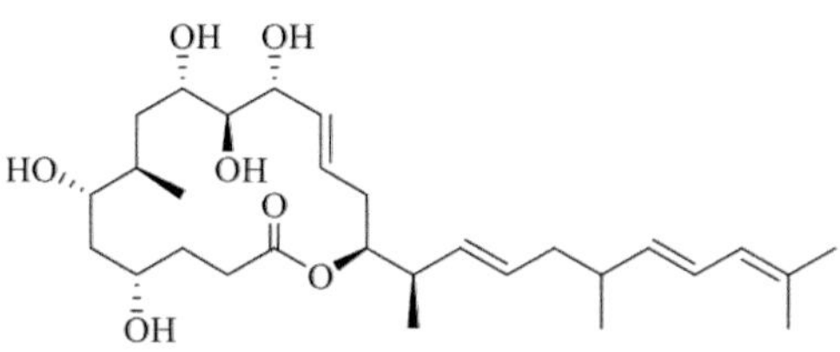

Iriomoteolide-4a

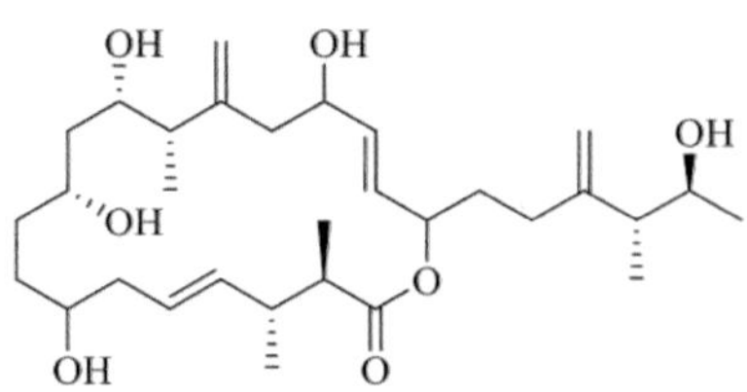

Iriomoteolide-5a.

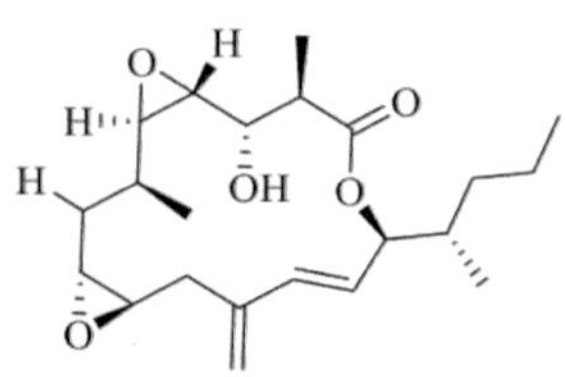

Iriomoteolide-9a

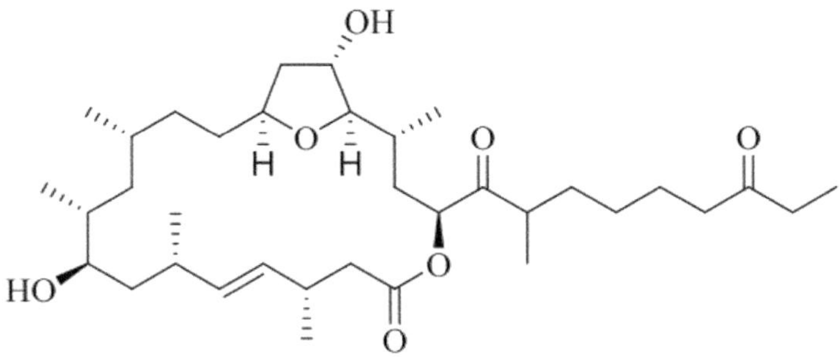

Iriomoteolide-10a

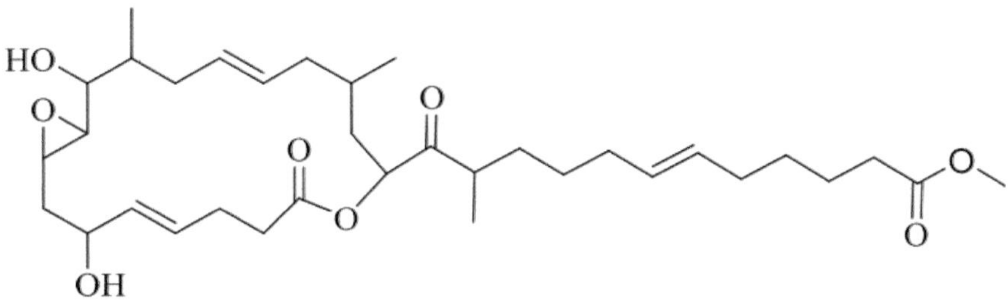

Iriomoteolide-11a

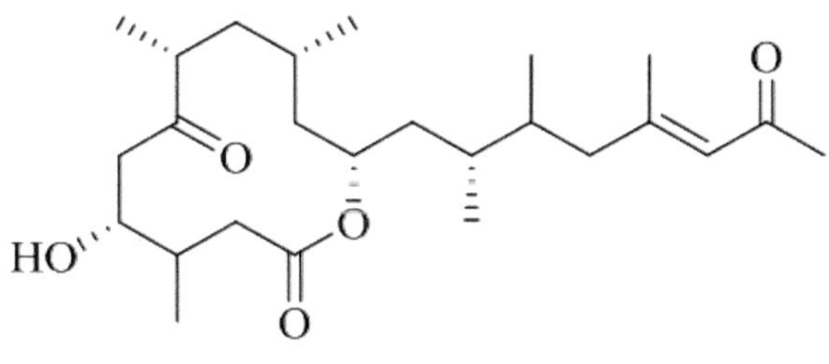

Iriomoteolide-12a

6.1.3 CYANOTOXINS OF MARINE BLUE-GREEN ALGAE (CYANOBACTERIA)

Blue-green algae (Cyanobacteria) and their blooms are normally seen in lakes, ponds, rivers, and brackish waters throughout the world. Strictly speaking, the occurrence of cyanobacterial species is rare in coastal waters though a few species are found in barackish and estuarine waters. These cyanobacteria produce cyanotoxins which are normally classified into four classes according to their toxicological target, viz., cytotoxins (hierridin, apratoxin, and lyngbouilloside) (*Cyanobium, Lyngbya,* and *Moorea*); neurotoxins (anatoxins and kalkitoxin) (microcystis, anabaena and lyngbya); hepatotoxins (microcystins and nodularins) (*Microcystis* and *Nodularia*); and dermatotoxins (lyngbyatoxins, aplysiatoxin and lipo-polysaccharide) (*Lyngbya, Leibleina, Moorea, Oscillatoria, Schizothrix,* and *Trichodesmium*) (Tang et al., 2019). Singh et al. (2011) reported that these marine cyanotoxins possess several bioactivities including antibacterial, anti-viral, cytotoxic, anti-protozoal, protease inhibitor, ca ion channel inhibitor, and so on.

6.1.3.1 CYTOTOXINS

The cytotoxin-producing marine cyanobacterial species and their therapeutic uses are given below.

***Cyanobium* sp. LEGE 06113**

Cytotoxicity: The polyketide toxin derived from this strain, viz., hierridin B has been reported to display selective cytotoxicity against human colon cancer cell line HT-29 with an IC50 value of 0.1 mM(Bajpai et al., 2018).

Lyngbya majuscula

Cytotoxicity: The apratoxin A of this species has been reported to demonstrate cytotoxic activity against KB oral epidermoid and LoVo colon cancer cell lines; and against U2OS osteosarcoma, HT29 colon adenocarcinoma and HeLa cervical carcinoma cell lines. Further, its apratoxin D exhibited similar activity against H-460 lung cancer cell line with an IC50 value of 2.6 nM (Costa et al., 2012). Furthermore, its small amides such as malyngamide C (8-O-acetyl-8-epi-malyngamide C) and 8-epi-malyngamide C were found to be cytotoxic against HT29 colon cancer cells with IC50 values of 5 and 15 μM, respectively (Bajpai et al., 2018).

Lyngbya sordida

Cytotoxicity: The cyclic depsipeptide, apratoxin D of this species has shown cytotoxicity against H-460 human lung cancer cells with an IC50 value of 2.6 nM (Costa et al., 2012).

***Lyngbya* sp.1**

Cytotoxicity: Apratoxin B and C derived from this unidentified species of *Lyngbya* exhibited cytotoxicity against KB oral epidermoid cancer and LoVo colon cancer cell lines (Costa et al., 2012).

***Lyngbya* sp.2**

Antitumor properties: A hybrid synthetic molecule combining both apratoxins A and E showed antitumor properties by downregulating receptor tyrosine kinases and VEGF-A in a colorectal tumor xenograft model (Qamar, et al., 2021).

***Lyngbya* sp.3**

Inducing autophagy: A synthetic analog of apratoxin A, viz., oxazoline has been reported to inhibit the function of Hsp90 (heat shock protein) and promote autophagy (Qamar, et al., 2021).

***Lyngbya* sp.4**

Antitumor effects: An apratoxin A analog, apratoxin S10 (Apra S10), has shown potential antitumor effects in the pancreatic cancer model by downregulating multiple receptor tyrosine kinases and inhibiting the secretion of cytokine (Qamar, et al., 2021).

Moorea bouilloni *(= Lyngbya bouilloni)*

Cytotoxicity and anticancer activity: A glycosidic macrolide toxin, viz., lyngbouilloside of this species was found to be moderately cytotoxic to neuroblastoma cells with an IC50 value of 17 μM (Cheney, 2016; Tan et al., 2002). Bajpai et al. (2018) reported that the depsipeptide toxin of this species, viz., apratoxin A has shown anticancer activity and as a clinical candidate, it is undergoing preclinical trials. Huang et al. (2016) reported that its apratoxin A showed dose-dependent anticancer activities in two human BxPC3 T1 pancreatic cancer and A549 lung cancer models. A dehydrated analog of apratoxin A, viz., apratoxin E of this species displayed significant cytotoxicity activity against HT29, HeLa, and U2OS cancer cell lines with IC50 values of 21, 72, and 59 nM, respectively. Further, its apratoxins F and G have also been reported to exhibit similar activity against H-460 cancer cell line with IC50 values of 2 and 14 nM, respectively (Costa et al., 2012).

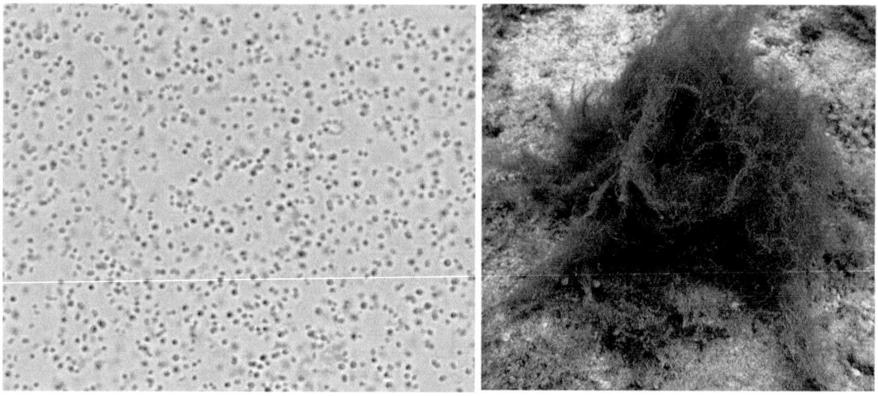

1 2

1: Cynobium sp.; *2. Lyngbya majuscula*

1. Own figure

2. *Image credit*: Philippe Bourjon. https://creativecommons.org/licenses/by-sa/4.0/

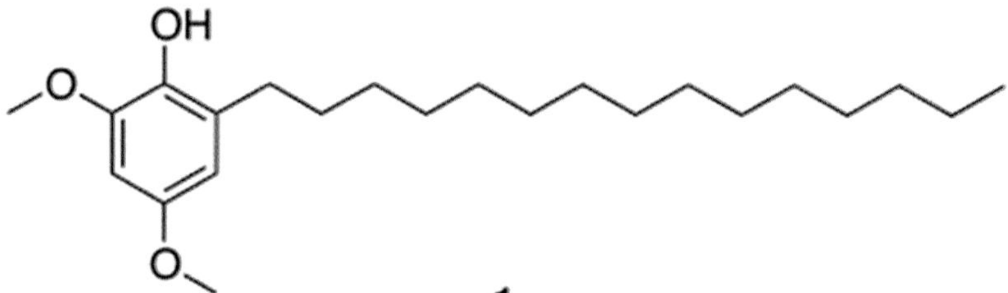

1

Hierridin B

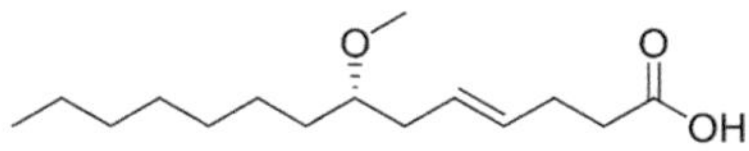

Lyngbic acid

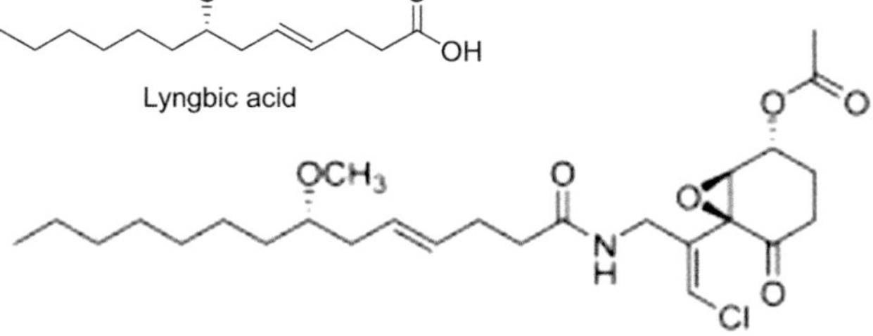

Malyngamide C

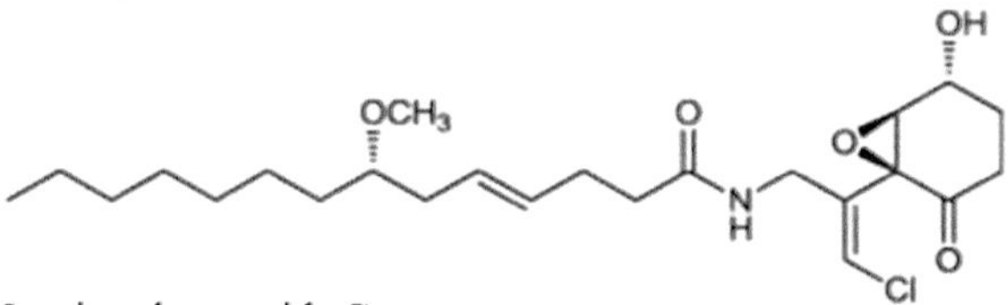

8-epi-malyngamide C

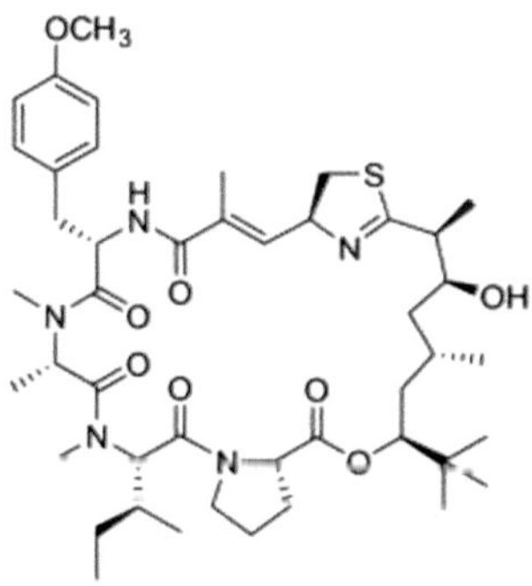

Apratoxin A.

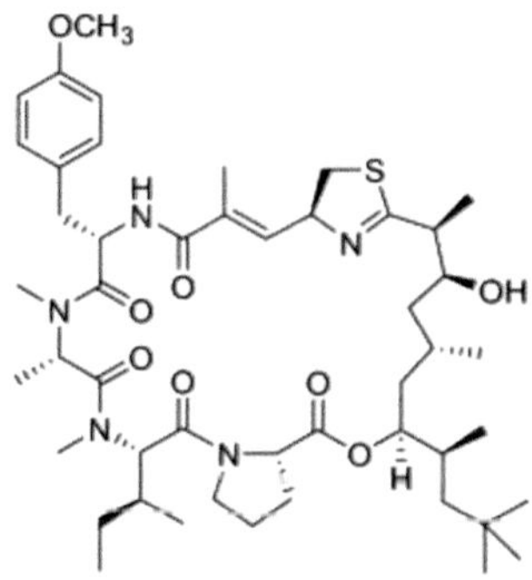

Apratoxin D

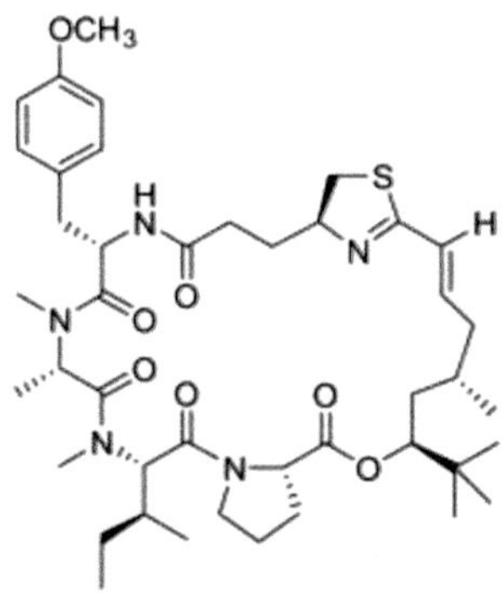

Apratoxin E.

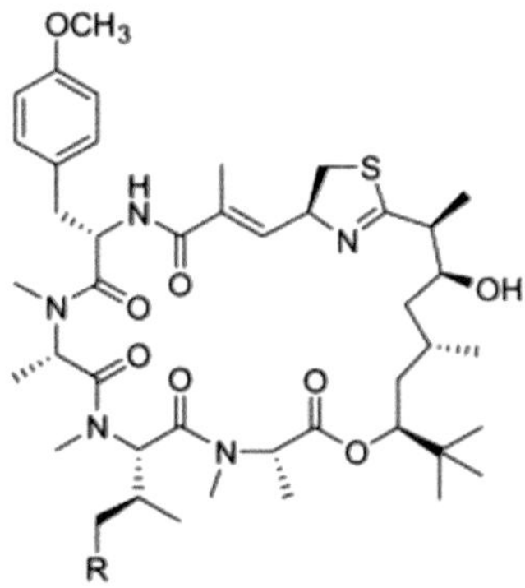

Apratoxin F (R=CH3); Apratoxin G (R=H).

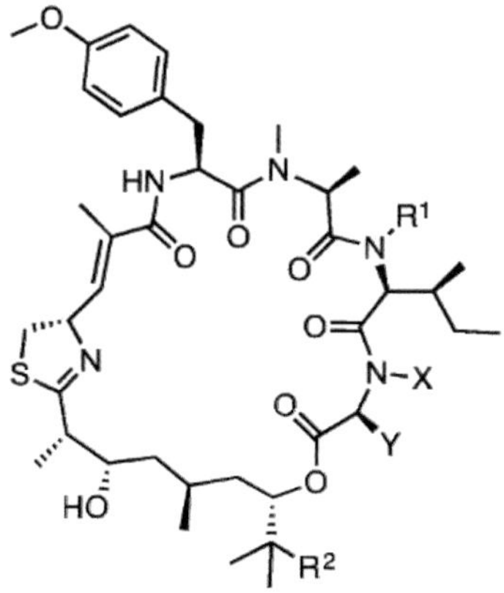

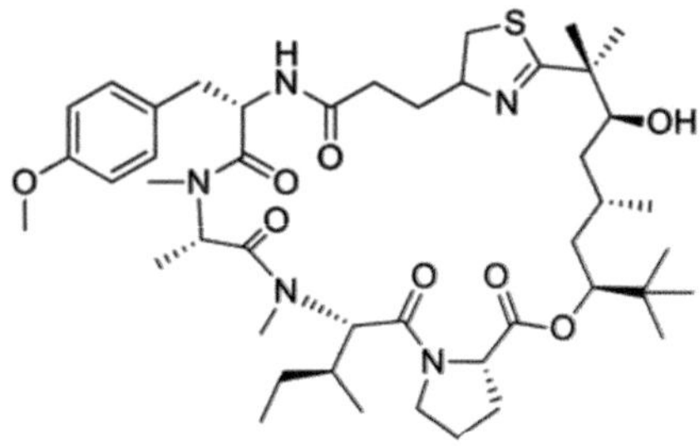

Apratoxin B (R1 ı.,R2=Me,X,Y= (CH2)3). Apratoxin S10

Apratoxin C. (R1=Me,R2=H,X,Y=(CH2)3)

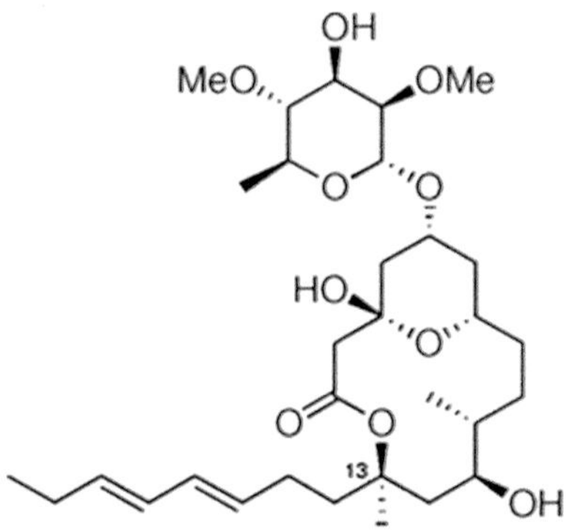

Lyngbouilloside

6.1.3.2 NEUROTOXINS

The neurotoxins of marine cyanobacteria include the bicyclic secondary amine, anatoxin-a; organophosphate, Anatoxin-a(s); carbamate alkaloids, viz., saxitoxins; and lipopeptide, kalkitoxin.

i) Anatoxin-a and Anatoxin-a(s)

Microcystis aeruginosa and Anabaena flos-aquae: The anaroxin-a is produced by the brackish-water cyanobacterial species, *Microcystis aeruginosa* and *Anabaena flos-aquae* (Yang 2007; Osterbauer and Dobbs, 2009).

Anatoxin-a is however produced by Anabaena flos-aquae (Moore et al., 1989).

Therapeutic Uses of Anatoxins

Both anatoxin-a and anatoxin-a(s) have been reported to possess anti-cholinesterase activity (Osterbauer and Dobbs, 2009). As cholinesterase inhibitors, anatoxins help treat glaucoma, a leading cause of blindness (Pope and Brimijoin, 2018). Tidgewell et al. (2010) stated that anatoxin-a(s) served as an irreversible inhibitor of acetylcholinesterase thereby reducing blood pressure and heart rate in rats dramatically.

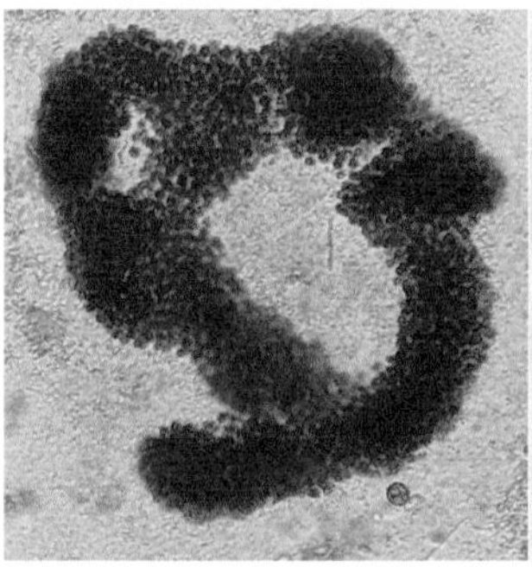

Anabaena flos-aquae
1

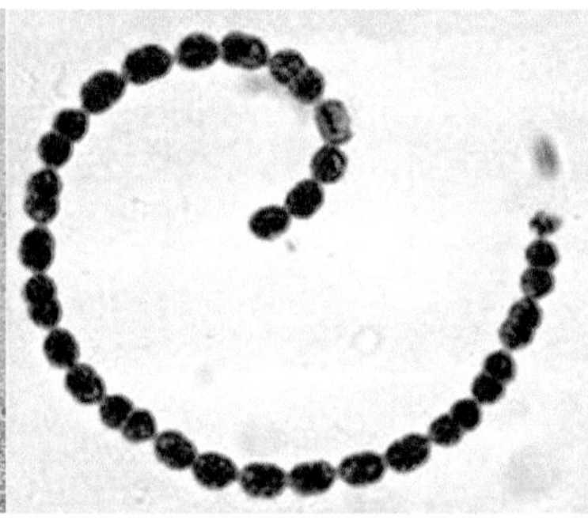

Microcystis aeruginosa
2

1. *Image credit*: Zihao Wang https://creativecommons.org/licenses/by/4.0/
2. *Image credit*: Environmental Protection Agency. https://creativecommons.org/publicdomain/zero/1.0/

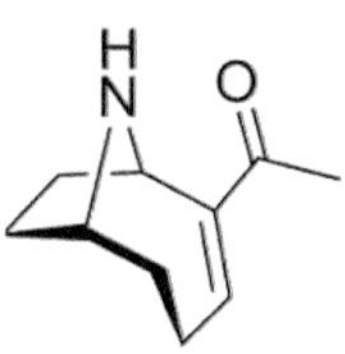

Anatoxin-a.

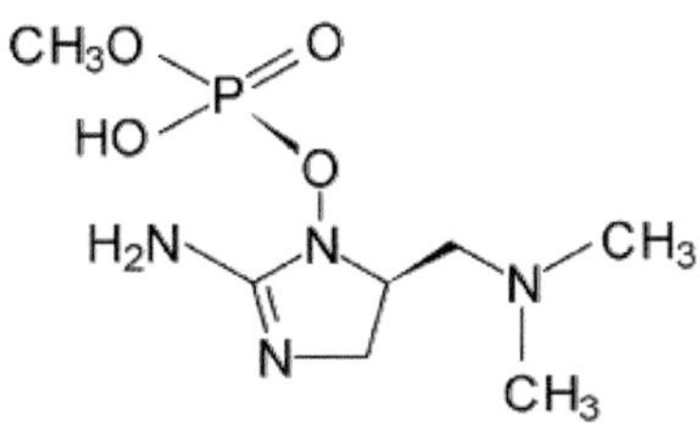

Anatoxin-a(s).

ii) Kalkitoxin

Kalkitoxin is a neurotoxic lipopeptide isolated from the cyanobacterial species, *Lyngbya majuscula* (Pradhan and Ki, 2022).

Therapeutic Uses of Kalkitoxin

Cytotoxicity: Pradhan and Ki (2022) reported that kalkitoxin displayed cytotoxicity against human HCT-116 colon cell lines. Further, this toxin is also found reducing hypoxia-induced initiation of HIF-1 in T47D breast tumor cells at an IC50 value of 5.6 nM.

Anti-metastatic activity: At a concentration of 10 mg/kg, kalkitoxin treatment was found to completely inhibit breast cancer metastasis to the lungs. In a mouse model, kalkitoxin has been reported to suppress osteoclast-soup-derived MDA-MB-231 cell invasion and subsequently induced osteolysis. Further, kalkitoxin inhibited CXCL5 and CXCR2 expression thereby suppressing the secondary growth of breast cancer cells in the lungs, bone, and brain. Keeping these in view, kalkitoxin could be developed as a novel therapeutic drug for the treatment of breast cancer metastasis (Shreshtha et al., 2023).

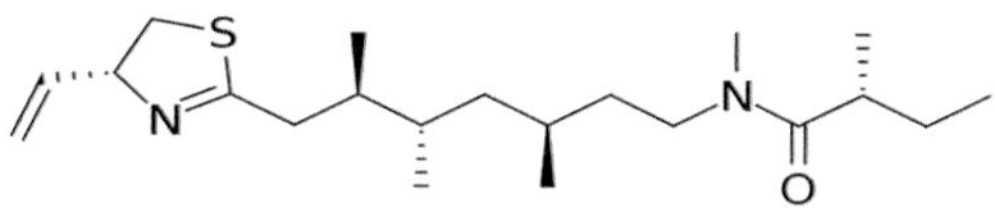

Kalkitoxin

6.1.3.3 HEPATOTOXINS

The hepatotoxins of marine cyanobacteria include mainly the cyclic heptapeptides, microcystins (or cyanoginosins); and cyclic pentapeptídes, nodularins (Zanchett and Oliveira-Filho, 2013). Although there are more than 65 variants of microcystins, the microcystin-LR and [D-Leu[1]] MC-LR-variants are well known (Ramos et al., 2015).

i) Therapeutic Uses of Microcystins

The cyanobacterial, brackish-water species, viz., *Microcystis aeruginosa has been reported to possess the hepatotoxin,* microcystins, the concentration of which may even exceed 2900 ppm (2.9 million ppb), though the WHO's limit is 1 ppb (0.001 ppm) (Miller et al., 2010). The accumulation of microcystins has been demonstrated in fresh- and saltwater corals, crustaceans, mussels, fish, and possibly humans.

Antimycobacterial and cytotoxicity: The MC-LR and [D-Leu[1]] MC-LR toxins of *Microcystis aeruginosa showed* antimycobacterial activity against the strains of *Mycobacterium tuberculosis with MIC values ranging between* 53 and 0.42 μM (Ramos et al., 2015).

Anticancer activity: A formulation from the microcystins has shown anticancer activity and it is patented (United States Patent-Patent No.: (45) Date of Patent: US 9,006,173 B2) (Monks et al., 2015).

Cytotoxicity: Microcystins have been reported to be potent inhibitors of the protein phosphatase families PP1 and PP2A. Microcystin-induced phosphatase inhibition resulted in potent cytotoxicity in the HeLa cell model (Monks et al., 2007).

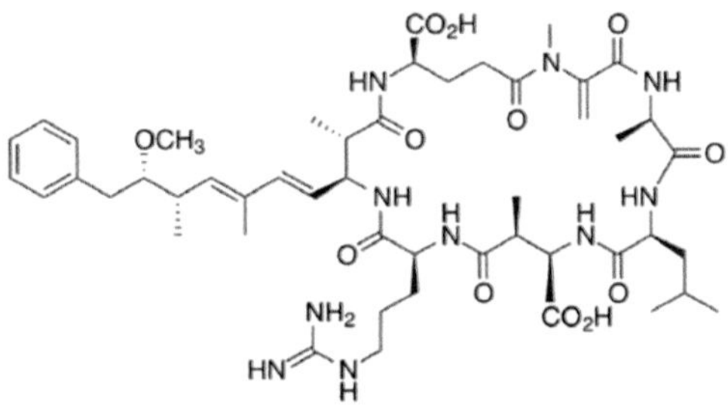

Microcystin-LR.

ii) Therapeutic Uses of Nodularins

Anticancer effects: The cyclic pentapeptide, nodularin produced by *Nodularia spumigena* has been reported to potently inhibit type 2A, (PP2A) protein phosphatase with an IC50 value of 0.026 nm. Further, nodulain-induced PP2A phosphatase inhibition and associated PLK1 (Polo Like Kinase 1) inhibition lead to the death of cancer cells by interfering with multiple stages of mitosis (Honkanen et al., 1991).

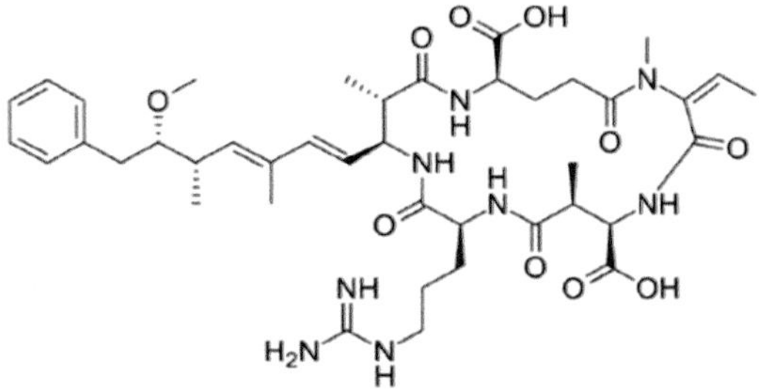

Nodularin

6.1.3.4 DERMATOTOXINS

The dermatotoxins of marine cyanobacteria include the alkaloid compounds, viz., lyngbyatoxin-a and aplysiatoxin; and lipopolysaccharides (LPS) (Zanchett and Oliveira-Filho, 2013).

6.1.3.4.1 Lyngbyatoxins and Their Therapeutic Uses

Cytotoxicity: The lyngbyatoxins produced by *Lyngbya majuscula* such as Lyngbyatoxin A and debromoaplysiatoxin. Lyngbyatoxin A was found to be cytotoxic in 24 h against A2058 human metastatic melanoma. Further, lyngbyatoxin A and ebromoaplysiatoxin were found to be cytotoxic against HL-60 C hHuman promyelocytic leukemia cells at the concentrations of 7 and 700 ng/mL, respectively (Sazdova et al., 2022).

Anti-HIV activity: The lyngbyatoxin A of *Lyngbya majuscula* has been reported to display anti-HIV activity (Tiwari and Tiwari, 2020).

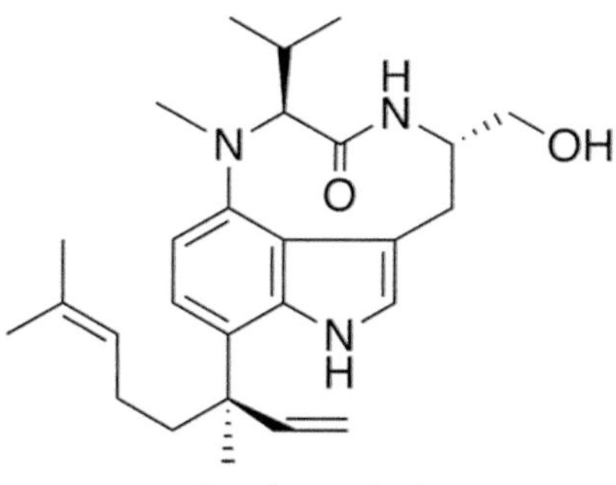

Lyngbyatoxin A

6.1.3.4.2 Aplysiatoxins (ATXs)

The aplysiatoxins (ATXs) are alkaloids and belong to the class of biologically active dermatotoxins with anti-proliferative activity and antiviral activity (Tang et al., 2019). The major aplysiatoxins are debromoaplysiatoxins, neo-debromoaplysiatoxins, and oscillatoxins

Therapeutic Uses of Aplysiatoxins (ATXs)

Lyngbya sp.

The debromoaplysiatoxin analogs, neo-debromoaplysiatoxin A and neo-debromoaplysiatoxin B, produced by *Lyngbya* sp. have been reported to display significant blocking activity against voltage-gated potassium channel Kv1.5 with IC50 values of 6.94 and 0.30 μM, respectively (Han et al., 2018). Further, the aplysiatoxins, Oscillatoxin E, Debromoaplysiatoxin and 30- methyloscillatoxin D derived from this species have shown selective blocking of Kv1.5 with IC_{50} values of 0.79, 1.28, and 1.47 μM, respectively. Kv1.5 has been believed to be a pivotal target for new treatment of atrial tachyarrhythmias involving minimal side effects (Tang et al., 2019).

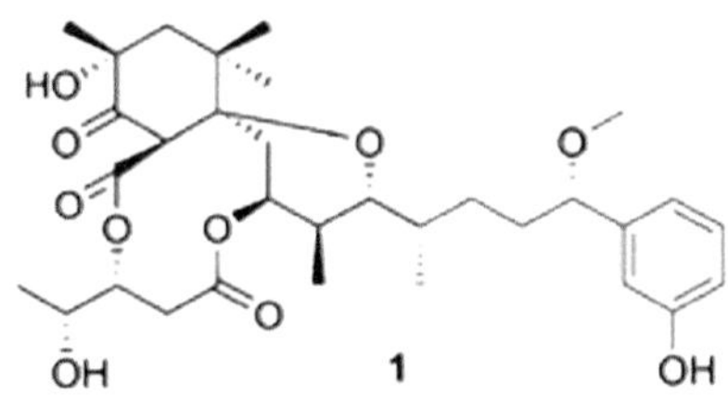

1

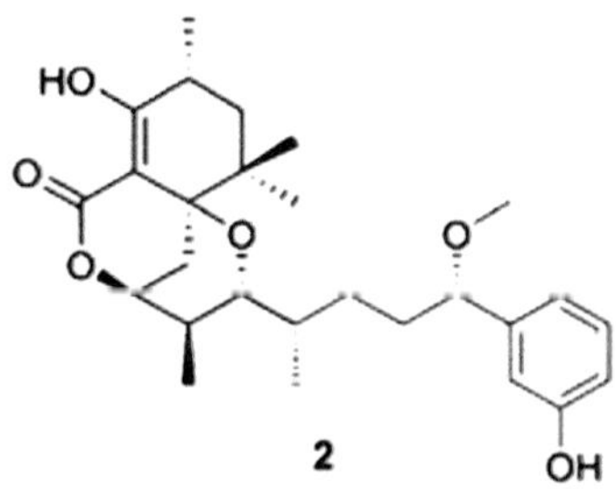

2

1: Neo-debromoaplysiatoxin A; 2: Neo-debromoaplysiatoxin B

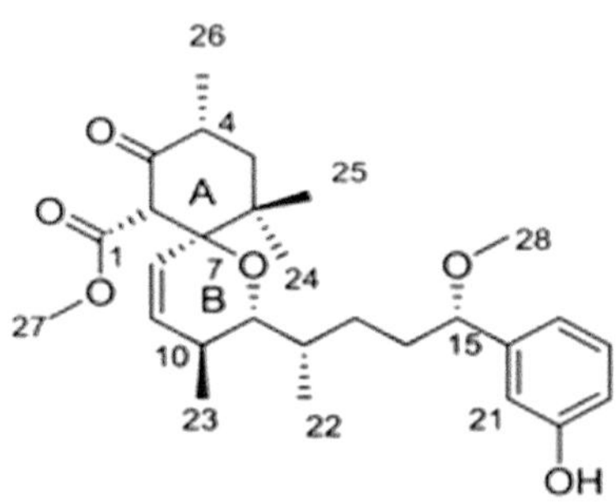

Oscillatoxin E.

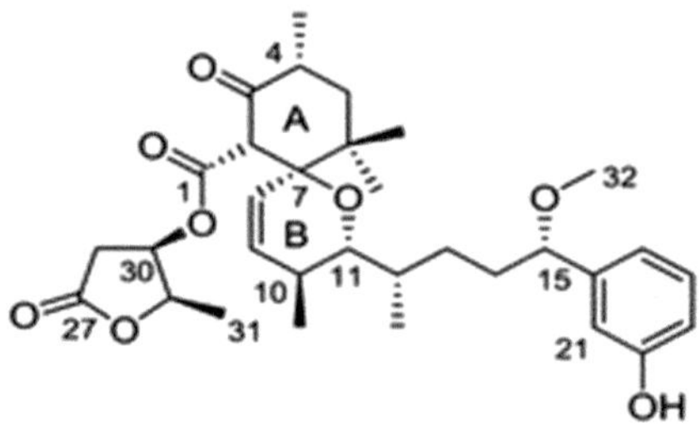

30-methyloscillatoxin D.

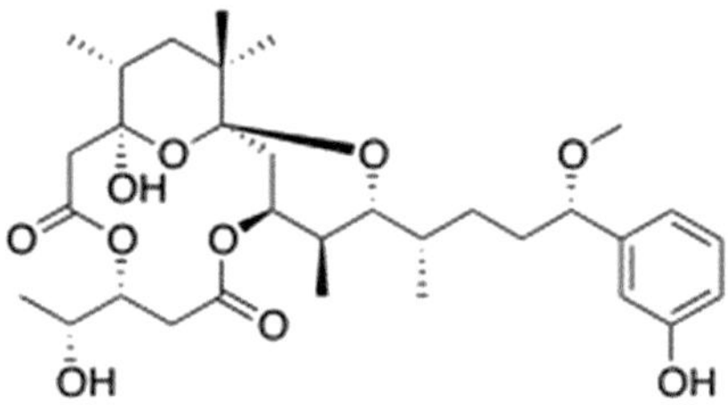

Debromoaplysiatoxin

Leibleinia gracilis *(= Lyngbya gracilis).*

The debromoaplysiatoxin produced by this species has shown activity against P-388 lymphocytic mouse leukemia (Mynderse et al., 1977).

Lyngbya majuscula

The debromoaplysiatoxin of this species has shown antineoplastic activity (Moore et al., 1984).

Kawaguchi et al. (2020) reported that the aplysiatoxin derivative, viz., neo-debromoaplysiatoxin of this species showed an inhibitory effect on K+ channel activity. Further, according to them, another derivative of aplysiatoxin of this speices, viz., neo-aplysiatoxin A displayed 85% cytotoxicity against L1210 mouse lymphoma cells. Furthermore, an aplysiatoxin analog, viz., oscillatoxin I, derived from this species showed 100% cytotoxicity against the L1210 mouse lymphoma cell line with an IC50 value of 4.6 μg/mL (Nagai et al., 2019).

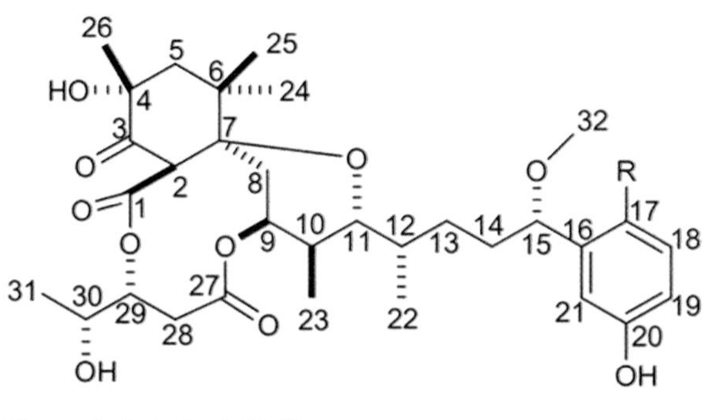

Neo-aplysiatoxin A: R=Br

Neo-debromoaplysiatoxin A: R=H

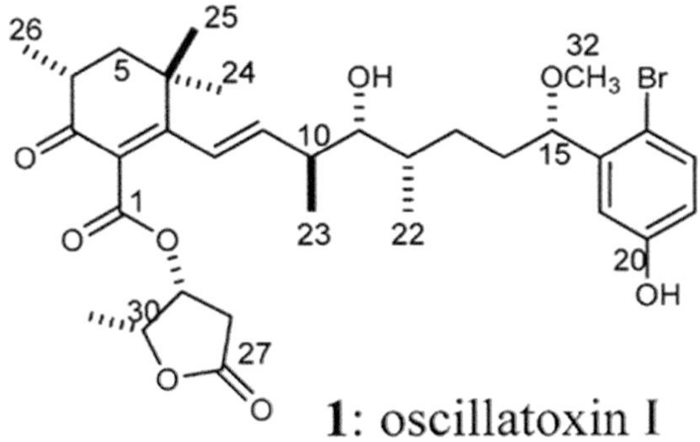

1: oscillatoxin I

Oscillatoxin I

Oscillatoria nigro-viridis and *Schizothrix calcicola*

The debromoaplysiatoxin of these species has shown antineoplastic activity (Moore et al., 1984).

Trichodesmium erythraeum

The aplysiatoxins, aplysiatoxin, debromoaplysiatoxin, and anhydrodebromoaplysiatoxin; and two new analogues, 3-methoxyaplysiatoxin and 3-methoxydebromoaplysiatoxin have been isolated from this species. Among these toxic compounds, debromoaplysiatoxin and 3-methoxydebromoaplysiatoxin displayed significant anti-CHIKV activity with EC50 values of 1.3 and 2.7 μM, respectively; and selectivity indices of 10.9 and 9.2, respectively (Gupta et al., 2014).

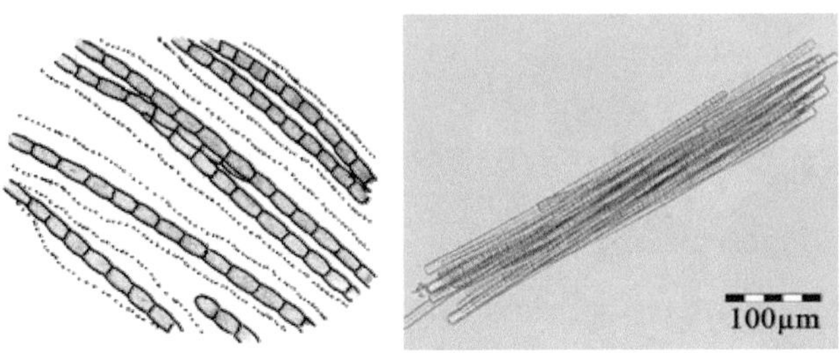

1 2

1. Schizothrix calcicole; 2. Trichodesmium erythraeum

1. Image credit: Freshwater Biological Association. https://creativecommons.org/licenses/by-sa/3.0/

2. Image credit: Reproduced with permission from Saud AlBusaidi

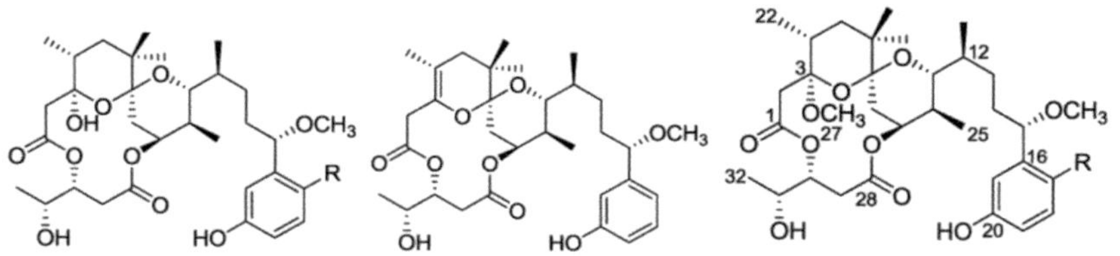

Debromoaplysiatoxin (R=H). Anhydrodebromoaplysiatoxin 3-methoxydebromoaplysiatoxin

6.1.3.5 LIPOPOLYSACCHARIDES (LPS)

Therapeutic Uses of Lipopolysaccharides (LPS)

Sazdova et al. (2022) reported that the cyanobacterial LPS in combination with benzylpenicillin inhibited cytokine production and enhanced antiinflammatory response in human meningioma cells and meningioma

Oscillatoria planktothrix

The "cyanobacterial product" (CyP), the most characterized cyanobacterial LPS from *Oscillatoria planktothrix* FP1 showed neuroprotective effect by its antagonist activity to toll-like receptor (TLR)-4. CyP ability to inhibit TLR-4 signaling resulted in an anticonvulsant effect in kainic acid-induced seizure (Tiwari and Tiwari, 2020). Further, the CyP of this species has been reported to inhibit cytokine production in human whole blood model and it is able to stimulate immune response in different ways (Gemma et al., 2016). Furthermore, the CyP of this species acted as cyanobacterial LPS antagonist for the human monocytic cell line (THP-1) (Sazdova et al., 2022).

6.1.3.6 MISCELLANEOUS TOXINS

Antitoxin (Bacterial toxin)

Synechocystis pevalekii

The antitoxin derived from this species has shown hypoglycemic and antidiabetic activity (Tiwari and Tiwari, 2020; Kopfmann et al., 2016). Mackieh et al. (2021) reported on other marine cyanobacterial toxins acting as blockers/activators of voltage-gated sodium channels (VGSCs) as shown in the following table.

TABLE 6.7 Other Marine Cyanobacterial Toxins Acting as Blockers/Activators of VGSCs.

Species	Toxin(s)
Leptolyngbya crossbyana	Crossbynol A
Lyngby majuscula (strain JHB)	Jamaicamide A-C
Lyngbya majuscula	Hoiamide A
Phormidium gracile	Hoiamide A
Lyngbya majuscula	Kalkitoxin
Leptolyngbya sp.	Palmyrolide A
Oscillatoria spp.	Palmyrolide A
Unidentified cyanobacterium 1	Hoiamide B
Unidentified cyanobacterium 2	Palmyramide A

Source: Mackieh et al. (2021).

6.2 TOXINS OF MARINE MACROALGAE (SEAWEEDS)

6.2.1 GREEN ALGAE

Toxins of Green Algae and Their Therapeutic Potential

The major toxins of marine blue-green algae include caulerpin, caulerpicin, caulerpenyne, cymopol, dopamine, and so on. The species of green algae producing these toxins and their therapeutic effects are given below.

i) Caulerpin

Several species of marine green algae such as *Caulerpa cupressoides, Caulerpa paspaloides, Caulerpa peltata, Caulerpa prolifera, Caulerpa scalpelliformis, Caulerpa racemosa,Caulerpa sertularioides, Codium decoratum, and Halimeda incrassata* have been reported to possess the bisindole alkaloid toxin, viz., caulerpin the concentration of which varies from 1 mg to 5.47 mg per dry gram of alga. Caulerpin has several bioactivities such as as antiviral, antioxidant, antifungal, acetylcholinesterase inhibitor (AChE), and antibacterial activity (Vest et al., 1983).

Caulerpa racemosa: Caulerpin of this species displayed antinociceptive (0.0945 μmol) and anti-inflammatory activity (100 μmol/kg) at the concentrations of 0.0945 and 100 μmol/kg, respectively (Rushdi et al., 2020).

Caulerpa ashmeadii: Caulerpin of this species exhibited antimycobacterial activity against Mycobacterium tuberculosis with an IC50 value of 0.24 μM. Further, this toxin has also been reported to exhibit cytotoxic activity against human leukemia (K562), human lung cancer (A549), human breast cancer (SK-BR-3), human liver cancer (Huh7), human cervical cancer (HeLa), and human colon cancer (HT29), cell lines with IC50 values of 4.67, 4.20, 3.71, 0.72 μM, 1.95 and 4.04, respectively (Rushdi et al. 2020).

Caulerpa cylindracea: The caulerpin of this species has been reported to possess antiviral and cytotoxic activities. In antiviral activity, this toxin inhibits the spike–ACE2 complex formation competitively; therefore, it can be used against SARS-CoV-2 virus infection. Its caulerpin showed cytotoxicity against breast cancer cell lines MDA-MB-231 and MCF7 with IC50 values of 11.9–14.7 and 13.8–15.7 uM, respectively (Erol et al., 2022).

1 *Caulerpa racemosa* **2** *Caulerpa ashmeadii* **3** *Caulerpa cylindracea*

1. *Image credit*: Nick Hobgood . https://creativecommons.org/licenses/by-sa/3.0/
2. *Image credit*: William Henry Harvey. https://creativecommons.org/publicdomain/zero/1.0/
3. *Image credit*: Bachir Bouiadjra B, Ghellai M, Daoudi M, Behmene IE, Bachir Bouiadjra MEA. https:// creativecommons.org/licenses/by/4.0/

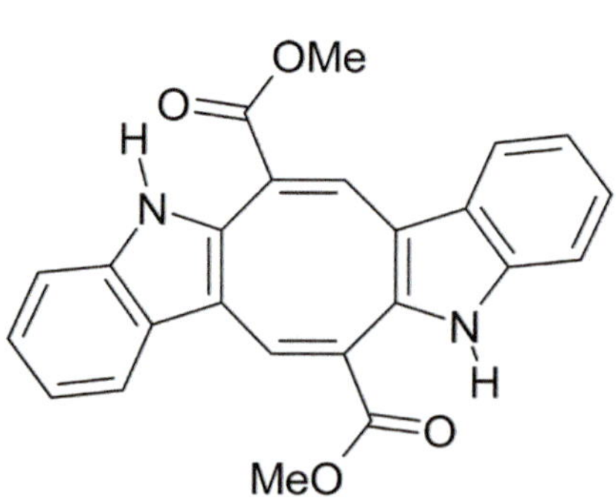

Caulerpin

ii) Caulerpicin

The toxin caulerpicin is made of a combination of ceramides and is present in the species of *Caulerpa* viz. *Caulerpa ashmeadii, Caulerpa racemosa* var. *uviferain, Caulerpa scalpelliformis,* and *Caulerpa sertularioides* and its concentration varies from 0.03 to 0.09 mg per dry gram of alga (Vest et al., 1983). Nielsen et al. (1982) reported that the caulerpicin of Caulerpa racemosa consists of a mixture of ceramides derived from 2S, 3R-sphinganine with C18(32%), C20(2%), C22(6%), C24(35%), and C26(25%) saturated fatty acid. These caulerpicins possess neurotrophic activity concerned with neuronal development, synaptic maturation, and plasticity through the activation of neurotrophic tyrosine.

1 2 3

1: Caulerpa scalpelliformis; 2: Caulerpa sertularioides; 3: Caulerpa serrulate

1. *Image credit*: Reproduced with permission from John Huisman

2. *Image credit*: Bob Peterson. https://creativecommons.org/licenses/by/2.0/

3. *Image credit*: Derek Keats . https://creativecommons.org/licenses/by/2.0/

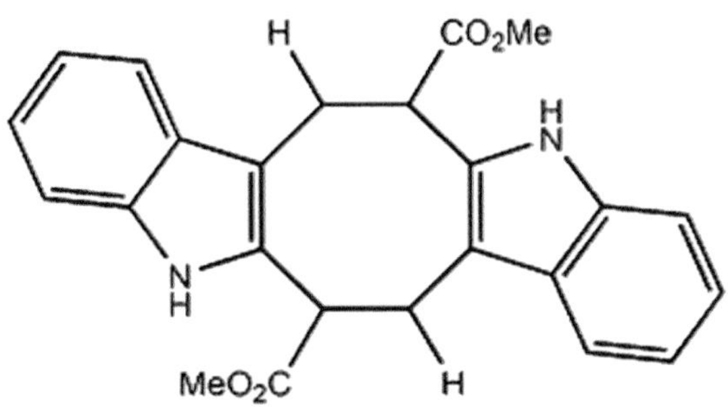

Caulerpicin

iii) Caulerchlorin and monomethyl caulerpinate

Caulerpa cylindracea*:* The monomethyl caulerpinate of this species has been reported to possess antiviral and cytotoxic activities. In antiviral activity, this toxin inhibits the spike–ACE2 complex formation competitively; therefore, it can be used against SARS-CoV-2 virus infection. Its caulerpin showed cytotoxicity against breast cancer cell lines MDA-MB-231 and MCF7 with IC50 values of 14–20.6 and 16.8–17.3 uM, respectively (Erol et al., 2022).

Caulerpa racemosa: The chlorinated bisindole alkaloid, caulerchlorin, and monomethyl caulerpinate of this species showed antifungal activity against *Cryptococcus neoformans* (32,609) with MIC80 values of 16 and 64 μg/mL, respectively (Quan et al., 2012).

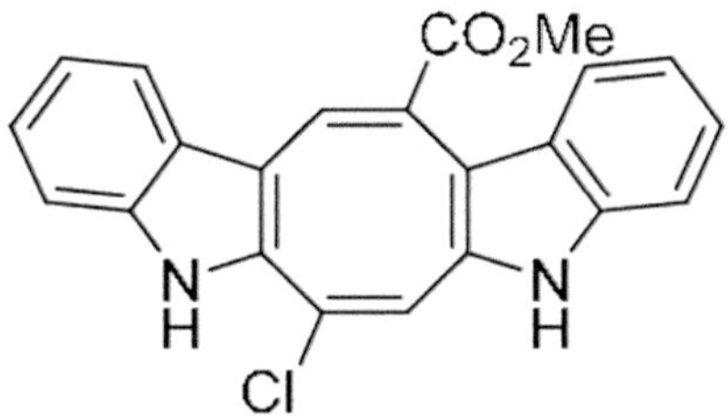

Caulerchlorin

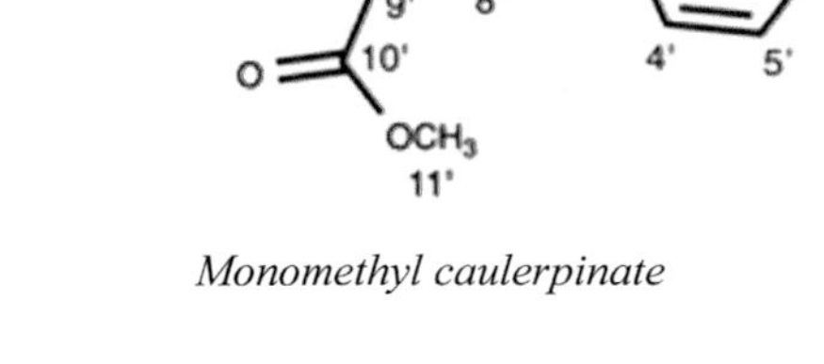

Monomethyl caulerpinate

iv) Caulersin

Caulerpa racemosa: The bisindole alkaloid, caulersin-containing extract of this species has shown significant protein tyrosine phosphatase-1B (PTP1B) inhibitory activity with an IC50 value of 5.86 μM. The PTP1B is believed to be a major negative regulator of the insulin (Rushdi et al., 2020).

Caulerpa serrulata: The caulersin of this species has shown antibacterial activity against *Escherichia coli, Staphylococcus aureus, Shigella* sp., *Salmonella typhi and Pseudomonas aeruginosa*. The values of maximum inhibition zone for Escherichia coli and Salmonella typhi were 21 and 10 mm, respectively (Rushdi et al., 2020).

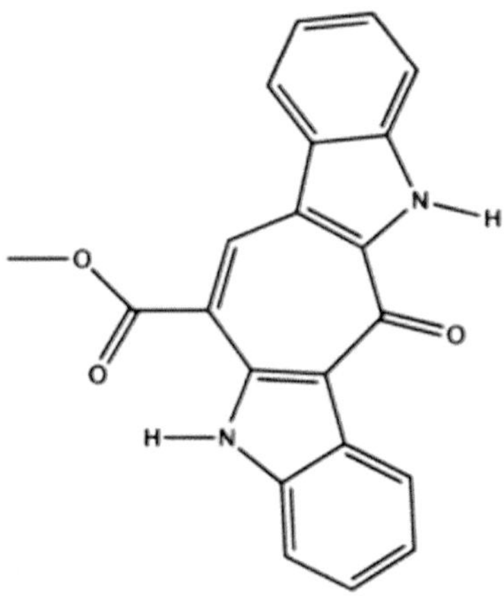

Caulersin

v) Caulerprenylols

Caulerpa racemosa: The prenylated p-xylenes, viz., caulerprenylol A and caulerprenylol B of this species have shown antifungal activity against *Candida glabrata, Trichophyton rubrum,* and *Cryptococcus neoformans* with MIC80 values between 4 and 64 μg/mL (Rushdi et al., 2020).

vi) Caulerpenyne

Caulerpenyne, an oxygenated sesquiterpene toxin is found in *Caulerpa prolifera* and *Caulerpa taxifolia* (Rushdi et al., 2020; Higa and Kuniyoshi, 2000).

Caulerpa prolifera: The caulerpenyne of this species has shown anticancerous, antitumour, and antiproliferative activities. Further, this compound has been found very active against pancreatic lipase and therefore, it possesses significant therapeutic values in the treatment of pancreatic cancer (Yadav and Palanisamy, 2019).

Caulerpa taxifolia: The caulerpenyne of this species was found to be cytotoxic against many cell lines, including neuroblastoma SK-N-SH cell line.

It induced the inhibition of the cell proliferation of this cell line with an IC50 value of 10 μM after 120 min incubation (Rushdi et al., 2020). Further, the caulerpenyne derivatives of this species have been reported as a potent source of anti-cancer drugs as they have cytotoxic effects against bone marrow cells (hematopoietic progenitors CFU-GM), melanocytes and keratinocytes, and immortalized keratinocytes (HaCaT and HESV) at IC50 values ranging between 6 and 24 μM(Rushdi et al., 2020).

4

5

6

4: Caulerpa taxifolia; 5: Ulvaria obscura; 6: Bryopsis plumosa

4. Image credit: Fish&Dive. https://creativecommons.org/licenses/by-sa/4.0/

5. Image credit: Reproduced with permission from Dr. Mandy Lindeberg,

6. Image credit: Javier. https://creativecommons.org/licenses/by/4.0/

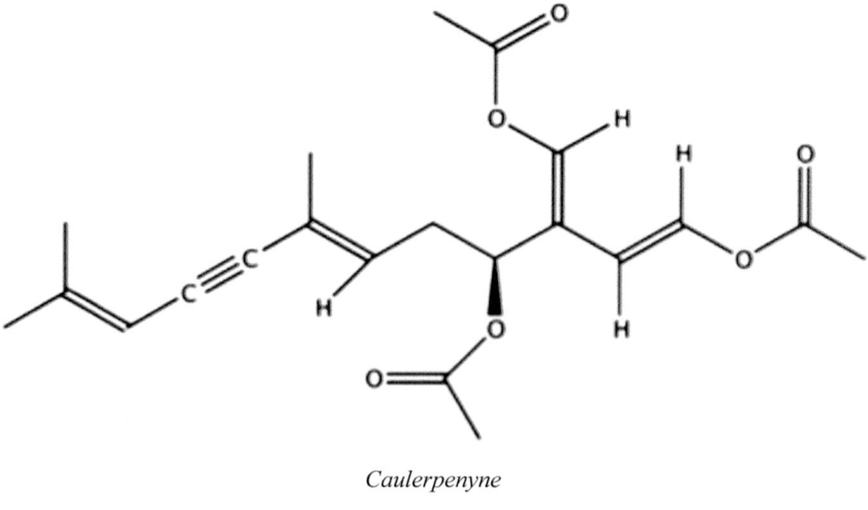

Caulerpenyne

vii) Cymopol and cyclocymopol

Cymopolia barbata: A monoterpene-bromohydroquinone toxin, viz., cymopol has been isolated from this species (Silva et al., 2021). This compound has shown both antioxidant and anti-inflammatory effects (Bousquet et al., 2020). Its DPPH antioxidant activity detected through dichlorodihydrofluorescein diacetate (DCFH-DA) recorded an IC50 value of 4.0uM. Both cymopol and cyclocymopol of this species have also been reported to possess antibiotic and antifungal properties (Bali, 1992).

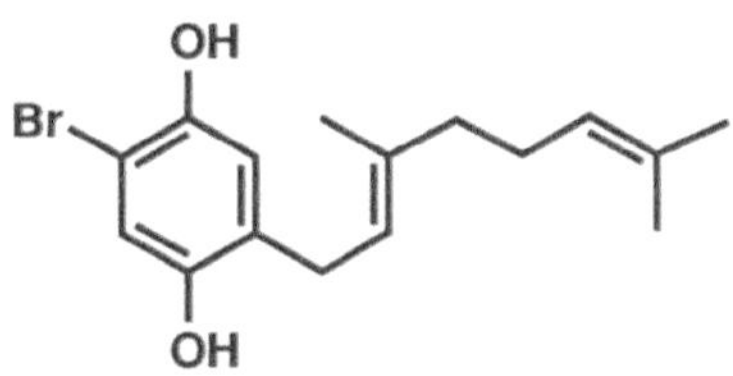

Cymopol

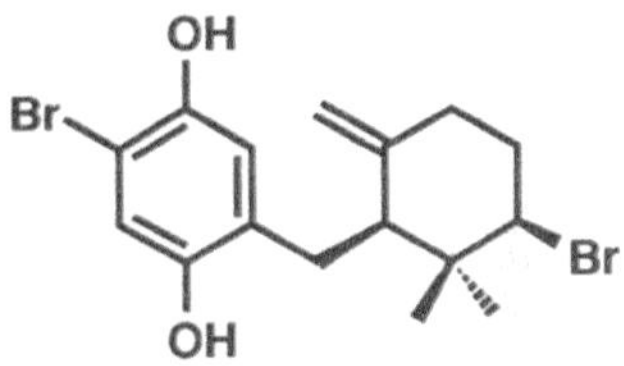

Cyclocymopol

viii) Dopamine

Ulvaria obscura: A hydroxytyramine toxin, dopamine, isolated from this species may treat low blood pressure, low heart rate, poor blood flow to vital organs, and cardiac arrest. Further, at low infusion rates of 0.5 to 2 µg/kg/min, dopamine acts on the visceral vasculature to produce vasodilation, including the kidneys, thereby increasing urinary flow. Furthermore, dopamine is also known to serve as a neurotransmitter in the brain as it plays a vital role in a variety of functions in the brain including memory, behavior and cognition, sleep, and movement. Deficiency of this toxin in the body is also believed to cause Parkinson's disease (https://www.news-medical.net/health/Dopamine-Therapeutic-Use.aspx).

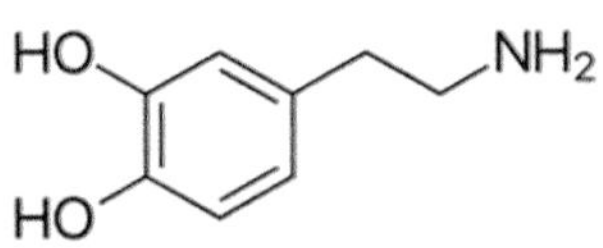

Dopamine

ix) Kahalalide F: It is a peptide toxin decorated with fatty acid and is isolated from the marine green alga, Bryopsis plumosa. This compound could be effective in the treatment of lung cancer and AIDS (Yadav and Palanisamy, 2019).

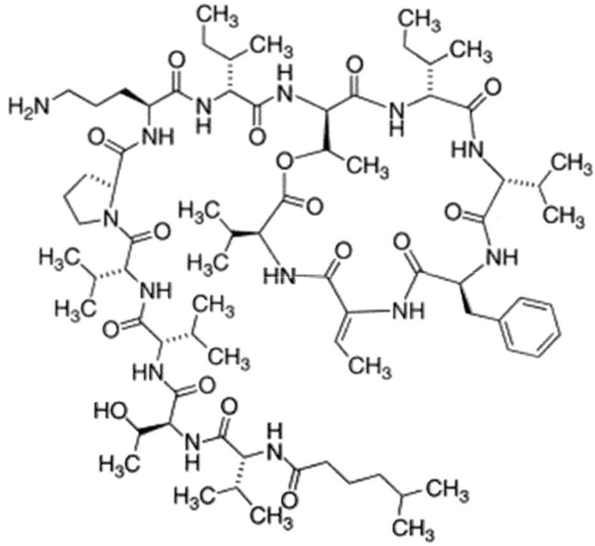

Kahalalide F

6.2.2 BROWN ALGAE

Among the brown algae *Cladosiphon okamuranus, Nemacystus decipiens,* and *Kjellmaniella crassifolia* have been reported to be toxic due to the presence of rich fucoidans, and sulfated polysaccharides (Cheney, 2016; Gideon and Rengasamy, 2008; Yasumoto,1993). Fucoidan may have potential liver toxicity and in a clinical study, 4 of 17 patients who took 6 g of fucoidan daily exhibited symptoms of diarrhea (Lin et al., 2020).

Therapeutic Uses of Fucoidans

Cladosiphon okamuranus and Kjellmaniella crassifolia

Fucoidans are well-known for their anticoagulant and anti-thrombotic activity. Fucoidans from *Cladosiphon okamuranus* and *Kjellmaniella crassifolia* were found to possess anti-inflammatory effects by inhibiting IL-6 production in CMT-93 cells with the down-regulation of NF-κB nuclear translocation. Therefore, fucoidan derived from these species may be useful as a dietary substance for patients with inflammatory bowel disease (Gideon and Rengasamy, 2008). Fucoidan are also known to display anti-cancer effects through cell cycle arrest, and induction of apoptosis (Lin et al., 2020).

Laminaria angustata var. *longissima, Undaria pinnatifida* and *Sargassum duplicatum*

Fucoidans isolated from *Laminaria angustata* var. *longissima* have been reported to stimulate the production of nitric oxide (NO), tumor necrosis factor-α (TNF-α), and interleukin-6 (IL-6) in RAW 264.7 cancer cells. Further, fucoidan from *Undaria pinnatifida* was found to induce IL-6, IL-8, and TNF-α production in human neutrophils. Furthermore, the fucoidan from *Sargassum duplicatum* has exerted immunomodulatory improving effects (Sinurat et al., 2021).

Dictyota bartayresii

The diterpene toxins, viz., pachydictyols (8ß-Hydroxypachydictyol A) isolated from this species exhibited cytotoxicity and antiviral activities (Yadav and Palanisamy 2019).

Sphaerotrichia divaricata

This alga is known to produce diethyl ether peroxides (Cheney, 2016) which are explosive materials (under the influence of light and air) causing skin, eye, and respiratory tract irritation; and on inhalation of its vapor may cause, nausea, narcosis, loss of consciousness, dizziness, and drowsiness. Further, it is also known to cause digestive tract irritation and central nervous effects by ingestion (https://sites.chemengr.ucsb.edu/~ceweb/faculty/scott/Chemical%20SOPs/DiethylEther.pdf). The therapeutic uses of diethyl ether peroxides are, however, unknown though diethyl ether is a potent anesthetic.

Cladosiphon okamuranus

Image credit: Reproduced with permission from Prof.Eduard Titlyanov and Dr.Tamara Titlyanova.

1 2

1: Laminaria angustata; 2: Undaria pinnatifida

1. Image credit: Reproduced with permission from Prof.Eduard Titlyanov and Dr.Tamara Titlyanova

2. Image credit: Nordisk familjebok .Public domain

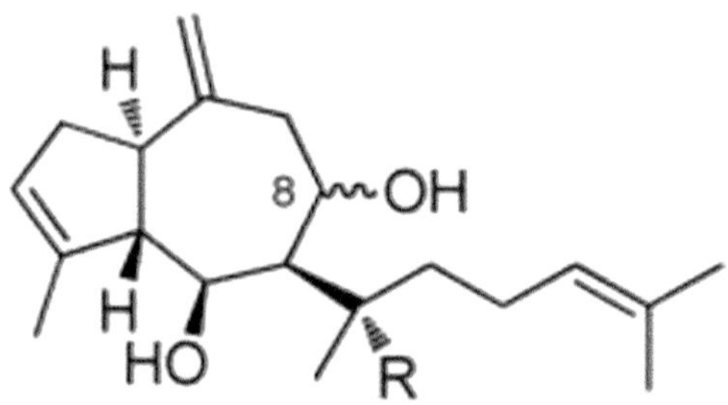

R=H 8 β

8β-Hydroxypachydictyol A

6.2.3 ***RED ALGAE***

Certain species of red algae have been reported to possess toxins including domoic acid and kainic acid, caulerpin and caulerpicin and aplysiatoxins. However, information on the therapeutic potential of such toxins is wanting. The species of such red algae and their toxins are given in the following table.

TABLE 6.8 Species of Red Algae and Their Toxins.

Source species	Toxins	Chemistry	References
Acanthophora specifera	Polycavernoside A	Glycosidic macrolide	Cheney (2016)
Alsidium corallinum	Domoic acid	Amino acid	(1)
Amansia glomerata	Caulerpin and caulerpicin	Bis-indole alkaloid; ceramides	Vidal et al. (1984)
Asparagopsis armata	Haloforms, haloacids, and haloketones	Halogenated compounds	Silva et al. (2021)
Caloglossa leprieurii	Caulerpin	Bis-indole alkaloid	Clinton et al. (2014)
Centroceras clavulatum	Domoic acid and kainic acid	Amino acids	Yadav and Palanisamy (2019)
Chondria armata	Domoic acid and kainic acid	Amino acids	Steele et al. (2022)
Digenea simplex	Domoic acid and kainic acid	Amino acids	Freile-Pelegrín and Tasdemir (2019)
Gracilaria coronopifolia	Aplysiatoxins and debromoaplysiatoxin	Alkyl phenols	Higa and Kuniyoshi (2000)
Gracilaria tsudai	Polycavernosides	Glycosidic macrolides	Higa and Kuniyoshi (2000)
Gracilaria verrucosa	Prostaglandin E2	Prostanoid(lipid)	Higa, and Kuniyoshi (2000)
Gracilaria sp	Yasumoto toxin	–	Paul (1993)
Halimeda incrassata	Caulerpin	Bis-indole alkaloid	Clinton et al. (2014)
Hydropuntia edulis	Prostaglandin E2, polycavernoside A and B	Prostanoid(lipid); glycosidic macrolides	Cheney (2016)
Hypnea concornis	Caulerpin	Bis-indole alkaloid	Clinton et al. (2014)
Hypnea musciformis	Sulphated polysaccharide	–	Yadav and Palanisamy (2019)
Laurencia majuscula	Caulerpin	Bis-indole alkaloid	Clinton et al. (2014)
Laurencia nidifica	Caulerpin and caulerpicin	Bis-indole alkaloid; ceramides	Vidal et al. (1984)
Palmaria palmata	Kainic acid	Amino acid	FAO and WHO (2022)
Plocamium lepitophyllum	Halogenated monoterpenes	–	Silva et al. (2021)

1) http://extoxnet.orst.edu/tics/domoic.asc.

KEYWORDS

- **therapeutic uses phycotoxins**
- **dinoflagellate toxin patenting**
- **therapeutic phytoplankton**
- **therapeutic seaweeds**
- **therapeutic applications of biotoxins**

CHAPTER 7

Therapeutic Potential of Marine Zootoxins: Invertebrate Toxins

ABSTRACT

This chapter deals with the therapeutic potential of marine zootoxins, namely sponge toxins, cnidarian toxins, TTX of flatworms, TTX of ribbon worms, annelid toxins, crustacean toxins, molluskan toxins, and echinoderm toxins.

Several species of marine invertebrates are known for their toxic secretions that are complex mixtures comprising poisons, venoms, and toxungens (Rodrigo et al., 2021). Invertebrates with poisons lack a delivery mechanism, that is, ingested, inhaled, or absorbed across the body surface; toxungenous invertebrates on the other hand deliver their toxins to the body surface of another animal without an accompanying wound; and venomous animals deliver their toxins to the internal tissues of another animal via the creation of a wound (Nelson et al., 2014).

7.1 THERAPEUTIC POTENTIAL OF TOXINS OF MARINE SPONGES

Halichondria okadai and *Halichondria melanodocia*

Toxic marine sponges are very rare and to our knowledge, only the species, *Halichondria okadai* and *Halichondria melanodocia*, have been reported to possess dinophysistoxin-1, member of the family of okadaic acid (Blanco, 2018). It is also worthy of mention here that okadaic acid was first isolated from these sponge species (https://www.sciencedirect.

Therapeutic Potential of Marine Biotoxins: From Poison to Pharmacy. Santhanam Ramesh, Ramasamy Santhanam, & Subbiah Balasundari (Authors)

DOI: 10.1201/9781003610991-7

com/topics/pharmacology-toxicology-and-pharmaceutical-science/dinophysistoxin-1).

Therapeutic uses of sponge toxins

Serine/theorine protein phosphatase inhibitor: The okadaic acid has been reported to inhibit the serine/theorine protein phosphatases PP1 and PP2A with IC50 values of 15–20 and 0.1 nM, respectively (https://www.cellsignal.com/products/activators-inhibitors/okadaic-acid/5934). Protein serine-threonine phosphatases are known to play a critical role in the control of cell functions.

Cytotoxicity: The dinophysistoxin-1 has been reported to display significant cytotoxicity in MCF-7, Neuro-2a, and NG108-15 cell lines (Soliño et al., 2015).

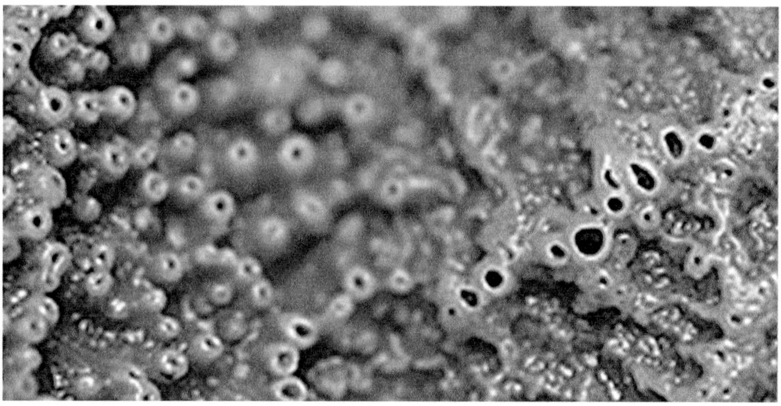

Halichondira sp.

Image credit: Minette Layne . https://creativecommons.org/licenses/by-sa/2.0/

7.2 THERAPEUTIC POTENTIAL OF TOXINS OF MARINE CNIDARIANS

Jellyfish (Phylum: Cnidaria) are commonly seen in several coastal areas around the world. They cause unpleasant reactions on recreating humans, due to the venom present in their specialized organelles called nematocysts. In the affected humans by the stings of these jellyfish inflammation and pain are the common symptoms.

Cnidarian venoms and their Bioprospecting

The venom of the jellyfish is constituted mainly by proteins: phospholipases A2, metallopeptidases, serinepeptidases, CRISPs, lectins, pore-forming toxins, and protease inhibitors (Arrruda et al., 2021). The crude venoms extracted from several species of marine cnidarians have a wide range of bioactivities such as neurotoxic, cytotoxic, cytolytic, and hemolytic, and the harmful effects of these venoms on humans include hypotension, diffused neurotoxicity, dermonecrosis, edema, motorial and respiratory problems, cardiovascular symptoms, and occasionally death. However, these toxins have attracted the attention of researchers as they are considered to be very useful molecular probes for the study and analysis of ion channels involved in immune responses, which are of biomedical interest. It is also worthy of mention here that Charles Robert Richet won the Nobel Prize in Physiology/ Medicine in the year 1913 for his research on anaphylaxis, a potentially life-threatening immune hypersensitivity reaction to an antigen, discovered using the toxin of a species of sea anemone viz. *Actinia sulcata* (D'Ambra and Lauritano, 2020). Various toxin compounds derived from the marine cnidarians have been reported to possess activities against leukemia, lung adenocarcinoma, glioblastoma, melanoma, colon carcinoma, lymphoma, ovarian adenocarcinoma, breast adenocarcinoma, glioblastoma, hepatoma, pheochromocytoma, prostate and pancreatic carcinoma, fibrosarcoma, cervix carcinoma, epidermoid carcinoma, renal and central nervous system cancer (D'Ambra and Lauritano, 2020). The venoms of marine cnidarians have yielded hundreds of lead drug compounds. But the composition of most of these venoms is not known in detail. All these venoms are believed to contain mainly a variety of proteinaceous compounds such as peptides, proteins, enzymes, and proteinase inhibitors, and non-proteinaceous compounds viz. purines, quaternary ammonium compounds, biogenic amines, and betaines (D'Ambra and Lauritano, 2020). Among the venoms isolated from the marine cnidarians, only one compound viz. dalazatide (= Stichodactila toxin, ShK-186), a 35-residue peptide produced by the sea anemone *Stichodactila helianthus*, has been reported to block neuronal and lymphocyte voltage-dependent potassium channels, and therefore it is usable in the treatment of autoimmune diseases, including neuroinflammatory diseases and is ready to undergo Phase 2 clinical trials (Serrano-Albarrás et al., 2014; https://multiplesclerosisnewstoday.com/ dalazatide-formerly-shk-186-multiple-sclerosis/).

The species of marine cnidarians and their toxins/compounds are given in the following tables.

TABLE 7.1 Marine Cnidarians Possessing Neurotoxins.

Class	Species	Toxin/compound
Hydrozoa	*Olindias sambaquiensis*	Metalloproteinases
-Do-	*Physalia physalis*	PpV9.4 and PpV19.3
Scyphozoa	*Aurelia aurita*	Proteolytic enzymes,
-Do-	*Chrysaora hysoscella*	Cationic protein
-Do-	*Cyanea capillata*	CcNT
-Do-	*Phyllorhiza punctata*	Phospholipase A2
Cubozoa	*Carybdea marsupialis*	CmNt
-Do-	*Carukia barnesi*	CbTX-I,II
-Do-	*Chiropsalmus quadrigatus*	CqTX-A
Anthozoa (Hexacorallia)	*Bunodosoma granulifera*	Bg II and III
-Do-	*Heteractis crispa (= Radianthus crispus; Radianthus macrodactyla)*	Hcr1b-2, Hcr1b-3, Hcr1b-4
-Do-	*Stichodactyla helianthus*	Sh1

Source: D'Ambra and Lauritano (2020), Santhanam (2020), Santhanam et al. (2019).

TABLE 7.2 Marine Cnidarians Possessing Cytotoxins.

Class	Species	Toxin/Compound
Hydrozoa	*Olindias sambaquiensis*	Oshem
-Do-	*Physalia physalis*	Collagenase
Scyphozoa	*Chrysaora hysoscella*	Cationic protein
-Do-	*Chrysaora quinquecirrha*	DNase
-Do-	*Cyanea capillata*	CcTX-1, Phospholipase A2,
-Do-	*Cyanea lamarckii*	ClGP-1, Phospholipase A2
-Do-	*Nemopilema nomurai*	Metalloproteinases
-Do-	*Pelagia noctiluca*	Proteinaceous macromolecules
-Do-	*Stomolophus meleagris*	Phospholipase A2
Cubozoa	*Chironex fleckeri*	CfTX-1,2
Anthozoa (Hexacorallia)	*Actinia villosa*	Avt120
-Do-	*Urticina coriacea*	U1

Source: D'Ambra and Lauritano (2020), Santhanam (2020), Santhanam et al. (2019).

TABLE 7.3 Marine Cnidarians Possessing. Hemolytic (Anemia Causing) Toxins.

Class	Species	Toxin/Compound
Hydrozoa	*Millepora platyphylla*	Milleporin-1
-Do-	*Olindias sambaquiensis*	Oshem1
-Do-	*Physalia physalis*	Collagenase, Elastases, PpV9.4, Physalitoxin
-Do-	*Tubularia larynx*	Phospholipase A2
Scyphozoa	*Aurelia aurita*	Proteolytic enzymes
-Do-	*Cassiopea andromeda, Cassiopea xamancha, Cyanea capillata, Cyanea capillata, Cyanea lamarckii, Rhizostoma nomadica, Stomolophus meleagris*	Phospholipase A2
-Do-	*Nemopilema nomurai*	Metalloproteinases
-Do-	*Pelagia noctiluca*	Proteinaceous macromolecules
-Do-	*Rhizostoma pulmo*	Rhizoprotease, Rhizolysin
Cubozoa	*Alatina moseri*	CaTX-A
-Do-	*Akatina alata*	CAH1, CaTX-A, CaTX-B
-Do-	*Carybdea marsupialis*	Haemolysin
-Do-	*Carybdea rastonii*	CrTX-I, II, III
-Do-	*Carukia barnesi*	Phospholipase A2
-Do-	*Chironex fleckeri*	Phospholipase A2; CfTX-A,B;
-Do-	*Chiropsalmus quadrigatus*	CqTX-A
Anthozoa (Hexacorallia)	*Actinia equina*	Equinatoxin-I, II, III, and IV
-Do-	*Actinia tenebrosa*	Tenebrosin-A, B, and C
-Do-	*Actinia villosa*	Avt-I
-Do-	*Anthopleura asiatica*	Bandaporin
-Do-	*Heteractis crispa (= Radianthus crispus; Radianthus macrodactyla)*	Actinoporin RTX-A, Actinoporin RTX-S II
-Do-	*Heteractis magnifica (= Radianthus magnifica; Radianthus paumotensis; Radianthus ritteri)*	Magnificalysins I and II; HMgIII
-Do-	*Metridium senile*	Metridin
-Do-	*Phyllodiscus semoni*	PsTX-60A; PsTX-60B
-Do-	*Stichodactyla helianthus*	Sticholysin-I and II
-Do-	*Urticina coriacea*	U1
-Do-	*Urticina piscivora*	Up-I

Source: D'Ambra and Lauritano (2020), Santhanam (2020), Santhanam et al. (2019).

TABLE 7.4 Marine Cnidarians Possessing. Cardiotoxins and Dermotoxins.

Class	Species	Toxin/Compound
Hydrozoa	*Physalia physalis*	PpV19.3
Scyphozoa	*Cyanea capillata*	Basic protein(s)
Cubozoa	*Chironex fleckeri*	CfTX-1,2
Anthozoa (Hexacorallia)	*Anemonia sulcata*	ATX-II
-Do-	*Anthopleura elegantissima*	Anthopleurin-C
-Do-	*Anthopleura xanthogrammica*	Anthopleurin-A and B
Scyphozoa	*Aurelia aurita*	Tetramine; Proteolytic enzymes*
-Do-	*Chrysaora hysoscella*	Cationic protein*
-Do-	*Cassiopea andromeda, Cassiopea xamancha*	Phospholipase A2*
-Do-	*Chrysaora quinquecirrha*	DNase*
-Do-	*Cyanea capillata*	Basic protein(s)*
-Do-	*Pelagia noctiluca*	Proteinaceous macromolecules*
Cubozoa	*Chironex fleckeri*	CTX-1,2*
-Do-	*Malo kingi*	MkTX-A; MkTX-B*

Source: D'Ambra and Lauritano (2020), Santhanam (2020), Santhanam et al. (2019).

**Dermotoxins.*

The therapeutic potential of species of the various classes of marine cnidarians is given below.

7.2.1 HYDROZOA

Aequorea victoria: A calcium-binding protein namely apoaequorin found in the venom of this luminescent jellyfish has been reported to possess anti-Alzheimer's properties by regulating calcium in the human brain and reduce memory loss and mental decline. Apoaequorin has almost a similar structure to human calcium-binding proteins and is designated as a dietary supplement that helps brain health and ageing-related memory loss. It has also been used off label in patients with multiple sclerosis. However, the FDA has not approved this compound as a therapy for memory loss or neurologic illnesses (Sudarshan et al., 2021). Further, apoaequorin has also been reported to theoretically treat or prevent age-associated dementia, Alzheimer's disease, chemotherapy-induced cognitive dysfunction ("chemobrain"), Parkinson's dementia, Creutzfeldt–Jakob disease (mad cow disease), and vascular dementia and Wernicke–Korsakoff syndrome (alcohol dementia) (https://www.verywellhealth.com/does-jellyfish-protein-work-for-memory-loss-2252421).

Physalia physalis: The toxins PpV9.4 and PpV19.3 of this species have been reported to possess antidiabetic properties with insulin secretion which is by increasing the cytosolic Ca2+ concentration. Both PpV9.4 and PpV19.3 toxins increased the fraction of secreting betacells in a 2.4- and a 2.1-fold increases, respectively (Diaz-Garcia et al., 2012).

Gonionemus vertens: An unidentified neurotoxin of this species has been reported to modulate the adhesion of macrophages. Adhesion of macrophages to material substrates plays an important role in biomedical applications, including wound healing and cancer treatment (Jayathilake and Gunathilake, 2020; https://dr.lib.iastate.edu/entities/publication/3f35276b-991c-4461-9e8d-3c6efa194b29). Macrophage is a type of white blood cell that surrounds and kills pathogenic microorganisms, removes dead cells, and stimulates the action of other immune system cells.

1 2 3

1. Aequorea victoria; 2. Physalia physalis; 3. Gonionemus vertens

Image credit: Sierra Blakely. https://creativecommons.org/publicdomain/zero/1.0/

Image credit: Biusch. https://creativecommons.org/licenses/by-sa/3.0/deed.en

Image credit: Reproduced with permission from Hans Hillewaert

7.2.2 SCYPHOZOA

Aurelia aurita: The peptide toxin, aurelin derived from the mesoglea of this moon jellyfish displayed significant activity against Gram-negative and Gram-positive bacteria (Jayathilake and Gunathilake, 2020). Sudarshan et al. (2021) reported that the antimicrobial activity of aurelin is membrane-mediated involving electrostatic interactions with membranes of bacterial cells.

The venom extract of the tentacles of this species has also shown anticoagulant activity. This extract displayed intense fibrinogenolytic activity, cleaving the α and β chains from the fibrinogen molecule, and it completely liquefied fibrin clots in 24 h (Sudarshan et al., 2021).

Cassiopea andromeda: The crude venom of this species was found to selectively induce ROS-mediated cytotoxicity by directly targeting mitochondria, isolated from cancer tissue of patients with breast adenocarcinomas (Jayathilake and Gunathilake, 2020).

Chrysaora quinquecirrha: The peptide isolated from the venom of this species displayed significant cytotoxicity against cervical cancer and alveolar epithelial carcinoma cell lines without affecting normal human lymphocytes. Further, the venom of this species has also been reported to exhibit significant immunomodulatory activity especially in higher concentrations (Jayathilake and Gunathilake, 2020). Sudarshan et al. (2021) reported that its venom showed antioxidant and antitumor properties against Ehrlich ascites carcinoma (EAC) tumor cells. Furthermore, its venom showed moderate antibacterial activity against pathogenic bacteria such as *Salmonella paratyphi*, *Klebsiella pneumoniae*, *Pseudomonas aeruginosa*, *Escherichia coli*, *Vibrio cholerae*, *Proteus mirabilis*, Proteus vul-garis and Klebsiella oxytoca, Salmonella paratyphi (Jayathilake and Gunathilake, 2020).

Cyanea capillata: A toxic protein (CcTX-1) from the fishing tentacle venom of this species has shown cytotoxicity on human hepatocyte cell line HepG2 (Lassen et al., 2011). Wang et al. (2013) reported that tentacle venom extract of this species could induce apoptosis and necrosis in NRK-52E cells.

Nemopilema nomurai: The venom (NnV) of this jellyfish has been reported to possess anticancer activity against human HepG2 cells. NnV induced potent cytotoxicity and apoptotic cell death (Lee et al., 2017). Sudarshan et al. (2021) reported that the venom of this Atlantic sea nettle has antioxidant activity. Further, according to these authors, the collagen extracts of this species stimulated the production of immunoglobulins and cytokines with a regulatory effect on the immune system. Furthermore, a glycoprotein derived from this species viz. oniumucin, had potential disease-modifying effects through the degeneration of articular cartilage in an osteoarthritis model.

Pelagia noctiluca

Anticancer and anti-inflammatory activities: Three fractions (F1, F2, and F3) of the venom of this species showed a dose-dependent cytotoxicity against human myelogenous leukemia (K562), human bladder carcinoma (RT112), and glioblastoma (U87) cell lines. These fractions also displayed a dose-dependent anti-inflammatory activity that was associated with nitric oxide (NO) production inhibition activities in IFN-γ/LPS-stimulated murine RAW 264.7 macrophages (D'Ambra and Lauritano, 2020).

Analgesic and anti-BuChE activities: The crude venom and two fractions of this species displayed analgesic and anticholinesterase activities. While the crude venom and fraction 1 possessed less significant inhibitory activity against analgesic and antibutyrylcholinestrasic models, fraction 2 possessed the highest analgesic and antibutyrylcholinestrasic effects (Ayed et al., 2012).

Rhizostoma pulmo: The crude venoms derived from the nematocyst and surrounding tissues of this species displayed significant cytotoxicity and killed V79 cells at the highest tested concentration within 2 h (Carli et al., 1996).

Rhopilema esculentum: Two toxic peptides, VKP (342 Da) and VKCFR (651 Da), of this jellyfish have been reported to display angiotensin converting-enzyme (ACE) inhibitor activity by protecting endothelial cells against oxidative damage which is through quenching the hydroxyl radical and enhancing antioxidant enzyme activity (Sudarshan et al., 2021).

Stomolophus meleagris: The toxic protein, smP90, derived from the nematocysts of this species displayed superoxide anion radical scavenging activity (Jayathilake and Gunathilake, 2020).

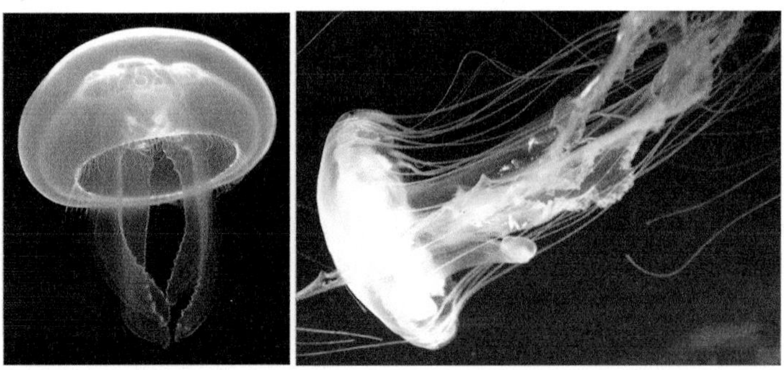

1 2

1: Aurelia aurita.; 2: Chrysaora quinquecirrha.

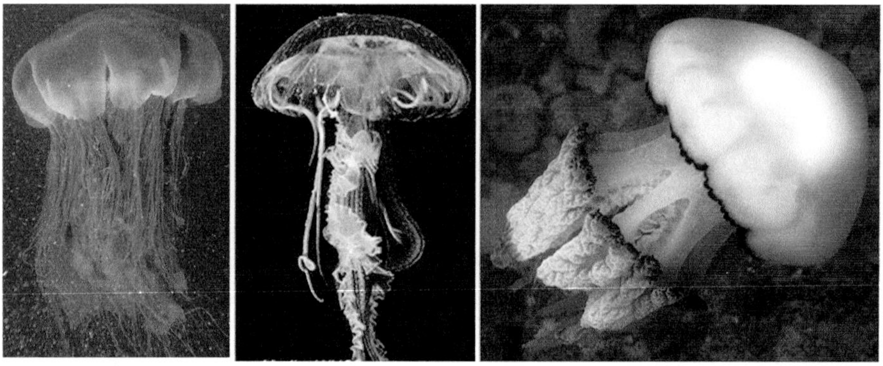

3 4 5

3: Cyanea capillata; 4: Pelagia noctiluca; 5: Rhizostoma pulmo

3. Image credit Dan Hershman . https://creativecommons.org/licenses/by/2.0/

4. Image credit: Reproduced with permission from L.P. Madin

5. Image credit: Ales Kladnik . https://creativecommons.org/licenses/by/2.0/

7.2.3 CUBOZOA

The cubozoans or box jellyfish are known for their venomosity and the species *Chironex fleckeri* is believed to be the world's most venomous animal (https://wchh.onlinelibrary.wiley.com/doi/10.1002/psb.1837). Very few species of this class of jelly fish are of therapeutic value and are listed below.

Chiropsalmus quadrumanus: The toxic extract containing peptides (characterized as cryptides, and 14 low molecular mass compounds) from this box jellyfish has been reported to possess antioxidant/neuroprotective activity. The aforesaid toxins increase neuritogenesis. Further, these toxins increase neurite outgrowth length and branching junction thereby amplifying the contact between SH-SY5Y neurons. Therefore, this extract is considered to be a promising tool for neuronal connection recovery that is essential for the treatment of neurodegenerative diseases (Arruda et al., 2021).

Chiropsoides quadrigatus *(= Chiropsalmus quadrigatus):* A mixture of toxic proteins derived from this species has shown cytotoxicity which is by its venom-induced breaking of DNA bands, leading to apoptotic cell death cancer cells. Further, the expression of its p53, a well-known tumor suppressor gene, was found to increase in human malignant glioblastoma cells (Sudarshan et al., 2021).

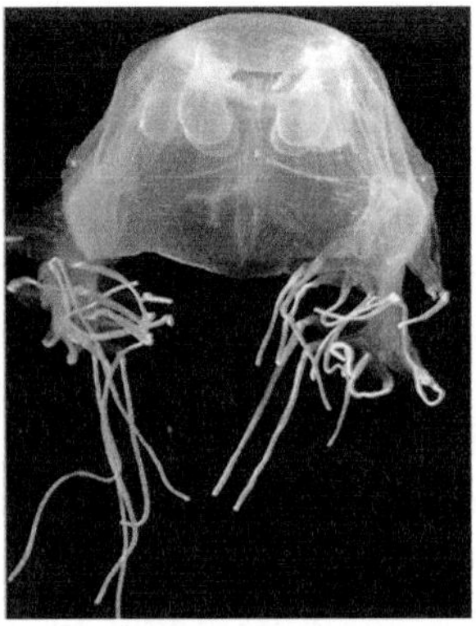

Chiropsalmus quadrumanus

Image credit: Reproduced with permission from Alvaro E.Megotto & Dr. Juliana Mozer Sciani

7.2.4 ANTHOZOA

The class Anthozoa includes the venomous animals like sea anemones, soft corals, hard corals, and sea pens. The toxins of these animals are seen in their nematocysts located on the tentacles, mucous coat that covers the animal body and acrorhagi and acontiae. The nematocyst toxins of some species of sea anemones such as *Anemonia sulcata* and *Phyllodiscus semoni* are believed to be more dangerous to humans. Persons stung by such nematocysts, local inflammations, including severe pain, redness, and edema, appear immediately. Though the composition of cnidarian venoms is not adequately known, about 250 of toxins have so far been identified (Honma and Shiomi, 2006). The majority of the sea anemone toxins are of three types, viz. pore-forming cytolysins (e.g., EqTx), neurotoxins (e.g., ATX II) acting on voltage-gated sodium channels, and neurotoxins (e.g., ShK and BgK) acting on voltage-gated Kv1 potassium channels (Honma and Shiomi, 2006). All these toxins are of great use therapeutically. Neurotoxins interacting with sodium channels could be used in the treatment of many diseases related to nerves, muscles, and cardiovascular besides forming as lead drug candidates for new anesthetics. On the other hand, the neurotoxins relating to potassium channel toxins could be potential use in the treatment of autoimmune diseases. Pore-forming toxins are believed to serve as new coagulants. The important sea anemone toxins and their major functions as reported by Wang et al. (2023) are given in the following table.

Sea anemone toxins and voltage-gated sodium channels (VGSCs/NaVs)

Sodium channels are the major proteins involved in electrical signaling in the nervous system, thereby involving in critical functions like heartbeat and brain activity. Voltage-gated sodium channels have been reported to be largely responsible for initiating and propagating actions in excitable cells, including nerve, muscle, and neuroendocrine cell types.

These VGSCs are very important as they help in maintaining homeostasis of living organisms. Cardiac, muscle, and neurological disorders often develop as a result of genetic abnormalities in VGSC genes (Mackieh et al., 2021). Several sodium channel subtypes exist, and types Nav1.2 and Nav1.6 are the major isoforms found in the brain. These different subtypes are known to exhibit differential drug sensitivity. As sodium channel modulators, sea anemone toxins either block, activate, or modify sodium ion channels. Thus, they are used in neurobiology and have larger implications in pharmacology. Sodium channel toxins from sea anemones that target VGSCs are of three types, viz. type 1–3 toxins. While type 1 and 2 toxins are composed of 46 to 49 amino acid residues, type 3 toxins are shorter peptides with 27 to 32 amino acid residues only. It is also interesting to note that the distribution of type 1 and 2 sodium channel toxins is more related to the taxonomical position of sea anemones. For example species of the family Actiniidae contain only type 1 toxins, and species of the family Stichodactylidae possess type 1 or 2 toxins or both. Apart from these 1–3 type toxins, two novel sodium channel toxins viz. calitoxins I and II with 46 amino acid residues are also seen in certain species of sea anemones. These calitoxins also act on voltage-gated sodium channels probably in a similar manner to the 1–3 type toxins. The species of sea anemones possessing 1–3 type sodium channel toxins as reported by Wang et al. (2023) are given in the following table.

The therapeutic uses of the toxins of the species of sea anemones (stony corals, Class: Hexacorallia) and soft corals (Class: Octocorallia) are given below.

7.2.4.1 HEXACORALLIA (STONY CORALS)

Actinia equina* and *Anemonia sulcata

Cytotoxicity: The toxin containing extracts of these species displayed cytotoxicity murine RAW 264.7 macrophages. At 24 h, the aqueous extract of *Actinia equina* exhibited the highest toxicity with an IC50 value of 0.629

TABLE 7.5 Toxins of Sea Anemones and Their Major Functions.

Species	Toxin	Function
Actineria villosa	Avt120; Avt I	High ATP degradation activity, hemolytic activity
Actinia equina	EqTx I, II, III, IV, V; Ae I	Hemolysis, platelet aggregation, degranulation
Actinia tenebrosa	Tenebrosin-A, B, C	Hemolytic activity
Anemonia sulcata	ATX II, III; AsKs; AsKC	Inactivation of sodium and potassium channels
Anthopleura asiatica	Bandaporin	Hemolytic activity
Anthopleura elegantissima	Anthopleurin-C	Modulating sodium channel
Anthopleura fuscoviridis	AFT-I, II	–
Anthopleura xanthogrammica	Anthopleurin-A; -B	Modulating sodium channel
Bunodosoma caissarum	Bc-III; Bc2	Activation of sodium channel subtypes 1.3 and 1.6.
Bunodosoma cangicum	Bcg-3	Induce hemolysis; prevent the combination of TTX and sodium channels.
Bunodosoma granuliferum	Bg II; Bg III; GRX; BgK	Neurotoxic; inhibit Kv1, -2, -3
Condylactis gigantea	Condylactis toxin	Hemolytic activity,
Epiactis prolifera	Epiatin A, B, C	–
Exocoelactis actinostoloides (= Parasicyonis actinostoloides)	PaTX; parasitoxin	Hemolytic activity
Homostichanthus duerdeni	HTX-I	Delays the inactivation of TTX-sensitive sodium current
Oulactis orientalis	Or-A	Low cytotoxicity and hemolytic activity
Phyllodiscus semoni	PsTX-60A; -B; -20A	Hemolytic activity
Pseudactinia varia,	Variolysin	Hemolytic activity

TABLE 7.5 *(Continued)*

Species	Toxin	Function
Radianthus crispa (= Radianthus macrodactylus; Heteractis crispa)	RTX-VI; RTX-A; RTX-S II	Cytotoxicity and hemolytic activity
Radianthus magnifica (= Radianthus paumotensis; Heteractis magnifica	RpI; RpII; RpIII; RpIV; HmK. HMgs	Modulating sodium channel
Stichodactyla gigantea (= Stoichactis kenti)	Gigantoxin I, II, III; keratin	Effect on A431 (human epidermoid carcinoma) cell line, hemolytic activity
Stichodactyla helianthus	Sticholysin I, II, III; ShK	Block Kv1.3 channel in jurkat T lymphocytes

Source: Wang et al. (2023).

mg/mL. On the other hand, at the highest concentration (1 mg/mL), the extract of *Anemonia sulcata* was found to be more toxic with time, causing a significant reduction (62.63%) in cell viability after 72 h. Further, incubation of cells with *Anemonia sulcata* (1 mg/mL) and *Actinia equina* (0.5 mg/mL) extracts leads to an increased activity of caspase-3, suggesting a caspase-dependent cell death in macrophages. These two extracts have also shown cytotoxicity of human gastric adenocarcinoma cells. However, in the same concentration range, *Actinia equina* extract was found to be more cytotoxic with an IC50 value of 0.365 mg/mL (D'Ambra and Lauritano, 2020-A-H). Giraldi et al. (197) reported that the cytolytic toxins (equinatoxins) of *Actinia equina* exhibited potent cytotoxicity on both Ehrlich carcinoma and L1210 leukemia cell lines with ED50 of a few ng/mL.

Antibacterial activity: A Na + channel blocking toxin viz. Neurotoxin 2 (ATX-II) derived from *Anemonia sulcata* has shown antibacterial activity. Further, its another neurotoxin namely Ueq 12–1 was found to inhibit the growth of a series of human pathogens including *Staphylococcus aureus* and *Corynebacterium glutamicum* (Jayathilake and Gunathilake, 2020).

Anti-inflammatory and immunomodulatory activities: The toxin homarine extracted from the tentacles of *Anemonia sulcata* demonstrated both anti-inflammatory and immunomodulatory activities (Jayathilake and Gunathilake, 2020).

Others: A cytolytic toxin (equinatoxin) of polypeptide named Ae I has also been isolated from this species. The therapeutic use of this toxin is yet to be known (Frazão et al., 2012).

Anemonia erythraea: It is known to produce three polypeptide toxins viz. AETX I, II, and III.

AETX I is made of 47 amino acid residues, is an analog of sea anemone type I toxins. On the other hand, AETX II and III are composed of 59 residues and are quite distinct from the known sea anemone polypeptide toxins. Interestingly, both toxins have sequence similarities with neurotoxins (Shiomi et al., 1997; Frazão et al., 2012). The therapeutic benefits of these toxins are not known.

Anemonia sulcata* and *Anemonia viridis: *Anemonia sulcata* has been reported to possess a variety of toxins such as sodium channel blocking toxins such as ATX-I, -III, -IV; potassium channel blocking toxins such as kalicludin-1,-2,-3; BDS-I,-II; kaliseptin; and other toxins like ATX-II and SA5II. On the other hand, *Anemonia viridis* showed the presence of only ATX-I, II, III, V. All these toxins have shown antimicrobial activities (Trapani et al., 2014; Frazão et al., 2012).

TABLE 7.6 1–3 Type Sodium Channel Toxins of Sea Anemones.

Species source	Toxin	Type of NaV Toxin
Actinia equina	Ae I	Type I
Anemonia erythraea	AETX I	-Do-
Anemonia sulcata	ATX Ia, Ib; ATX II, ATX V	-Do-
Anemonia viridis	ATX II and II	
Antheopsis maculata	Am III	-Do-
Anthopleura elegantissima	ApC; APE 1 to APE 5	-Do-
Anthopleura fuscoviridis	AFT I and II	-Do-
Anthopleura xanthogrammica	ApA and ApB; PCR1–2, 2–1, 2–5, 2–10, 3–6, and 3–7	-Do-
Anthopleura sp.	Hk2	-Do-
Bunodosoma caissarum	Bc III	-Do-
Bunodosoma cangicum	Cangitoxin	-Do-
Bunodosoma granuliferum	Bg II and III	-Do-
Condylactis gigantea (= Condylactis Passiflora)	Cp I and II	-Do-
Halcurias sp.	Halcurin	-Do-
Radianthus crispa (= Radianthus macrodactylus; Heteractis crispa)	Rc I	-Do-
Stichodactyla gigantea	Gigantoxin II	-Do-
Radianthus crispa (= Radianthus macrodactylus; Heteractis crispa).	RTX I, II, III, IV and V	Type II
Radianthus magnifica (= Radianthus paumotensis; Radianthus (Heteractis) paumotensis	Rp II, III	-Do-

TABLE 7.6 *(Continued)*

Species source	Toxin	Type of NaV Toxin
Stichodactyla gigantea	Gigantoxin III	-Do-
Stichodactyla helianthus	Sh I	-Do-
Anemonia viridis	ATXIII	Type III
Dofleinia armata	ATXIII; Da I and II	-Do-
Entacmaea quadricolor (= Entacmaea ramsayi; Entacn aea actinostoloides)	ATXIII; Er I; Ea I(PaTx)	-Do-

Source: Mackieh et al. (2021).

***Antheopsis maculata*.:** The peptide toxins AmI, II, and III (with 26, 46, and 47 amino acid residues respectively) have been isolated from this species. Among them AmIII is an analogue of the well-known type 1 sea anemone sodium channel toxins. The therapeutic properties of these toxins are not known (Honma and Shiomi, 2006).

***Anthopleura elegantissima*:** Its toxins such as APE 1-1, APE 2-1, and APE 5-3 produced a positive inotropic effect (increased contractility) in mammalian heart muscle, although these toxins differ in their potency (Bruhn et al., 2001).

***Anthopleura xanthogrammica*:** Its toxin A (ApA), an effective cardiotonic polypeptide, has been reported to play an important role in the heart (Frazão et al., 2012).

***Calliactis parasitica*:** This species has been reported to possess tetrodotoxin (TTX), the therapeutical benefits of which have been elaborated already (Rossignoli et al., 2023).

Palythoa caribaeorum

Antitumor and antiparasitic activities: The peptide toxins of this species showed major inhibitory effects on the tumoral cell lines namely glyoblastoma cell line U251 and human lung cancer cells SKLU-1 with inhibition values of 52.61 and 41.5%, respectively. Further, two toxic extracts of this species exhibited antiparasitic activity against the flagellated protozoan viz. Giardia intestinalis in a dose-dependent manner with IC90 values of 116 and 603 μg/mL, respectively (Lazcano-Perez et al., 2018).

Neuroprotective activity: A novel Kunitz-like peptide (PcKuz3) derived from this species has the potential to treat neurodegenerative diseases. Further, its another type of kunitz peptide was displayed neuroprotective activity against 6-hydroxydopamine (Jayathilake and Gunathilake, 2020).

***Palythoa toxica*:** The palytoxin of this species has been reported to modulate bone cell activity.

In mouse calvarias cultured ex vivo, palytoxin proved to be a potent inducer of bone resorption at low doses. Therefore, palytoxin has the potential for the treatment of diseases where bone formation is increased (Iduarte and Juárez, 2020).

Palythoa variabilis *(= Protopalythoa variabilis):* A novel toxic peptide (PpVα) of this species has been reported to have neuroprotective activity by suppressing the 6-hydroxydopamine (6-OHDA) (= Oxidopamine) which is known to induce neurotoxicity. Further, the above toxin also prevented the 6-OHDA-induced excessive reactive oxygen species (ROS) generation.

Excess ROS is known for its oxidative stress/damage which is often implicated in various disease states such as neurodegeneration, aging, atherosclerosis, diabetes, and cancer (Jayathilake and Gunathilake, 2020).

Radianthus crispa *(= Heteractis crispa; Radianthus macrodactylus).*

Anti-inflammatory activity: The Kunitz-type polypeptide toxins viz. HCRG1 and HCRG2 have been reported to show anti-inflammatory activity by reducing the secretion of the pro-inflammatory mediators tumor necrosis factor-α (TNF-α), interleukin 6 (IL-6), and proIL-1β expression in lipopolysaccharide (LPS)-activated macrophages. Further, a HCGS 1.20 recombinant polypeptide toxin isolated from this species displayed anti-inflammatory activity by inhibiting the lipopolysaccharide-stimulated increase in the concentration of nitric oxide in RAW 264.7 mouse macrophages (D'Ambra and Lauritano, 2020).

Antihistamine and analgesic activities: Two recombinant peptides isolated from this species acted as Kunitz-type serine protease inhibitors rHCGS1.19 and rHCGS1.36 and displayed antihistamine activity at 10 μM by inhibiting the increase in the calcium ion concentration in murine bone marrow-derived macrophages developed by histamine at 62 and 84%, respectively. Further, at a concentration of 0.5 mg/kg, HCGS1.36 and HCGS1.10 exhibited analgesic effects in the thermal pain stimulation model (D'Ambra and Lauritano, 2020; Monastyrnaya et al., 2019).

Antidiabetic activity: A β-Defensin-like peptide toxin viz. magnificamide isolated from the mucus of this species showed antidiabetic activity by inhibiting human saliva alpha-amylase (AMY2A) with Ki's equal to 7.7 nM. This peptide is therefore considered to be a potential drug candidate for diabetes treatment (D'Ambra and Lauritano, 2020).

Antibacterial activity: Jayathilake and Gunathilake (2020) reported that its cytolytic actinoporins, cardio stimulatory proteins, and cytolysins have shown antibacterial activity against pathogenic bacteria including *Staphylococcus aureus* and *Salmonella typhi*.

Others: This species has also been reported to accumulate palytoxin, the therapeutic uses of which have already been given elsewhere.

Stichodactyla helianthus

Antidiabetic activity: The glycosidase toxin, helianthamide isolated from this species exhibited antidiabetic activity by serving as an excellent source of human pancreatic α-amylase inhibitors with possible applications in the control of blood sugar levels in the management of diabetes mellitus patients (D'Ambra and Lauritano, 2020).

Antiparasitic activity: Two toxins viz. sticholysine I and II derived from this species have shown antiparasitic activity against the flagellate protozoan viz. *Giardia duodenalis* (Jayathilake and Gunathilake, 2020).

Kv1.3 (voltage-gated potassium (K) channel) inhibitor: A 37-amino acid synthetic peptide viz. dalazatide (= ShK-186 and SL5) derived from this species has been reported to inhibit Kv1.3. Inhibition of Kv1. 3 improves the insulin sensitivity by increasing the amount of GLUT4 at the plasma membrane. Kv1.3 is therefore identified as a pharmacologic target for the treatment of diabetes (Xu et al., 2004; Tarcha et al., 2017).

Stichodactyla mertensii: Pore-forming toxins (PFTs) namely cytolysins and cardio stimulatory proteins, cytolytic actinoporin of this species have been reported to display antimicrobial properties against human pathogens such as *Staphylococcus aureus*, *Salmonella typhi*, *Escherichia coli*, and *Mycobacterium tuberculosis* (Jayathilake and Gunathilake, 2020).

Urticina crassicornis: The pharynx extract of this species yielded neurotoxin (BDS)-like antimicrobial peptides (AMPs)-Crassicorin-I and its putative homolog (Crassicorin-II). Among these toxins, Crassicorin-I exhibited antimicrobial activity against *Bacillus subtilis*, *Escherichia coli*, and *Salmonella enterica* (Kim et al., 2017).

Urticina eques: The neurotoxin, Ueq 12^{-1}, derived from this species displayed antibacterial activity against human pathogens like Corynebacterium glutamicum, *Staphylococcus aureus* (Jayathilake and Gunathilake, 2020).

1 2 3

1: Actinia equina; 2: Anemonia sulcata; 3: Anemonia viridis; 4: Anthopleura elegantissima; 5: Anthopleura xanthogrammica; 6: Calliactis parasitica; 7: Palythoa caribaeorum; 8: Palythoa toxica; 9: Stichodactyla helianthus; 10: Stichodactyla mertensii; 13: Urticina eques

1. Image credit: Attrattorestrano. https://creativecommons.org/licenses/by-sa/3.0/deed.en

2. Image credit: Attrattorestrano. https://creativecommons.org/licenses/by-sa/3.0/deed.en

3. Image credit: Diego Delso. https://creativecommons.org/licenses/by-sa/4.0/

4. Image credit: Ed Bierman. https://creativecommons.org/licenses/by/2.0/deed.en

5. Image credit: Stan Shebs. https://creativecommons.org/licenses/by-sa/3.0/

6. Image credit: Diego Delso, https://creativecommons.org/licenses/by-sa/4.0/

7. Image credit: Jean Bize. https://creativecommons.org/licenses/by-sa/4.0/deed.en

8. Image credit: iNaturalist. https://creativecommons.org/licenses/by-sa/4.0/

9. Image credit: Linél. https://creativecommons.org/licenses/by-sa/3.0/deed.en

10. Image credit: Brocken Inaglory. https://creativecommons.org/licenses/by-sa/3.0/

11. Image credit: Ed Bierman. https://creativecommons.org/licenses/by/2.0/

7.2.4.2 OCTOCORALLIA (SOFT CORALS)

The toxins of soft coral species have been reported to possess several bioactivities including antitumor, antimalarial, anti-inflammatory, antimicrobial, and antituberculosis as shown in the following table.

Fairly elaborate therapeutic studies have been made on the toxins of certain species of soft corals and such species and their bioactivities are given below.

Isis hippuris: The sesquiterpene toxins derived from this species showed anticancer activity against human colon adenocarcinoma HT-29 and mouse lymphoma P388 with the lowest IC50 value of 0.000005 μg/mL (D'Ambra and Lauritano, 2020).

Leptogorgia punicea: The polyoxygenated steroids namely punicinols A, B, C, D, and E produced by this species exhibited cytotoxicity. Among these toxins, punicinols A and B were found to be most active against against human lung cancer A549 cells with IC50 values of 9.7 and 9.6 μM, respectively (D'Ambra and Lauritano, 2020).

Leptogorgia setacea: The diterpene toxin, epoxypukalide-induced β-cell proliferation was found to be mediated by ERK1/2 activation targeting cyclin D2 and cyclin E. Based on these findings, it is suggested that epoxypukalide may be of great use in the treatment of diabetes (López-Acosta et al., 2013).

Leptogorgia virgulata: Its homarine and/or a homarine analog showed antimicrobial activity against *Vibrio harveyii* and *Micrococcus luteus* (Jayathilake and Gunathilake, 2020).

Litophyton arboretum: Its unidentified toxin showed anticancer activity against human cervical cancer cell lines and leukemia cancer cell lines (Jayathilake and Gunathilake, 2020).

Stragulum bicolor: The macrolide toxins viz. Amphidinolide (AMP)-PX1, -2, -3, and Stragulin A derived from this species have shown anticancer activity against Melanoma (A2058), Ovarian Adenocarcinoma (A2780), and Neuroblastoma (SK-N-BE-2) cell lines. Among these toxins, AMP-PX1 displayed potent activity against A2780 cell line with an IC50 value of 4.3 μM, and Stragulin A against A2058 cell line with an IC50 value of 0.18 μM (Nuzzo et al., 2019).

TABLE 7.7 The Toxins of Soft Coral species.

Species Source	Toxin	Chemistry	Bioactivity	References
Antillogorgia acerosa (= *Pseudopterogorgia acerosa)*	Bis(pseudopterane) amine	Dialkylamine	Antitumor	Khoei (2015)
Antillogorgia bipinnata (= Pseudopterogorgia bipinnata	Bipinnapterolide B	Terpenoid	Antituberculosis	-Do-
-Do-	Caucanolide A and D	Diterpenoids	Antimalarial	-Do-
Antillogorgia elisabethae (= *Pseudopterogorgia elisabethae)*	Pseudopterosin X	-Do-	Antimicrobial	-Do-
-Do-	Ileabethoxazole	-Do-	Antituberculosis	-Do-
-Do-	Homopseudopteroxazole	-Do-	-Do-	-Do-
-Do-	Caribenols A and B	Terpenoids	-Do-	-Do-
-Do-	Elisapterosin B	Diterpenoid	-Do-	-Do-
-Do-	Aberrarone	-Do-	Antimalarial	-Do-
Antillogorgia kallos (= *Pseudopterogorgia kallos)*	Bielschowskysin	-Do-	Antimalarial and antitumor	-Do-
Antillogorgia rigida (Pseudopterogorgia rígida)	Curcuphenol	Terpenoid	Antimicrobial	-Do-
Antillogorgia sp *(Pseudopterogorgia* sp*)*	Secosterols	Sterol	Antitumor and anti-inflammatory	-Do-
Briareum stechei (= Briareum excavata)	Briaexcavatin E	Diterpenoid	Anti-inflammatory	-Do-
-Do-	Briaexcavatolides [illegible] and P	-Do-	Antitumor	-Do-
-Do-	Briarellin D, K, and L	-Do-	Antimalarial	-Do-
Eunicea fusca	Fuscisides	-Do-	Anti-inflammatory	
Eunicea sp.	Sesquiterpenoids	–	Antimalarial	-Do-

TABLE 7.7 *(Continued)*

Species Source	Toxin	Chemistry	Bioactivity	References
Isis hippuris	Suberosenol B	Terpenoid	Antitumor	-Do-
-Do-	Polyoxygenated steroids	Steroid	-Do-	-Do-
-Do-	A –nor-hippuristanol	-Do-	-Do-	-Do-
-Do-	Isishippuric acid B	-Do-	-Do-	-Do-
-Do-	Sesquiterpenes	–	Anticancer	D'Ambra and Lauritano (2020)
Junceella fragilis	Frajunolides B and C	Terpenoids	Anti-inflammatory	Khoei (2015)
Leptogorgia punicea	Punicinols A, B, C, D	Steroids	Cytotoxic	D'Ambra and Lauritano (2020)
Leptogorgia setacea	Epoxypukalide	Diterpene	Antidiabetic	López-Acosta et al. (2013)
Leptogorgia virgulata	Homarine	Cyclic amine alkaloid	Antimicrobial	Jayathilake and Gunathilake (2020)
Litophyton arboretum	Toxin	–	Anticancer	-Do-
Pinnigorgia flava (= *Euplexaura flava)*	Butenolide	Lipid	Anti-inflammatory	Khoei (2015)
Ptilosacrcus gurneyi	Ptilosarcone	–	Neuroprotective	Wratten et al. (1977)
Stragulum bicolor	Amphidinolide PX1,2,3; stragulin	Macrolides	Cytotoxic	Nuzzo et al. (2019)

1: Antillogorgia bipinnata; 2. Briareum sp.; 3. Junceella fragilis; 4. Leptogorgia punicea; 5. Leptogorgia virgulata; 6. Litophyton arboretum; 7. Stragulum bicolor

1. *Image credit*: Public domain
2. *Image credit*: Haplochromis . https://creativecommons.org/licenses/by-sa/3.0/
3. *Image credit*: Anne Hoggett . https://creativecommons.org/licenses/by/3.0/
4. *Image credit*: Reproduced with permission from South Carolina Department of Natural Resources (SCDNR)
5. *Image credit*: Public domain
6. *Image credit*: Derek Keats. https://creativecommons.org/licenses/by/2.0/
7. *Image credit*: Reproduced with permission from Dr. Ralf T. S. Cordeiro

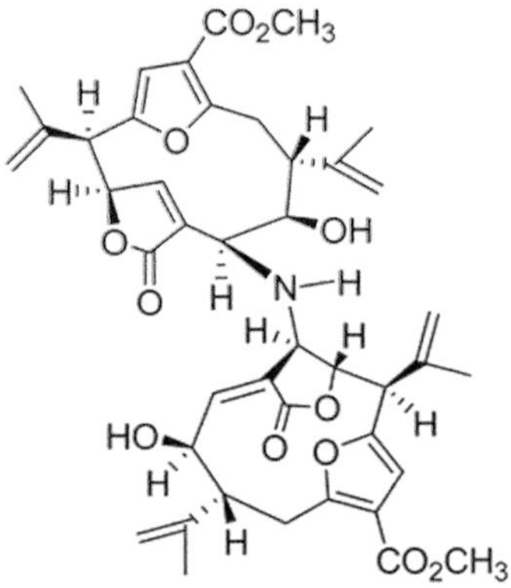

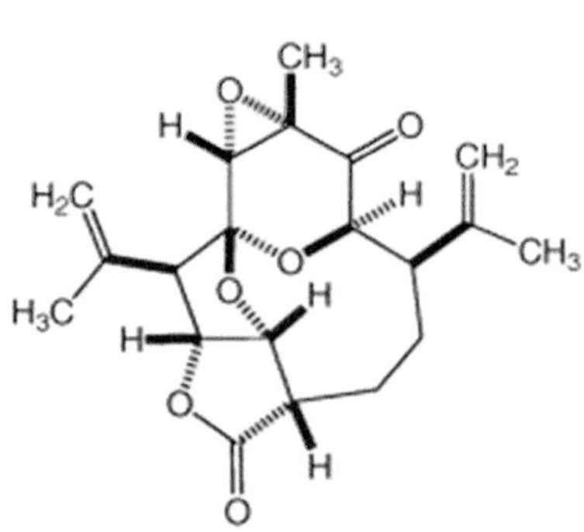

Bis(pseudopterane) amine *Bipinnapterolide B*

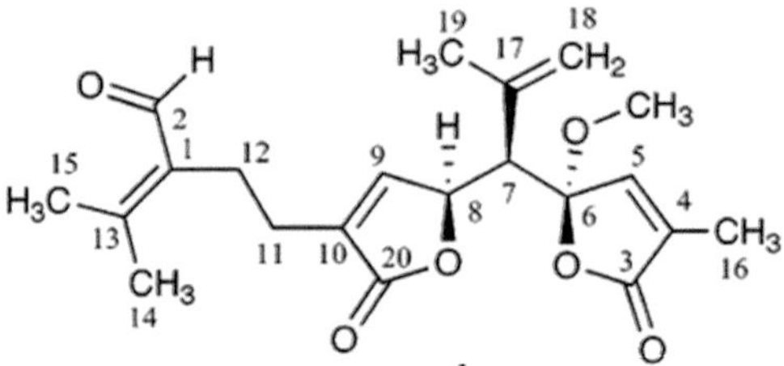

1

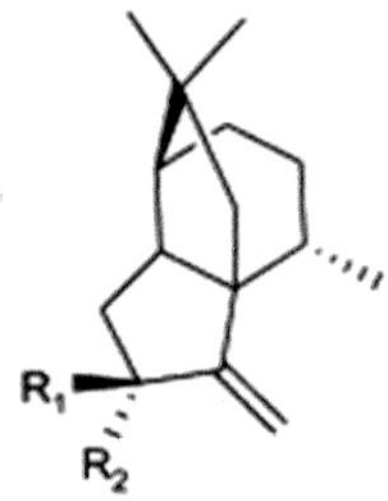

Caucanolide A. *Suberosenol B*(**R1=H; R2=OH**)

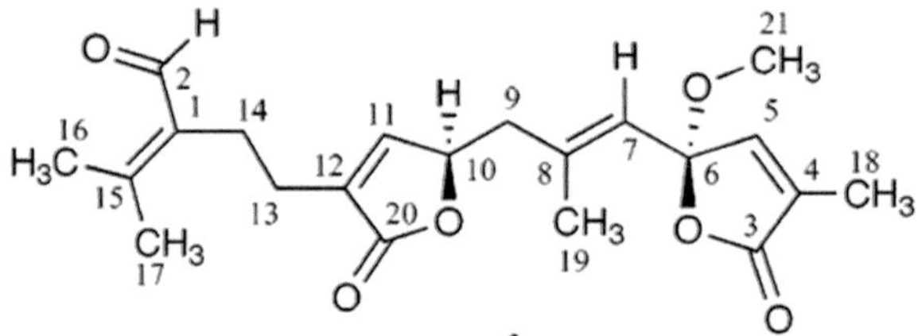

4

Caucanolide D

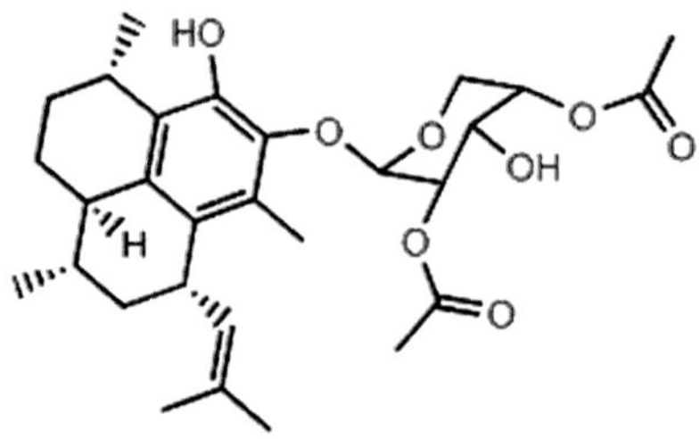

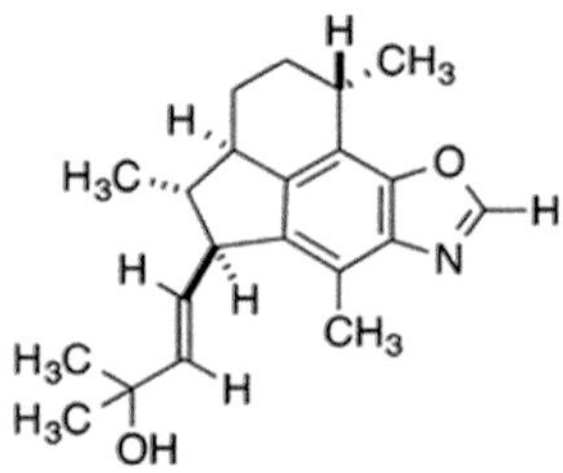

Pseudopterosin X. Ileabethoxazole

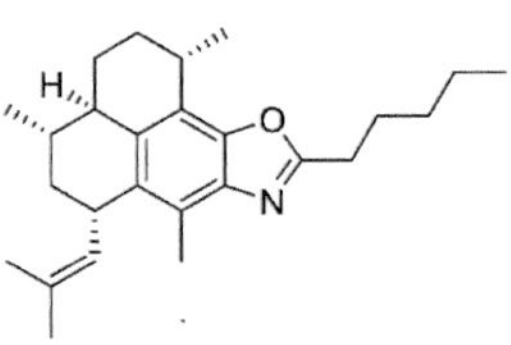

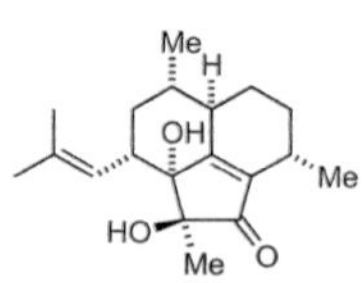

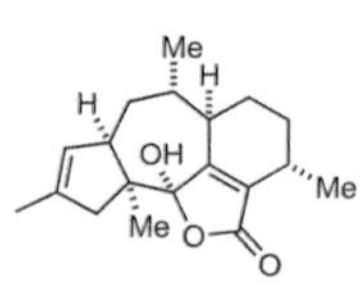

Homopseudopteroxazole. Caribenol B. Caribenols A

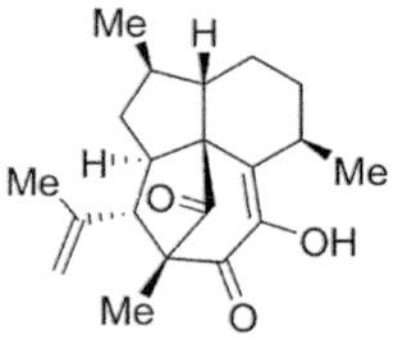

Elisapterosin B.

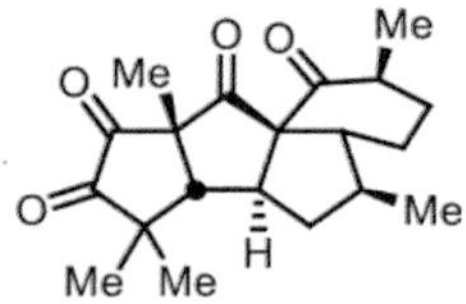

Aberrarone.

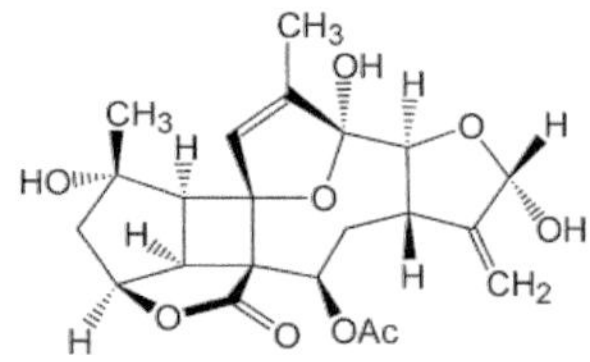

Bielschowskysin

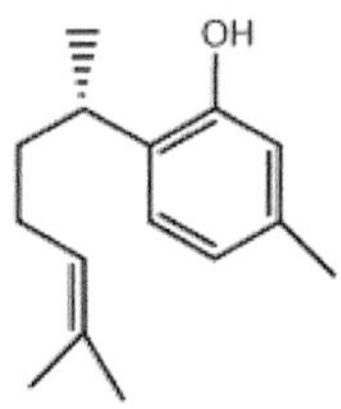

Curcuphenol.

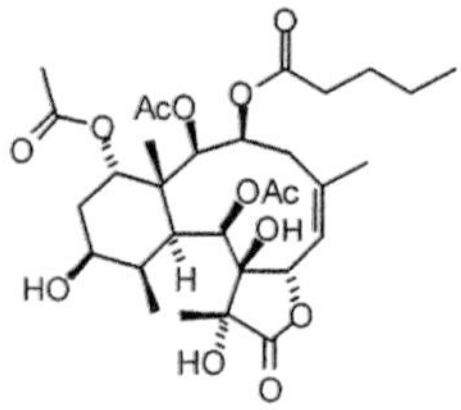

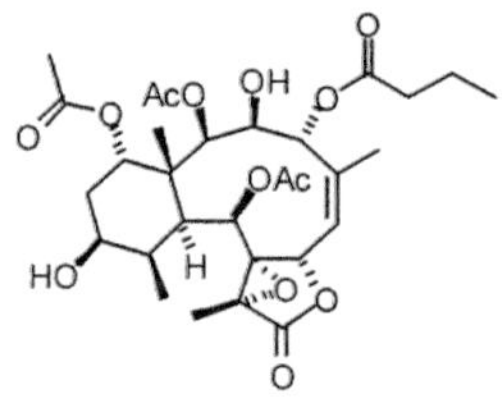

Briaexcavatolides L,P

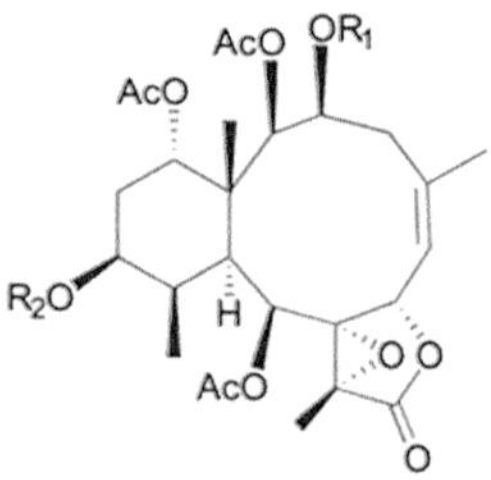

2: R_1 = Ac, R_2 = $CO(CH_2)_2CH{=}CH_2$
3: R_1 = Ac, R_2 = $COCH_2CH(CH_3)_2$

Briaexcavatin E

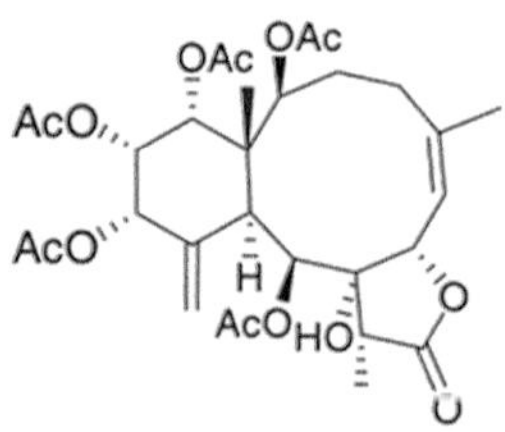

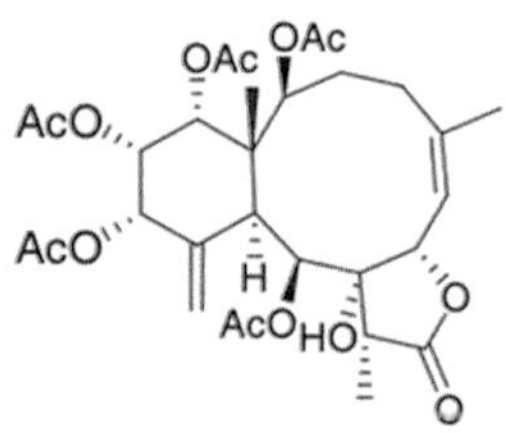

Frajunolide B C

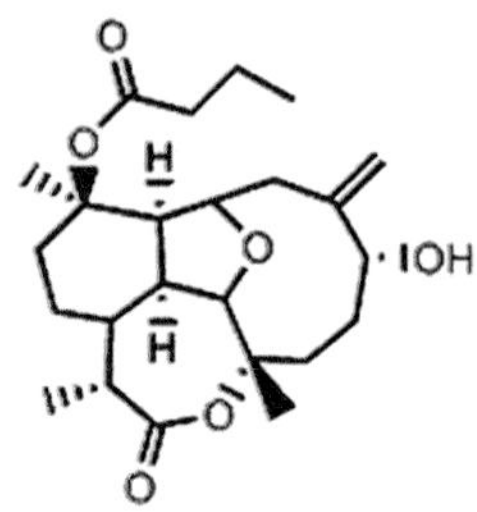

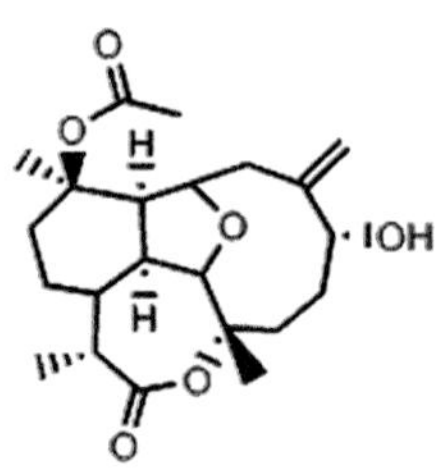

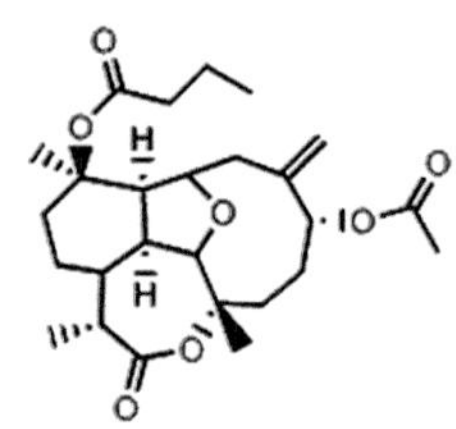

Briarellin D K L

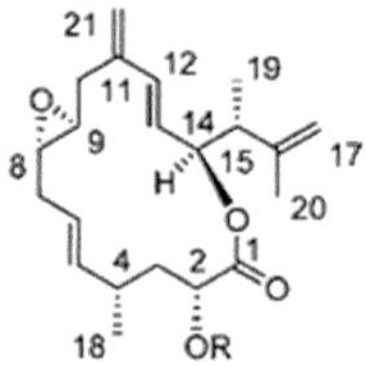

R=H; R=Ac

Amphidinolide PX1, -PX2

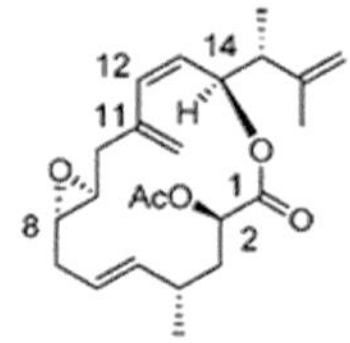

AMP PX3.

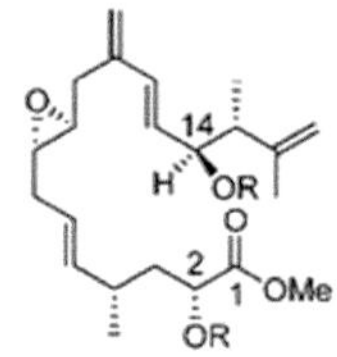

Stragulin A

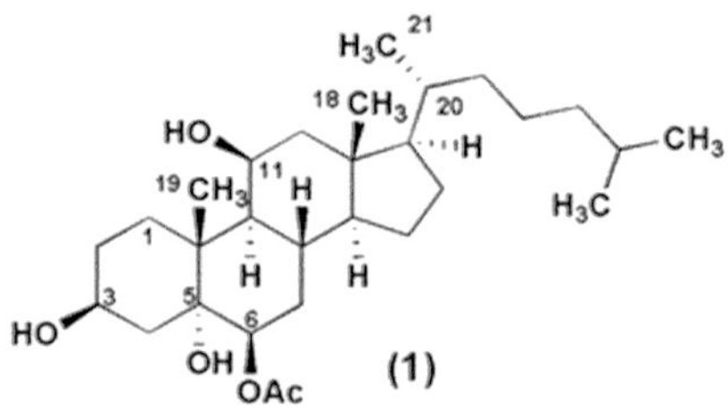

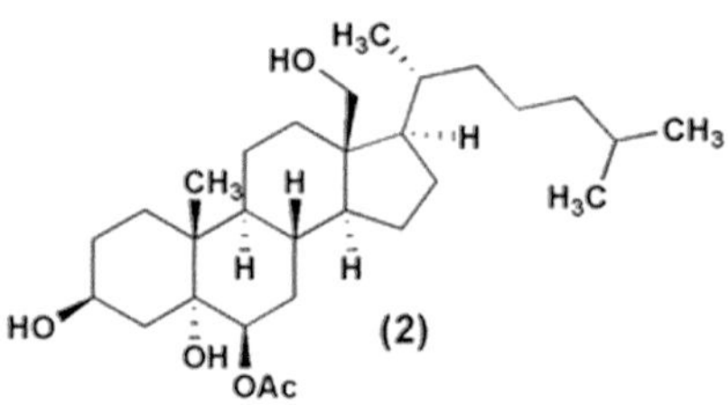

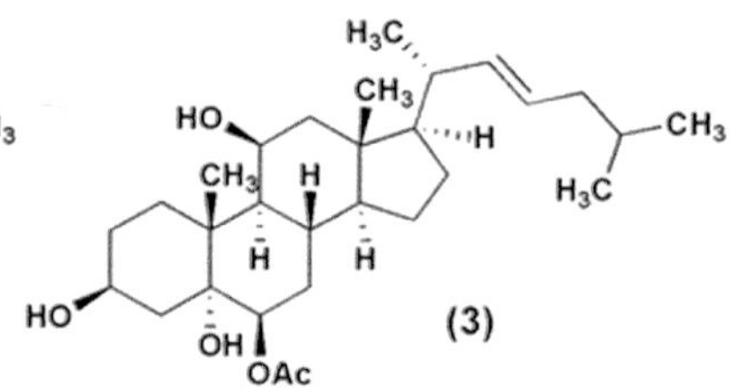

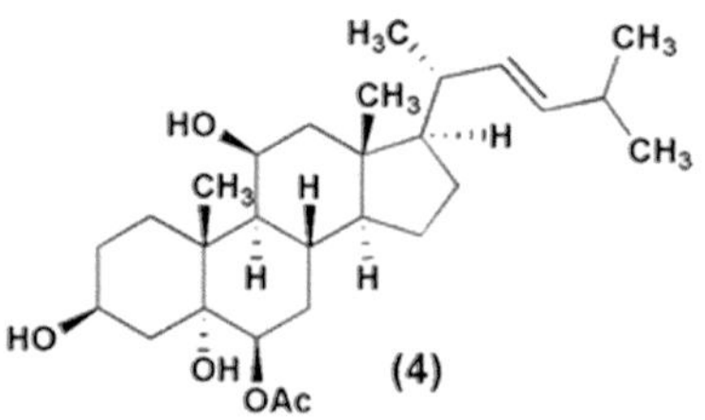

1-4: Punicinols A, B, C, D.

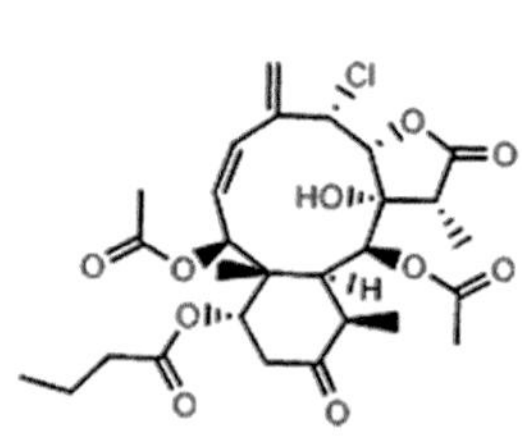

Ptilosarcone

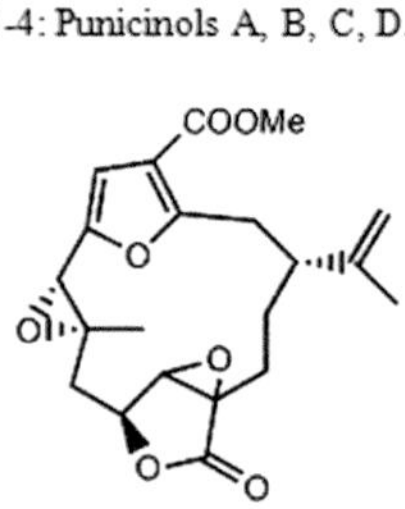

Epoxypukalide

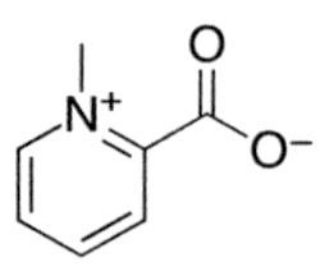

Homarine.

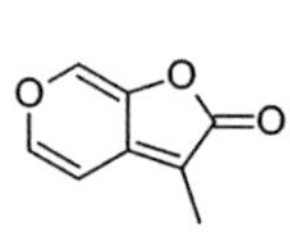

Butenolide

7.3 THERAPEUTIC POTENTIAL OF TTX OF MARINE FLATWORMS

The marine flatworm species such as the acoel flatworm viz. Amphiscolops sp., and polyclad flatworms such as *Planocera multitentaculata*, *Planocera reticulata* and *Stylochoplana* sp. have been reported to accumulate toxins collected from their prey. Fruther, these worms are also considered to be important suppliers of such toxins to other organisms including fish through the marine food chain.

***Planocera multitentaculata*, *Planocera reticulata*, *Stylochoplana* sp:** These worms have been reported to accumulate TTX, the concentration of which ranged from 468 and 3634 μg (Ueda et al., 2018). The therapeutic potential of TTX collected from these species is yet to be known.

***Amphiscolops* sp.:** Kobayashi, et al. (1988) reported that the toxin-producing dinoflagellate *Amphidinium* sp. was isolated from this Okinawan flatworm. The toxins of *Amphidinium* sp. viz. amphidinolide A, B have shown antitumor activity as already discussed (Kobayashi, et al., 1988).

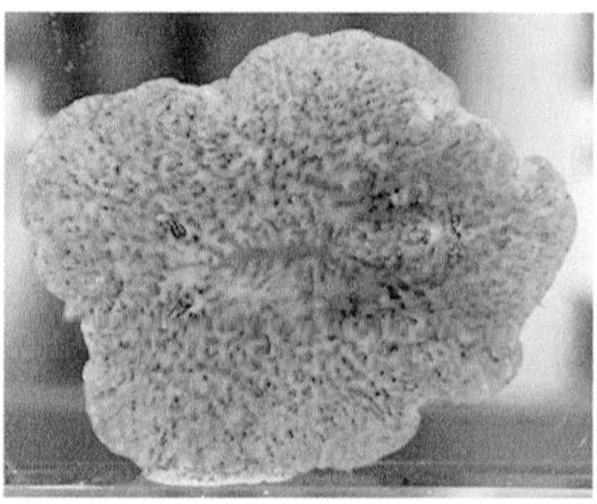

1

2

1. Planocera reticulata; 2. Amphiscolops sp.

2. Image credit: Reproduced with permission from Mario Munaretto, Biologiamarina.org

4. Image credit: Reproducd with permission from Dr. Scott Johnson

7.4 THERAPEUTIC POTENTIAL OF TTX OF MARINE RIBBON WORMS (NEMERTINE WORMS)

Among the marine ribbon worms (nemertine worms), *Cerebratulus lacteus, Notospermus geniculatus, and Yininemertes pratensis* have been reported to produce peptide toxins.

Cerebratulus lacteus: This species has been reported to produce a polypeptide cytotoxin viz. cytotoxin A-III; the therapeutic potential of this toxin is yet to be known (Luo et al., 2018).

Notospermus geniculatus: This species is also known to produce cytotoxin A-III which possesses antithrombotic, neuroprotective, and anticancer properties corresponding to their inhibitory activities of Serine protease, ion channel, and metalloproteinases. The important targets of these are blood clot, modulation of VGSC, and breast cancer progression respectively (Luo et al., 2018). Göransson et al. (2019) reported that this species produces several toxins with bioactivities as given in the following table.

TABLE 7.8 Toxins with Bioactivities Produced by *Notospermus geniculatus.*

Toxin	Function
Acidic phospholipase A2 EC-I	Inhibits platelet aggregation
Delta-actitoxin-Amc1a	Inhibits VGSC
Mu-theraphotoxin-Hhn2a 4	Blocks TTX-sensitive VGSC
Snaclec agkicetin-C subunit beta	Antithrombotic action
Snaclec bothroinsularin subunit alpha	Inhibits platelet aggregation
Snaclec dabocetin subunit alpha	Inhibits platelet aggregation

Source: Göransson et al. (2019).

Yininemertes pratensis: This estuarine hetero-nemertean worm has been reported to produce neurotoxic tetrodotoxin (TTX), 5,11-dideoxy TTX (m/z 288), and 11-norTTX-6(S)-01 (m/z 305.97) which possess cytotoxicity (Kwon et al,. 2017).

7.5 THERAPEUTIC POTENTIAL OF TTX OF CHAETOGNATHA

Among the marine chaetognath arrow worms, venomous species such as *Eukrohnia hamata*, *Pseudosagitta lyra* (= *Flaccisagitta lyra*), *Parasagitta elegans*, and *Zonosagitta nagae*, have been reported to possess TTX. The common marine bacterial species *Vibrio alginolyticus* associated with these arrow worms is believed to be responsible for the TTX of the aforesaid arrow worms. Though the therapeutic uses of TTX of other marine animals are fairly known, the TTX of chaetognaths needs further study to identify its therapeutic potential (Thuesen and Kogure, 1989).

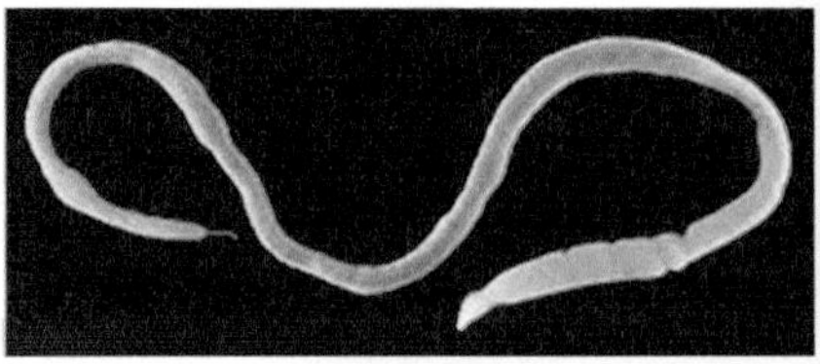

1

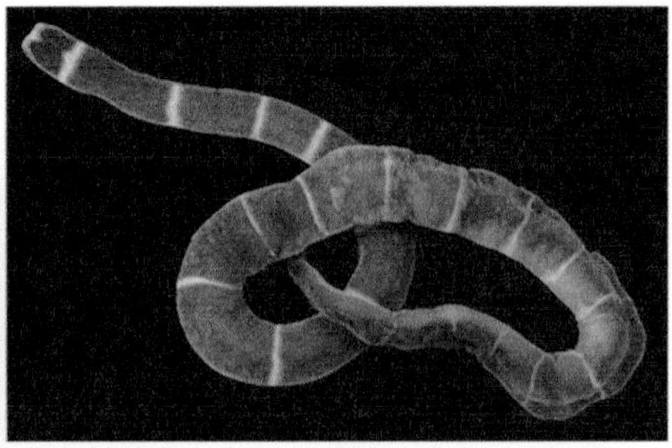

2

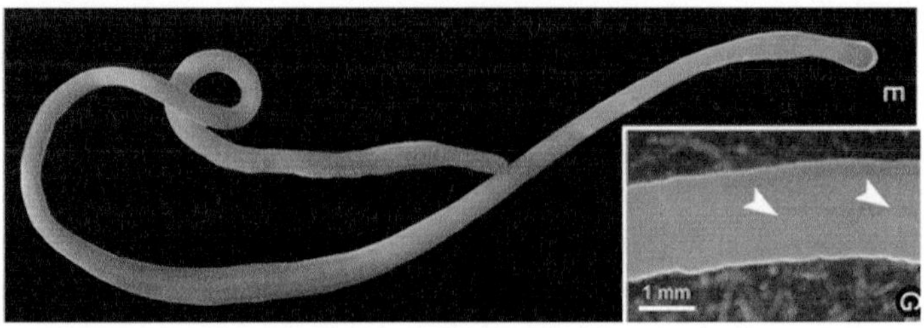

3

1. Cerebratulus lacteus; 2. Notospermus geniculatus; 3. Yininemertes pratensis

1. Image credit: Eric A. Lazo-Wasem. https://creativecommons.org/publicdomain/zero/1.0/

2. Image credit: Reproduced with permission from Denis Riek

3. Image credit: Reproduced with permission from Hiroshi Kajihara

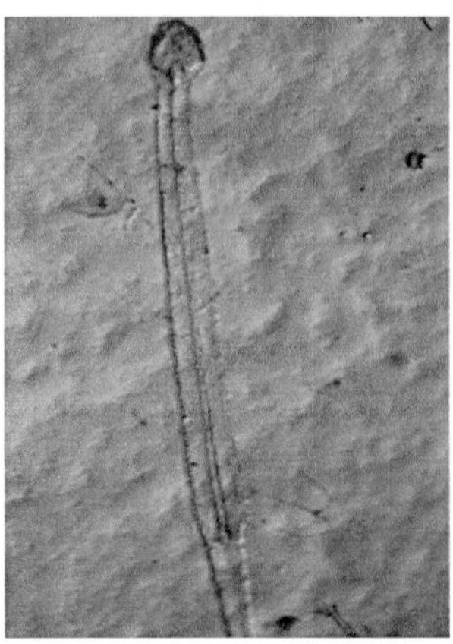

1. Parasagitta elegans

1. Image credit: MarinFaunistik . https://creativecommons.org/licenses/by-sa/4.0/

7.6 THERAPEUTIC POTENTIAL OF TTX OF MARINE ANNNELIDA

There are approximately 20,000 described annelid species that are a promising but often neglected group for venomics research. Although the salivary secretions of the potent neurotoxins of glycerid bloodworms have been widely studied, venom in most of the marine annelid worms remains largely uncharacterized. The toxins of marine annelid species and their probable therapeutic potential are given below.

Arenicola marina: The peptide toxins viz. arenicins produced by the coelomocytes of this lugworm have shown antimicrobial and cytotoxic activities.

Antimicrobial activity: At micromolar concentrations, these arenacins have been reported to demonstrate significant antimicrobial activity toward a wide range of Gram-positive and Gram-negative bacteria, as well as of fungi (Umnyakova et al., 2018). Ovchinnikova et al. (2004) reported that both arenicins 1 and 2 were found to be active against Gram-negative *Escherichia coli*, strain ML-35p and fungus *Candida albicans*, strain 820 with MIC values of 4 and 4.5 L g/mL, respectively. Sandvang et al. (2008) stated that its arenacin-3 showed very potent activity against *Klebsiella pneumoniae, Salmonella enterica*, *Pseudomonas aeruginosa*, and *Escherichia coli* with MIC values of < 1 mg/mL. Edwards et al. (2022) reported that its arenacin-3 displayed antibacterial activity against the strains of *Escherichia coli* ATCC 25922, *Klebsiella pneumoniae* ATCC 13883, *Acinetobacter baumannii* ATCC 19606, *Pseudomonas aeruginosa* ATCC 27853, *Bacillus subtilis* ATCC 6051, and *Staphylococcus aureus* ATCC 43300 with MIC values of 0.5, 1, 0.5, 1, 8, and >32 μg/mL, respectively.

Cytotoxicity: Rodrigo et al. (2021) stated that its arenacin displayed cytotoxicity toward cancer cell line with proapoptotic effects. Further, its arenacin-3 exhibited anticancer activity against mammalian HEK-293 cells (Edwards et al., 2022).

Eulalia viridis: A fluorescent proteinaceous toxin found in the skin mucus viz. phyllotoxins showed neuromuscular effect. Further, its purified mucus toxins that are mainly cysteine-rich proteins exhibited cytotoxicity against MCF7 human adenocarcinoma breast cancer cell line with an IC50 value of 0.154 μM. Its higher toxicity toward this cell line can bring bioprospecting for antiumor agents (Rodrigo et al., 2018). Ceuvas et al. (2018) reported that its phyllotoxins had antimicrobial activity by strongly inhibiting the marine pathogenic bacterium *Vibrio fischeri*.

***Eulalia* sp.:** The purified toxin-containing extract of this species displayed higher antiproliferative effect toward MCF7 and A2780 cancer cell lines with IC50 values of 0.15 and 0.08 μg/μL, respectively. However, this extract displayed moderate activity with other cancer cell lines viz. HCT, A549, and K562 with IC50 values of 0.24, 0.46, and 0.37 μg/μL, respectively (Rodrigo et al., 2021).

Eurythoe complanata, Hermodice carunculata,* and *Paramphinome jeffreysii: A total of 13 distinct toxin classes have been identified from these toxic annelid species. These toxins belong to functional categories viz. protease inhibitors, neurotoxins, proteolytic enzymes, and other proteins. It is also reported that both *Eurythoe complanata* and *Hermodice carunculata* had the most diverse toxin homologs viz. Kazal-domain and Kunitz-type protease inhibitors. On the other hand, neurotoxins including agatoxin, ctenitoxin, hainantoxin, and latrotoxin homologs were reported to be the second most diverse expressed putative toxins of *Paramphinome jeffreysii* (Verdes et al., 2018).

Glycera convoluta: Its toxin, GLTx, has been reported to act as selective blockers of Cav2.2 in human embryonic kidney cells (HEK293). Based on this activity, GLTx may represent a promising tool with great potential to explain the role of Cav2.2 in a wide range of neuronal processes in the human central nervous system (Meunier et al., 2002).

Glycera dibranchiata, Glycera tridactyla,* and *Glycera fallax: The venom of these species is a complex and heterogeneous cocktail including neurotoxins (glycerotoxin) and pore-forming toxin. The protein venom and Cavv2.2 agonist viz. glycerotoxin is known to interact with N-type calcium channels (Cav2.2) and upregulate their activity reversibly (Cabral, 2021; Ceuvas et al., 2018; Richter et al., 2017).

Hediste diversicolor: The peptide toxin, hedistin, of this species has been reported to show antibacterial activity against a large range of bacteria including methicillin-resistant *Staphylococcus aureus* and *Vibrio alginolyticus* (Cabral, 2021).

Lumbrinereis heteropoda: A non-proteinaceous neurotoxic alkaloid viz. nereistoxin, produced by this species, has been reported to display neuromuscular effects (Ceuvas et al., 2018).

Nicomache (Nicomache) minor: The peptide toxin viz. nicomicin has been reported to show cytotoxicity toward cancer cell lines with proapoptotic effects (Rodrigo et al., 2021).

1. Arenicola marina; 2. Eulalia viridis; 3. Eurythoe complanate; 4. Glycera dibranchiate; 5. Glycera fallax; 6. Glycera tridactyla; 7. Hediste diversicolor; 8. Hermodice carunculata; 9. Lumbrinereis sp.; 10. Paramphinome jeffreysii

1. Image credit: Auguste Le Roux. https://creativecommons.org/licenses/by/3.0/deed.en

2. Image credit: Arne Nygren/Sjøfartsmuseet Akvariet Gøteborg. https://creativecommons.org/licenses/by-sa/4.0/

3. Image credit: Brocken Inaglory. https://creativecommons.org/publicdomain/zero/1.0/

4. Image credit: Public domain

5. Image credit: Eric A. Lazo-Wasem - Gall L . https://creativecommons.org/publicdomain/zero/1.0/deed.en

6. Image credit: Diego Delso. https://creativecommons.org/licenses/by-sa/4.0/

7. Image credit: Arne Nygren/Sjøfartsmuseet Akvariet Gøteborg. https://creativecommons.org/licenses/by-sa/4.0/

7.7 THERAPEUTIC POTENTIAL OF TTX OF MARINE ARTHROPODS

Among marine arthropods, the crustaceans (Subphylum: Crustacea) such as copepods, krill, crabs, and lobsters, and horse-shoe crabs (Subphylum: Chelicerata) have been reported to possess TTX.

7.7.1 COPEPODS

Marine copepods may be either planktonic or parasitic. The TTX-containing marine copepods include the species *Caligus fugu* (= *Pseudocaligus fugu*) and *Taeniacanthus* sp. which are parasitic on certain species of pufferfish. Further, in these parasitic copepods, the TTX is not endogenously produced and is derived by its symbiotic bacterium *Roseobacter* (Lago et al., 2015).

7.7.2 KRILL

Krill (means "whale food") are small crustaceans (Order: Euphausiacea) and found in all the world's oceans. These organisms are believed to be a key potential vector for the transfer of the neurotoxin, domoic acid (DA) to higher trophic level organisms. DA in the tissue of North Pacific krill, *Euphausia pacifica*, has been reported to vary between 0.1 and 44 µg DA eq/g tissue. The source of DA in krill is their food viz. DA-rich diatom species viz. *Pseudo-nitzschia australis* and this was evident by the presence of *Pseudo-nitzschia australis* frustules in the digestive tract of *Euphausia pacifica.* Although the therapeutic potential of DA is well known and discussed elsewhere, venomic research in krill is still wanting (Bargu et al., 2002).

7.7.3 CRABS

7.7.3.1 XANTHID CRABS

Several species of xanthid crabs (Family: Xanthidae) are known to possess TTX and STX. For example, the species *Demania cultripes, Demania toxica, Demania reynaudi, Lophozozymus incisus, Lophozozymus pictor*, and *Atergatopsis germaini* have been reported to have TTX and less toxic. The average toxicity of these species varied from 59 to 611 mouse units (MU). In these species, in addition to TTX, TTX derivatives such as 11-oxoTTX, deoxyTTXs, 4,9-anhydroTTX, and 4-epiTTX have also been detected qualitatively. In Atergatis floridus,11-oxoTTX and 11-norTTX-6(R)-ol, in addition to TTX, were isolated and identified. The source of TTX in this species has been reported to be symbiotic bacterium, *Vibrio* strain. Species like *Zosimus aeneus, Atergatis floridus,* and *Platypodia granulosa.* that inhabit coral reefs have been reported to contain STX and are generally highly toxic. Nine components, neoSTX, hyneoSTX, STX, hySTX, dcSTX, GTX2, GTX3, dcGTX2, and dcGTX3, were also detected in these xanthid crab species (Tsai et al., 2006; Zhang et al., 2021; Lagos et al., 2015).

7.7.3.2 ROCK CRAB

Among the rock crabs (Family: Cancridae), *Cancer irroratus* has been reported to possess elevated levels of domoic acid (DA) toxin. However, the therapeutic potential of the DA of this species is yet to be known (https://cadivingnews.com/domoic-acid-advisory-issued-for-consumers-of-lobsters-and-rock-crabs/).

7.7.4 LOBSTERS

Among the lobsters, the Southern Rock Lobster, *Jasus edwardsii* (Family: Palinuridae), has been reported to accumulate dangerous levels of paralytic shellfish toxin, STX. The dinoflagellate blooms of *Alexandrium catenella* are believed to be responsible for the toxicity of this lobster. During blooms, the cells of this dinoflagellate have been reported to be highly toxic (9.9 pg STX.2HCl $cell^{-1}$. The accumulation STX in the hepatopancreas of this lobster species was found to be 6.65 mg STX.2HCl eq/kg (mean). It is worthy of mention here that during non-bloom periods, these lobsters had low concentrations of PST in the hepatopancreas on receipt (mean 0.03 mg STX.2HCl eq/kg) (Turnbull et al., 2021). The therapeutic potential of STX of lobsters is yet to be known.

7.7.5 HORSE-SHOE CRABS

The species of horseshoe crabs such as *Limulus polyphemus, Carcinoscorpius rotundicauda, Tachypleus gigas*, and *Tachypleus tridentatus* have been reported to possess tetrodotoxin (TTX). The average TTX in both

Carcinoscorpius rotundicauda and *Tachypleus tridentatus* was found to range from 6.2–8.0 μg/kg to 3.8–8.4 μg/kg, respectively (Zhu e al., 2022). The source of TTX and anh-TTX in *Carcinoscorpius rotundicauda* is its associated bacterial species, *Vibrio alginolyticus* (Kungsuwan et al., 1988).

Antimicrobial activity of toxins of horseshoe crabs: Liu et al. (2018) reported that the peptide toxin, tachyplesin III, of a horse-shoe crab species displayed antibacterial activity against *Escherichia coli, Acinetobacter baumannii, Klebsiella pneumoniae,* and *Pseudomonas aeruginosa.*

The peptide toxin, tachyplesin, derived from *Tachypleus tridentatus* was found to inhibit the growth of both Gram-negative and -positive bacteria at low concentrations. The range of values of inhibitory concentrations recorded for the Gram(−) bacteria, *Salmonella typhimurium* LT2(S), *Salmonella typhimurium* SL1102(Re), *Salmonella minnesota* 1114W(S), and *Salmonella minnesota* R595(Re) were 3.1–6.3, 0.8–1.6, 3.1–6.3, and 3.1–12.5 ug/mL, respectively. The corresponding values for Gram(+) *Staphylococcus aureus* 209P and Gram(+) *Staphylococcus aureus* ATCC 25923 were found to be 3.1 and 6.3 ug/mL, respectively (Nakamura et al., 1988).

Miyata et al. (1989) reported that the peptides tachyplesin I and II, and polyphemusins I and II isolated from the hemocytes of the American horseshoe crab, *Limulus polyphemus* inhibited the growth of not only Gram-negative and Gram-positive bacteria but also fungi, such as *Candida albicans* M9. The values of minimum inhibitory concentrations (ug/mL) recorded for the aforesaid toxins are given in the following table.

7.8 MARINE MOLLUSKAN TOXINS AND THEIR THERAPEUTIC POTENTIAL

7.8.1 MARINE BIVALVES

Bivalve mollusks (shellfish) are invertebrates and are characterized by the presence of an external two-part hinged shell which encloses the soft parts. Marine bivalve mollusks play important roles in the marine ecosystems by serving as habitats and prey for a variety of sea lives. This diverse group of species, estimated at around 9200, inhabits virtually the entire world's oceans. Among the marine bivalves, about 180 species (including mussels, oysters, scallops, cockles, and clams) have long been a part of the diet of coastal human populations (Santhanam, 2018). As these shellfish are filter feeders, they are often contaminated by ingesting harmful, toxin-containing phytoplankton as their food. Such toxins are known to be accumulated actively by these shellfish and concentrated in their digestive organ, hepatopancreas. Although some toxins are accumulated very regularly by specific bivalve species in some areas, prediction of contamination levels in general remains very challenging. Consumption of contaminated shellfish with high levels of certain biotoxins viz. domoic acid (DA), okadaic acid (OA), tetrodotoxin (TTX), and pectenotoxin (PTX) can lead to serious and potentially fatal illnesses such as: paralytic shellfish poisoning (PSP), amnesic shellfish poisoning ASP), and diarrhetic shellfish poisoning (DSP). Though the therapeutical potential of these toxins produced by marine microalgae viz. diatoms, dinoflagellates, and cyanobacteria is fairly well known, the pharmacological benefits of such toxins accumulated by the bivalve shellfish through food chain are yet to be known. The PSP, DSP, and ASP toxins in marine bivalve shellfish examined, and the accumulation of biotoxins by their various species are given in the following tables.

TABLE 7.9 Antimicrobial Activity of Tachyplesins and Polyphemusins (MIC, ug/mL).

Strain of Bacteria/Fungi	Tachyplesin I	Tachyplesin II	Polyphemusin I	Polyphemusin II
Salmonella typhimurium LT2 (G -)	3.1	3.1	3.1	6.3
*Salmonella typhi*murium 1102 (G -)	0.8–1.6	1.6	3.1	3.1
Salmonella minnesota 1114W(G-)	3.1	3.1	6.3	12.5
Salmonella minnesota R595(G-)	1.6	1.6	3.1	3.1
Escherichia coli K12 (G-)	1.6–3.1	3.1	6.3	12.5
Pseudomonas aeruginosa(G-)	12.5	–	–	–
Staphylococcus aureus 209P (G+)	3.1	1.6–3.1	6.3	6.3
Staphylococcus aureus ATCC 25923 (G+)	6.3	6.3	6.3	12.5
Bacillus subtilis (G+)	3.13	–	–	–
Candida albicans M9	3.1	3.1	6.3	6.3
Cryptococcus neoformans IMF 40040	1.56	–	–	–

Source: Miyata et al. (1989).

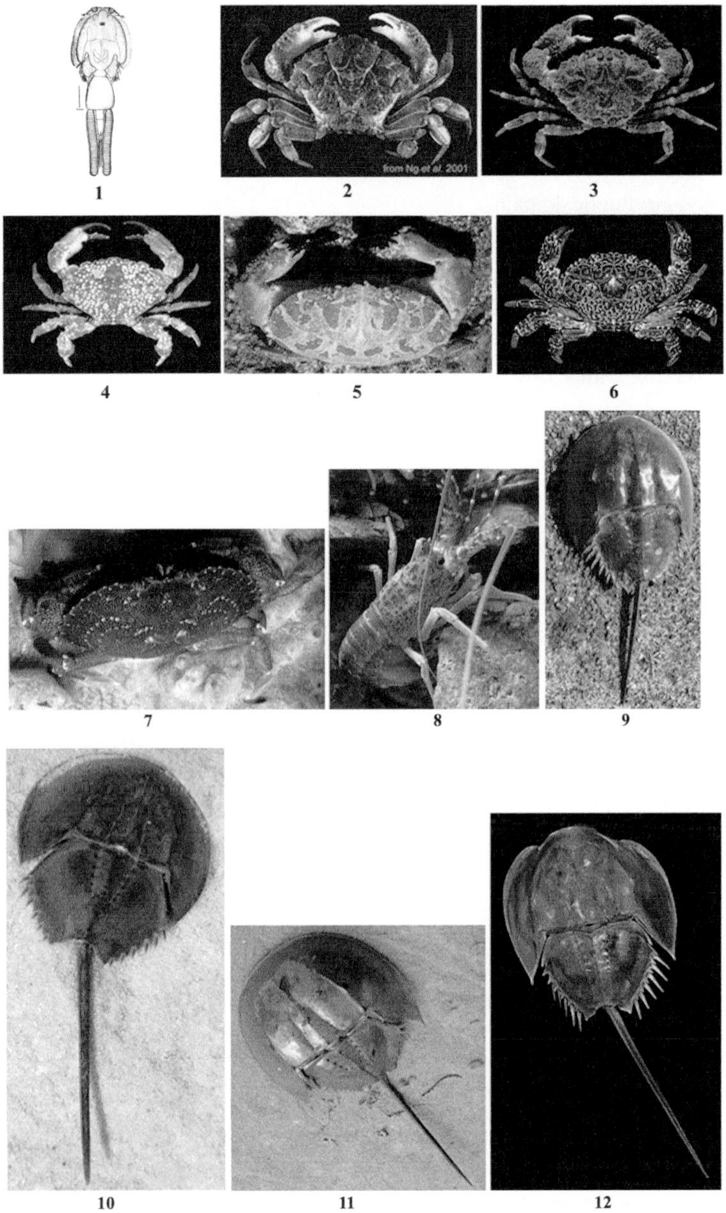

1. Caligus fugu; 2. Demania cultripes; 3. Demania reynaudi; 4. Lophozozymus Pictor; 5. Atergatopsis germaini; 6. Zosimus aeneus; 7. Cancer irroratus; 8. Jasus edwardsii; 9. Limulus polyphemus; 10. Carcinoscorpius rotundicauda; 11. Tachypleus gigas; 12. Tachypleus tridentatus;

1. *Image credit*: Public domain
2. *Image credit*: Reproduced with permission from Ondrej Radosta
3. *Image credit*: Reproduced with permission from Ondrej Radosta
4. *Image credit*: Reproduced with permission from Ondrej Radosta
5. *Image credit*: Reproduced with permission from Dr. Scott Johnson
6. *Image credit*: Reproduced with permission from Ondrej Radosta)
7. *Image credit*: Public domain
8. *Image credit*: Stemonitis . https://creativecommons.org/licenses/by-sa/3.0/
9. *Image credit*: Kaldari. Publicdomain
10. *Image credit*: Amada44 . https://creativecommons.org/licenses/by/3.0/
11. *Image credit*: Shubham Chatterjee . https://creativecommons.org/licenses/by-sa/3.0/
12. *Image credit*: Didier Descouens . https://creativecommons.org/licenses/by-sa/4.0/

TABLE 7.10 PSP, DSP, and ASP Toxins in Marine Bivalve Shellfish Examined.

Species/Common Name	PSP Range (ug/kg)	PSP Average (ug/kg)	DSP Range (ug/kg)	DSP Average (ug/kg)	ASP Content (mg/kg)
Mytilus edulis (Blue mussel)	52–109	73	66–93	79	0.1; 0.1; 0.5
Tapes philippinarum (Japanese carpet shell)	54–77	63	67–112	84	0.2
Crassostrea gigas (Pacific oyster)	50–129	72	65–112	77	0.2
Ensis directus (Razor clam)	51–82	64	72–85	79	–
Cerastoderma edule (Common cockle)	52–97	58	70–81	75	–
Glycymeris glycymeris (Dog cockle)	52–64	55	77–101	89	–
Mercenaria mercenaria (Hard clam)	55–91	67	68–81	72	0.1
Pecten maximus (Great scallop)	54–784	467	–	–	0.2; 0.9; 2.7

Source: Madejska and Osek (2021).

TABLE 7.11 Domoic Acid (DA) Accumulation in Marine Bivalve Shellfish.

Species	DA Content (Max)	References
Acanthocardia tuberculata	2110 μg/kg	Kvrgić et al. (2022)
Aequipecten opercularis	810 μg/kg	Kvrgić et al. (2022)
Anadara antiquata	1000 μg/kg	Kvrgić et al. (2022)
Argopecten irradians	60 ug/g	Bogan (2006)
Argopecten purpuratus	103 mg/kg	Álvarez et al. (2020)

TABLE 7.11 *(Continued)*

Species	DA Content (Max)	References
Atrina vexillum,	1000 μg/kg	Kvrgić et al. (2022)
Callista chione	4900 μg/kg	Kvrgić et al. (2022)
Cerastoderma edule,	3710 μg/kg	Kvrgić et al. (2022)
Chama lazarus	1000 μg/kg (ca)	Kvrgić et al. (2022)
Chama limbula (= Chama iostoma,)	1000 μg/kg (ca)	Kvrgić et al. (2022)
Donax trunculus	53,000 μg/kg	Kvrgić et al. (2022)
Ensis sp.	2520 μg/kg	Kvrgić et al. (2022)
Flexopecten glaber (= Flexopecten proteus)	1657 μg/kg	Kvrgić et al. (2022)
Hyotissa hyotis	1000 μg/kg (ca)	Kvrgić et al. (2022)
Isognomon isognomum.	8 ug/g	Takata et al. (2009)
Mactra quadrangularis (= Mactra veneriformis)	1.99 mg/kg	Choi et al. (2009)
Magallana gigas (= Crassostrea gigas)	1730 μg/kg	Kvrgić et al. (2022)
Mesodesma donacium	–	Kvrgić et al. (2022)
Mizuhopecten yessoensis	90 μg/kg	Kvrgić et al. (2022)
Modiolus modiolus	–	Stewart et al. (1998)
Mya arenaria	–	Stewart et al. (1998)
Mytilus californianus	–	Bogan (2006)
Mytilus edulis	1040 μg/kg	Kvrgić et al. (2022)
Mytilus galloprovincialis	6549 μg/kg	Kvrgić et al. (2022)
Mytilus trossulus	100 μg/kg	Kvrgić et al. (2022)
Ostrea edulis	1040 μg/kg	Kvrgić et al. (2022)
Pecten jacobaeus	1657 μg/kg	Kvrgić et al. (2022)
Pecten maximus	2000 ug/g	Bogan (2006)
Perna canaliculus	–	Bogan (2006)
Perna viridis	1000 μg/kg (ca)	Kvrgić et al. (2022)
Peronidia venulosa	3.02 mg/kg	Choi et al. (2009)
Placopecten magellanicus	–	Stewart et al. (1998)
Placuna ephippium (= Placuna sella)	1000 μg/kg (ca)	Kvrgić et al. (2022)
Ruditapes sp.	2580 μg/kg	Kvrgić et al. (2022)
Siliqua patula	154 μg/g	Choi et al. (2009)
Solen sp.	2520 μg/kg	Kvrgić et al. (2022)
Spondylus sinensis	54 ug/g,	Takata et al. (2009)
Spondylus squamosus	88,160 μg/kg	Kvrgić et al. (2022)
Spondylus versicolor.	25, 39 ug/g	Takata et al. (2009_

1. Acanthocardia tuberculata; 2. Aequipecten opercularis; 3. Anadara antiquata; 4. Argopecten irradians; 5. Argopecten purpuratus; 6. Atrina vexillum; 7. Callista chione.; 8. Cerastoderma edule; 9. Chama lazarus; 10. Chama limbula (= Chama iostoma); 11. Crenomytilus grayanus; 12. Donax trunculus.

1. *Image credit*: https://creativecommons.org/licenses/by-sa/3.0/deed.en
2. *Image credit*: Didier Descouens - https://creativecommons.org/licenses/by-sa/4.0/
3. *Image credit*: H.Zell. https://creativecommons.org/licenses/by-sa/3.0/deed.en
4. *Image credit*: Eric A. Lazo-Wasem. Public domain
5. *Image credit*: James St. John. https://creativecommons.org/licenses/by/2.0/deed.en
6. *Image credit*: SSR2000 . https://creativecommons.org/licenses/by-sa/3.0/
7. *Image credit*: Tigerente -.https://creativecommons.org/licenses/by-sa/4.0/
8. *Image credit*: H. Zell. https://creativecommons.org/licenses/by-sa/3.0/deed.en
9. *Image credit*: H. Zell. https://creativecommons.org/licenses/by-sa/3.0/deed.en
10. *Image credit*: Reproduced with permission from Dr. Bret Raines
11. *Image credit*: PICRYL. Public domain
12. *Image credit*: Hans Hillewaert. https://creativecommons.org/licenses/by-sa/4.0/deed.en

TABLE 7.12 Other Toxins Accumulated by Marine Bivalve Shellfish.

Species	Toxin	Content (Max)	References
Argopecten irradians	OA	–	Blanco (2018)
Cerastoderma edule,	OA	–	Blanco (2018)
Crenomytilus grayanus	OA	-	Blanco (2018)
Donax trunculus	OA	–	Blanco (2018)
Magallana gigas (= Crassostrea gigas)	TTX; PTX	253 μg/kg; 0.026 mg/kg	Antonelli et al. (2022) Boundy, et al. (2020)
Mercenaria mercenaria.	TTX	172 μg/kg	Antonelli et al. (2022)
Mesodesma donacium	PTX2 and PTX2sa	–	Blanco (2018)
Mimachlamys crassicostata (= Mimachlamys nobilis).	PTX2		Blanco (2018)
Mizuhopecten yessoensis (= Patinopecten yessoensis)	TTX; PTX2	8000 μg/kg	Antonelli et al. (2022)
Moerella donacina (= Tellina donacina).	TTX	–	Rossignoli et al. (2023)
Mytilus edulis	TTX	541 μg/kg	Antonelli et al. (2022)
Mytilus galloprovincialis	OA; DTX2	–	Mudadu et al. (2021); Blanco (2018)
Mytilus unguiculatus (= Mytilus coruscus)	TTX	4.4 μg/kg	Antonelli et al. (2022)
Ostrea chilensis	PTX	0.043 mg/kg	Boundy et al. (2020)
Ostrea edulis	OA; TTX	80 μg/kg	Antonelli et al. (2022)
Paphies australis	TTX	800 μg/kg	Antonelli et al. (2022)
Pecten fumatus	OA	–	Blanco (2018)
Pecten maximus	OA	–	Blanco (2018)

TABLE 7.12 *(Continued)*

Species	Toxin	Content (Max)	References
Pecten novaezealandia	PTX	0.031 mg/kg	Boundy, et al. (2020)
Perna canaliculus	OA; TTX; PTX	160 µg/kg; 0.056 mg/kg	Blanco (2018); Antonelli et al. (2022); Boundy et al. (2020)
Pinna bicolor	OA	–	Blanco (2018)
Ruditapes decussatus	TTX	<10 µg/kg	Antonelli et al. (2022)
Ruditapes philippinarum (=Tapes philippinarum)	TTX	32 µg/kg	Réveillon et al. (2021)
Saccostrea glomerata (= Saccostrea commercialis)	TTX	140 µg/kg	Antonelli et al. (2022)
Sinonovacula constricta.	TTX	16 µg/kg	Antonelli et al. (2022)
Spisula solida	OA	–	Blanco (2018)
Venus verrucosa	TTX	176.5 µg/kg	Antonelli et al. (2022)

OA, Okadaic acid; DTX, Dinophysistoxin; TTX, Tetrodotoxin; PTX, Pectenotoxin.

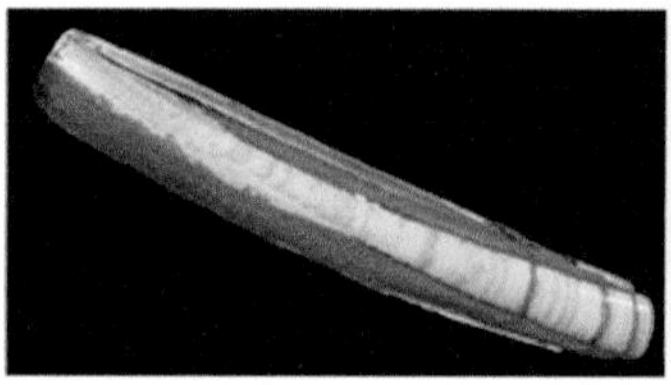

13 14 15

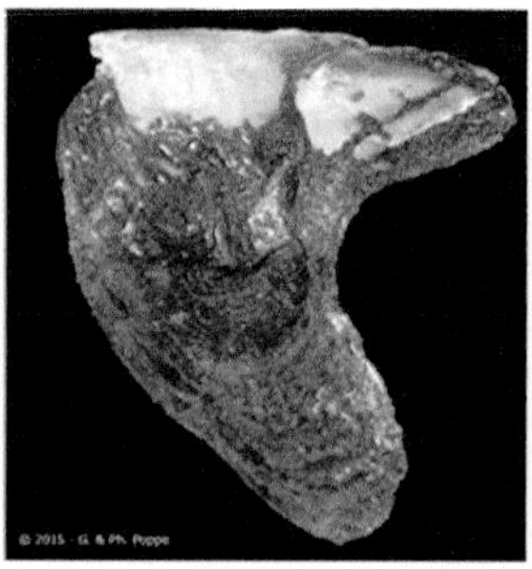

16 17 18

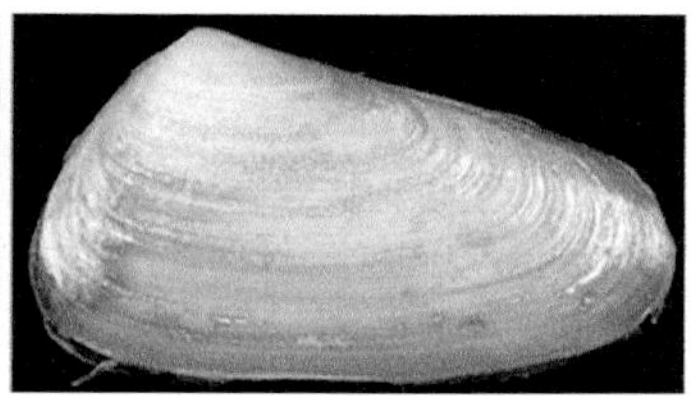

19 20 21

13. Ensis directus.; 14. Flexopecten glaber (= Flexopecten proteus); 15. Glycymeris glycymeris; 16. Hyotissa hyotis; 17. Isognomon isognomum; 18. Mactra quadrangularis (= Mactra veneriformis); 19. Magallana gigas (= Crassostrea gigas); 20. Mercenaria mercenaria; 21. Mesodesma donacium.

13. Image credit: Hans Hillewaert. https://creativecommons.org/licenses/by-sa/4.0/deed.en

14. Image credit: Didier Descouens . https://creativecommons.org/licenses/by-sa/4.0/

15. H. Zell. https://creativecommons.org/licenses/by-sa/3.0/deed.en

16. Image credit: Daderot . https://creativecommons.org/publicdomain/zero/1.0/deed.en

17. Image credit: Reproduced with permission from Guido & Philippe Poppe

18. Image credit: Reproduced with permission from Guido & Philippe Poppe

19. Image credit: Llez. https://creativecommons.org/licenses/by-sa/3.0/deed.en

20. Image credit: GFDL. https://creativecommons.org/licenses/by-sa/3.0/

21. Image credit: Reproduced with permission from Guido & Philippe Poppe

30 31 32

22. *Mimachlamys crassicostata (= Mimachlamys nobilis).*, 23. *Modiolus modiolus;* 24. *Moerella donacina (= Tellina donacina); 25. Mya arenaria; 26. Mytilus edulis;* 27. *Mytilus galloprovincialis.; 28. Mytilus trossulus; 29. Mytilus unguiculatus (= Mytilus coruscus); 30. Ostrea chilensis.; 31. Ostrea edulis; 32. Paphies australis.*

22. *Image credit*: Daderot . https://creativecommons.org/publicdomain/zero/1.0/deed.en

23. *Image credit*: H. Zell . https://creativecommons.org/licenses/by-sa/3.0/

24. *Image credit*: Hans Hillewaert. https://creativecommons.org/licenses/by-sa/4.0/deed.en

25. *Image credit*: Kirsten Poulsen. https://creativecommons.org/licenses/by-sa/3.0/deed.en

26. *Image credit*: Pallbo. Public domain

27. *Image credit*: H. Zell. https://creativecommons.org/licenses/by-sa/3.0/deed.en

28. *Image credit*: Reproduced with permission from Guido & Philippe Poppe

29. *Image credit*: Jin-Long Yang, Dan-Dan Feng, Jie Liu, Jia-Kang Xu, Ke Chen, Yi-Feng Li, You-Ting Zhu, Xiao Liang, Ying Lu, https://creativecommons.org/licenses/by-sa/4.0/

30. *Image credit*: Philippi. Public domain

31. *Image credit*: Jan Johan ter Poorten. https://creativecommons.org/licenses/by-sa/3.0/deed.en

32. *Image credit*: Reproduced with permission from Guido & Philippe Poppe

33. Pecten fumatus; 34. Pecten jacobaeus; 35. Pecten maximus; 36. Pecten novaezealandia; 37. Perna viridis.; 38. Pinna bicolor; 39. Placopecten magellanicus; 40. Placuna ephippium (= Placuna sella); 41. Ruditapes decussatus.

33. Image credit: Naturalis Biodiversity Center. Public domain

34. Image credit: Andreas Tille . https://creativecommons.org/licenses/by-sa/4.0/deed.en

35. Image credit: H. Zell. https://creativecommons.org/licenses/by-sa/3.0/deed.en

36. Image credit: GrahamBould. Public domain

37. Image credit: H. Zell. https://creativecommons.org/licenses/by-sa/3.0/

38. Image credit: Fuuke. https://creativecommons.org/licenses/by-sa/4.0/

39. Image credit: Eric A. Lazo-Wasem. Public domain

40. Image credit: Philipsson. Public domain

41. Image credit: Jan Johan ter Poorten. https://creativecommons.org/licenses/by-sa/3.0/

42. Ruditapes philippinarum (=Tapes philippinarum); 43. Saccostrea glomerata (= Saccostrea commercialis); 44. Siliqua patula; 45. Sinonovacula constricta; 46. Solen sp.; 47. Spisula solida; 48. Spondylus sinensis.; 49. Spondylus squamosus; 50. Spondylus versicolor; 51. Venus verrucose.

42. Image credit: Public domain

43. Image credit: Pelagic. https://creativecommons.org/licenses/by-sa/4.0/

44. Image credit: James St. John. https://creativecommons.org/licenses/by/2.0/deed.en

45. Image credit: Kakichi Mitsukuri. Public domain

46. Image credit: GFDL. https://creativecommons.org/licenses/by-sa/3.0/

47. Image credit: H. Zell. https://creativecommons.org/licenses/by-sa/3.0/

48. Image credit: Reproduced with permission from Guido & Philippe Poppe

49. Image credit: Víctor Cebollada. https://creativecommons.org/licenses/by-sa/3.0/

50. Image credit: James St. John. https://creativecommons.org/licenses/by/2.0/

51. Image credit: Hans Hillewaert. https://creativecommons.org/licenses/by-sa/4.0/

7.8.2 MARINE GASTROPODS

Marine gastropod mollusks inhabit in all possible habitats, including estuaries, rocky, and sandy shores. They are the largest subgroup within about 130,000 available and 100,000 described species, and are comprising about 60,000 shelled and 13,000 seaslug species. It is also worthy of mention here that about 180 species of marine gastropod species are edible and are of nutritional and commercial importance (Santhanam, 2018). Among the marine gastropods, only a few species including *Bolinus brandaris* have been reported to accumulate the ASP-causing potent neurotoxin domoic acid (DA) at >20 mg/kg (Papiol et al., 2013). On the other hand, several species of marine gastropods are known to accumulate TTX, the importance of which has already been dealt with. Amazingly, the species of *Conus* are unique as they produce the peptide conotoxins that are very important owing to their diversity and therapeutics. The species of marine gastropods of world seas accumulating TTX are given in the following table.

Conotoxins

There are about 1000 known species of *Conus* inhabiting mainly in tropical coral reefs and all are venomous. Each one of these species has been reported to produce an abundant, diverse, and unique armada of disulphide-rich peptide toxins called conotoxins which are mainly used for defence and prey capture (Twede et al., 2009). It is believed that the venom of each *Conus* species may have 100–200 toxins.. Though the total pool of wild-type distinct conotoxins may be about 100,000, only about 3000 conotoxins have so far been discovered and only about 200 of them have been functionally characterized. The major pharmacological properties of these conotoxins include neuroprotective, cardioprotective, antinociceptive, and antiepileptic, and thus have pharmacological applications in cancer, neuromuscular, and psychiatric disorders.

Though several conotoxins have entered clinical trials for their analgesic properties, one FDA-approved conotoxin (MVIIA (ziconotide) is presently used in the clinic to treat intractable chronic pain (Twede et al., 2009). The components of conotoxins have shown inherent exquisite selectivity which can be utilized both as unique pharmacological tools and as templates for drug design.

Conopeptides and Conotoxin families

The conotoxins are made of small peptides with mostly 10–40 amino acid residues in length. While the shortest conotoxin, conophanmus-V, possesses 8 amino acid residues, the longest conopeptide, viz. con-ikot-ikot, has 86 amino acid residues. The molecular weight of conopeptides has been reported to range from 847 to 9445 Da, and their isoelectric point is from very acidic to basic. The first conopeptide sequenced and confirmed by chemical synthesis was αconotoxin-GIC. It belonged to *Conus geographus* and is consisting of 16 amino acid residues with two disulphide bonds. Further, conopeptides containing one or no disulfide bonds are named very differently. Such disulfide-poor conopeptides are designated by combining a class name, followed by one- or two-letter code of the name of the species. For example, such conopeptides of *Conus striatus, Conus radiatus,* and *Conus geographus* are designated as conopressin-S, conotryphan-R, and contulakin-G, respectively. On the other hand, conopeptides with two or more disulfides are called disulfide-rich conopeptides or "conotoxins." A total of 26 conopeptide superfamilies are so far known. However, certain cysteine-poor conopeptides do not cluster in any of the known superfamilies. Examples of such conopeptides are contulakin and conantokin which belong to superfamilies B and C, respectively (Gao et al., 2017). According to Dao et al. (2017), superfamilies of conotoxins are designated as A, M, O, P, S, and T and each of such superfamilies will have its own conotoxin families or characteristics as given below:

- A-Superfamily: It consists of four toxin families viz. α-, αA-, κA-, and ρ-conotoxins
- M-Superfamily: It has three families viz. ψ-, μ-, and κM-conotoxins
- O-Superfamily: It is the largest group, with about 500 peptide sequences (excluding precursor sequences), in five distinct families viz. μO-, δ-, ω-, κ-, and γ-conotoxins
- P-Superfamily: This family consists of nine peptide sequences with the family name "spastics"
- S-Superfamily: It has been reported to target muscle-type nAChR and it has seven peptides

T-Superfamily: It includes both τ- and χ-conotoxin families and is consisting of ~ about 110 peptide sequences.

TABLE 7.13 TTX Content Of Marine Gastropods.

Species	TTX (µg/kg)
Aplysia depilans	<5.5
Babylonia japonica	11,000
Buccinum undatum.	7.4
Calliostoma sp.	<10
Charonia lampas lampas.	67–315,000
Charonia lampas	390,000
Gibbula umbilicalis	63
Monodonta lineata	90
Nassarius castus	2600
*Nassarius conoidalis****	7000–220,000
Nassarius glans	2500–2100
*Nassarius livescens**	1000
Nassarius nitidus	1350
Nassarius papillosus.	110,000
Nassarius pullus	9340
Nassarius scalaris	28,000
Nassarius sinarum.	37,000–190,000
Nassarius siquijorensis.	680
Natica lineata	144,000
Natica vitellus	4500
*Niotha clathrate****	–
Oliva annulata.	3400
Oliva hirasei	4800–5600
Oliva miniacea.	3600
Oliva mustelina.	3200
Oliva ornata.	4600
Onchidella celtica	12
Pleurobranchaea maculata	626–850,000
Polinices didyma	1.3
Polinices tumidus.	14,000
Pugilina ternotoma	–
Rapana rapiformis.	28,000
Rapana venosa ventosa	2600
Turbo smaragdus	<200
Tutufa lissostoma	140,000
*Umborium suturale***	–
Zeuxis castus-like	2600

Source: Antonelli et al. (2022); *Chen et al. (2002); **Assumcao et al. (2017); ***Whitelaw et al. (2020); ***Produced by symbiotic bacteria Vibrio and Pseudomonas (Lagos et al., 2015).

Conotoxins isolated from the venoms of *Conus* spp.

As on date, the conotoxins of major species of Conus have been characterized and the details are given in the following table:

TABLE 7.14 Conotoxins of Conus spp.

Species	Conotoxin
Conus asiaticus	Asi3a, Asi14a
Conus australis.	α-AusIA
Conus bandanus	BnIA, BnIIID
Conus betulinus	bt5a
Conus brunneus	α-BruIB
Conus cancellatus.	as25a, as25b
Conus capitaneus	αD-Cp
Conus caracteristicus	Ca11a, Ca11b
*Conus catus.**	CVID (AM336)
Conus consors	CnIIIC
Conus ermineus	α-EII_B
Conus fulgetrum.	fu6a
Conus geneis	ge6b, ge6c
Conus generalis.	α-GeXXA GeXXVIIA
Conus geographus	α-GVIIIB
*Conus geographus**	Contulakin-G (CGX 1160). Conantokin-G (CGX-1007)
Conus imperialis	Im10A, MfVIA, im23a, im23b
Conus limpusi	α-LsIA
Conus literatus	lt3a, lt9a, lt9b
Conus longurionis	Lo1a, Lo6/7a, Lo6/7b
*Conus magus**	MVIIA (Prialt®)
*Conus marmoreus**	MrIA (Xen-2174)[a]., MrVIB (CGX-1002)
Conus mustelinus	αD-Ms
Conus parius	pr3a, pr3b, pr6a, pr6b, pr6c, pr6d
Conus pictus	pc16a
Conus princeps	PiVIIA
Conus purpurascens	PIC, PIC[O7], p21a
Conus quercinus	qc16a
Conus radiatus	κ-RIIIJ

TABLE 7.14 *(Continued)*

Species	Conotoxin
Conus regius	RegIIA, α-RgIB
Conus regularis	RsXXIVA
*Conus stercusmuscarum***	μ-SIIIA
Conus tessulatus	ts14a
*Conus victoriae.**	Vc1.1 (ACV-1)
Conus virgo	vi6a

Source: Twede et al. (2009), *Layer and McIntosh (2006), **Mir et al. (2016).

Therapeutic potential of conotoxins

Voltage-gated ion channels

The ω-conotoxins, one of the well-studied conotoxins, have been reported to block specific voltage-gated calcium channels. Conotoxins targeting sodium ion channels may either prevent conductance (e.g., μ- and μO-conotoxins) or block sodium ion channel inactivation (e.g., δ-conotoxins). Further, conotoxins targeting voltage-gated potassium channels belong to different families and all target a single subfamily of potassium channels namely Shaker (or Kvl) channels (Mir et al., 2016).

Ligand-gated ion channels

Most of the ligand-gated ion channel superfamily inhibitors are nicotinic acetylcholine receptor antagonists. Among such inhibitors, α-conotoxins, the first characterized conopeptides, are known to act upon nicotinic acetylcholine receptors as competitive antagonists. Conotoxins targeting nicotinic acetylcholine receptor are of two subtypes viz. neuromuscular (α-conotoxin GI, α-conotoxin MI, α-conotoxin EI, α-conotoxin PIVA, α-conotoxin EIVA, α-conotoxin OIVB, α-conotoxin PelVB, and ψ-conotoxinPIIIE), and neuronal (α-conotoxin MII, α-conotoxin PIA, α-conotoxin ImII, α-conotoxin BuIA). σ-conotoxin GVIIIA target 5HT3 receptors. Further, conotoxins that target glutamate receptors (NMDA subclass: conantokin-G and conantokin-L; AMPA subclass: con-ikot-ikot) are also known (Mir et al., 2016).

Clinical status of conopeptides

Among the species of *Conus, Conus catus. Conus geographus, Conus magus, Conus marmoreus, Conus princeps, Conus stercusmuscarum*, and *Conus victoriae* have provided potential conopeptides of therapeutic importance. The preclinical and clinical status of these conopeptides is given in the following table.

7.8.3 CEPHALOPODS

Cephalopod mollusks (Phylum: Molluska; Class: Cephalopoda), the exclusively marine animals, are known for their highly diversified body shapes and sizes; advanced visual and nervous systems; and camouflage abilities and metabolic rates. Considering the number of species (845), Cephalopoda is the third richest class of the Phylum Molluska. These animals have worldwide distribution and are available from shallow to deep waters. They are also known for their complex behavior combined with their capacity to produce venom and enzymes which are present in their salivary glands. These cephalopods have received little attention as sentinel species in comparison to other mollusks, such as bivalves. While the coleoid cephalopods (Subclass: Coleoidea or Dibranchiata) which includes squid, cuttlefish, and octopuses are characterized by an internal shell, ink sac, and arm hooks, the nautilids (Subclass: Nautiloidea) lack an ink sac, arm hooks, suckers, and they have an external conch. Several species of cuttlefish, squid, and octopus are of great food value. However, certain species of blue-ringed octopuses of the genus Hapalochlaena found in tide pools and coral reefs in the Pacific and Indian oceans, from Japan to Australia, are known to present an envenoming risk to humans. Of the 12 documented bites by *Hapalochlaena,* six patients did not experience serious symptoms. Two fatalities due to *Hapalochlaena* have been reported in Australia, the first in Darwin in 1954, and the second at Camp Cove near Sydney in 1967. Further, several envenomings required life-preserving artificial ventilation (Jacups et al., 2008).

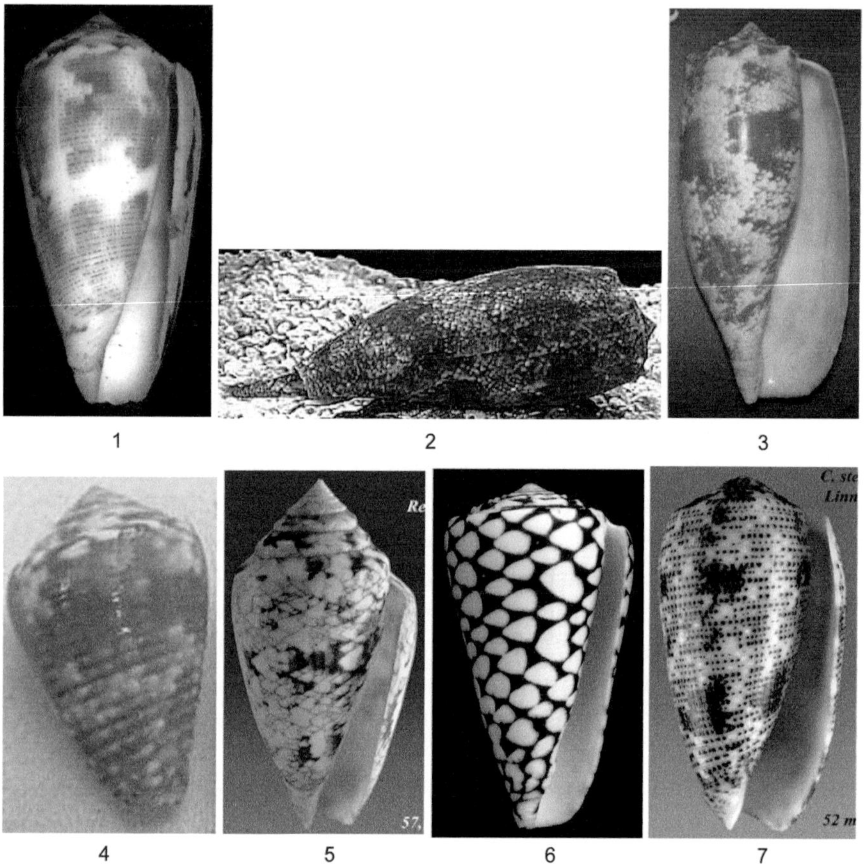

1: Conus magus; 2: Conus geographus; 3: Conus catus; 4: Conus victoriae; 5: Conus marmoreus; 6: Conus princeps; 7: Conus stercusmuscarum

1. Image credit: Richard Parker. https://creativecommons.org/licenses/by/2.0/

2. Image credit: Kerry Matz. Public domain

3. Image credit: Jan Delsing. Public domain

4. Image credit: Almed2. https://creativecommons.org/licenses/by-sa/3.0/

5. Image credit: H. Zell. https://creativecommons.org/licenses/by-sa/3.0/deed.en

6. Image credit: Shellnut. https://creativecommons.org/licenses/by-sa/3.0/

7. Image credit: Almed2. https://creativecommons.org/licenses/by-sa/3.0/

TABLE 7.15 Conopeptides-Based Therapeutics and Their Status.

Source Species	Conopeptide	Commercial Name	Mode of Action	Symptoms	Clinical Status
Conus magus	ω-MVIIA	Ziconotide, Prialt	N-type calcium channels/ blocker	Intractable pain	Phase IV; FDA approved
Conus geographus	Contulakin-G	CGX-1160	Neurotensin receptor/agonist	Neuropathic pain	Phase Ib; for SCI -sciatica patients
Conus catus.	ω-CVID	AM-336	N-type calcium channels/ blocker	Neuropathic pain	Phase IIa; for cancer patients
Conus geographus	Conantokin-G	CGX-1007	NMDA receptor (glutamate receptor) /antagonist	Intractable epilepsy	Phase I
Conus victoriae	α-Vc1.1.	ACV-1	Niconitic acetylcholine receptors/antagonist	Neuropathic pain	Phase II
Conus marmoreus	χ-MrIA	Xen2174	Norepinephrine transporter/ inhibitor	Neuropathic pain	Phase IIa; for cancer patients
Conus princeps	κ-PVIIA	CGX-1051	K-channels/blocker	Myocardial Infarct; Cardioprotection	Preclinical
Conus marmoreus	μO-MrVIB	CGX-1002	Sodium channels/subtype selective blocker	Neuropathic pain	Preclinical
Conus stercusmuscarum	μ-SIIIA	PEG-SIIIA	Sodium channels/blocke	Inflammatory pain	Preclinical

Source: Mir et al. (2016).

Cephalopod toxins

The toxins of coleoid cephalopods appear to be complex mixtures of proteins and small molecules and are of two kinds, viz. accumulated toxins and native toxins. The accumulated toxins include domoic acid (DA), saxitoxins (STXs), and palytoxins (PLTX) and the coleoids are believed to be the vectors of harmful algal bloom toxins (Lopes et al., 2013). The native toxins or the posterior salivary gland (PSG)-produced toxins of coleoids include tetrodotoxin, tachykinins, and cephalotoxins; biogenic amines like histamine, putrescine tyramine, cadaverine, dimethylamine, triimethylamine, serotonin, and octopamine; and a diverse range of enzymes like serine proteases, phospholipase A2, hyaluronidases, and chitinases (Cooke et al., 2017).

Harmful Algal Bloom (HAB)-toxins in coleiod cephalopods

The maximum levels of HAB toxins viz. DA, STXs, and PLtX accumulated in the tissues/organs of the different coleoid species are presented in the following table.

TABLE 7.16 HAB-Toxins in Coleoid Cephalopods.

Toxin	Species	Max. level
Domoic acid	*Sepia officinalis*	241.7 μg DA/g
-Do-	*Doryteuthis opalescens*	0.37 μg DA/g
-Do-	*Dosidicus gigas*	0.23 μg DA/g
-Do-	*Octopus vulgaris*	166.2 μg DA/g
-Do-	*Eledone moschata*	127 μg DA/g
-Do-	*Eledone cirrhosa*	18.8 μg DA/g
Saxitoxin	*Dosidicus gigas*	4.83 μg STX eq/g
-Do-	*Octopus vulgaris*	35 μg STX eq/g
-Do-	*Octopus (Abdopus)* sp.	2.46 μg STX eq/g
Palytoxin	Octopod (species unknown)	971 μg/kg OVTX-a and 115 μg/kg PlTX

Source: Lopes et al. (2013).

Therapeutical applications of cephalopod toxins

Among the various toxins of the cephalopod mollusks, the cephalotoxins, a group of so-called proteinaceous toxins (complex glycoproteins) produced by the salivary glands of coleoids, that is, squids, cuttlefishes, and octopuses, has been considered a promising group for the bioprospecting of novel compounds. It is also suggested that these cephalotoxins possess antiproliferative properties, and they may be of great use in the development of novel antibiotics and anticancer drugs (Gonçalves and Costa, 2021).

1

2

3

4

5

1. Acanthosepion esculentum; 2. Ascarosepion latimanus; 3. Acanthosepion pharaonis; 4. Alloteuthis subulata; 5. Doryteuthis pealeii.

1. Image credit: harum.koh. https://creativecommons.org/licenses/by-sa/2.0/

2. Image credit: Nick Hobgood. https://creativecommons.org/licenses/by-sa/3.0/deed.en

3. Image credit: portioid . https://creativecommons.org/licenses/by-sa/4.0/

4. Image credit: Pierre Corbrion, https://creativecommons.org/licenses/by-sa/4.0/deed.en

5. Image credit: Public domain

TABLE 7.17 Toxins Produced by Cuttlefish and Squid.

Group	Species	Toxin(s)	Type of Toxin	References
Cuttlefish	*Acanthosepion esculentum (= Sepia esculenta).*	SE-Cephalotoxin	Proteinaceous, neurotoxin	Gonçalves and Costa (2021)
-Do-	*Ascarosepion latimanus (= Sepia latimanus)*	phospholipase A2	Enzyme	Almeida et al. (2020)
-Do-	*Acanthosepion pharaonis (= Sepia pharaonis.)*	phospholipase A2	Enzyme	Almeida et al. (2020)
-Do-	*Rhombosepion prashadi (= Sepia prashadi.)*	Unidentified toxin	–	Gonçalves and Costa (2021)
-Do-	*Sepia officinalis*	Cephalotoxin.	Proteinaceous, neurotoxin	Gonçalves and Costa, 2021
Squid	*Alloteuthis subulata (= Loligo subulata)*	Histamine, putrescine, tyramine, and cadaverine	Biogenic amines	Prester et al. (2010)
-Do-	*Doryteuthis pealeii (= Loligo pealei; Doryteuthis (Amerigo) pealeiia)*	Unknown cytolytic protein	–	Gonçalves and Costa (2021)
-Do-	-Do-	Phospholipase A2	Enzyme	Alberghina and Gould (1993)
-Do-	*Illex illecebrosus.*	Carboxypeptidase	Enzyme	Raksakulthai et al. (2001)
-Do-	*Sepioteuthis australis*	Phospholipase A2	Enzyme	Almeida et al. (2020)
-Do-	*Todarodes pacificius*	Dimethylamine, trimethylamine	Biogenic amines	Gou et al. (2010)
-Do-	*Uroteuthis chinensis (= Loligo formosana)*	Serine protease	Enzyme	Singh and Benjakul (2017)
-Do-	*Uroteuthis (Aestuariolus)* noctiluca *(= Loliolus noctiluca)*	Phospholipase A2	Enzyme	Almeida et al. (2020)

TABLE 7.18 Toxins Produced by Octopuses.

Species	Toxin(s)	Type of Toxin	References
Eledone cirrhosa (= *Eledon aldrovandi*).	Eledoisin	Tachykinin	Silva et al. (1972)
Eledone moschata	Eledoisin	Tachykinin	Gonçalves and Costa (2021)
-Do-	Histamine, putrescine, tyramine, and cadaverine	Biogenic amines	Prester et al. (2010)
Enteroctopus dofleini.	Neurotoxin	–	Gonçalves and Costa (2021)
Hapalochlaena fasciata	TTX	Neurotoxin	Wu et al. (2014)
Hapalochlaena lunulata	TTX	Neurotoxin	Wu et al.,2014
Hapalochlaena maculosa. (= *Octopus maculosus*)	Tetrodotoxin,* anhydrotetrodotoxin, 4-epitetrodotoxin	Neurotoxin	Hwang et al. (1989)
-Do-	Maculotoxin	TTX-like	Jacups et al. (2008)
-Do-	Hapalotoxin	Lethal toxin	Jacups et al. (2008)
-Do-	Histamine, tyramine, serotonin and phenolic amines;	Biogenic amines	Jacups et al. (2008)
-Do-	Hyaluronidase, phospholipase A2, protein NP2	Enzymes	Almeida et al. (2020); Jascups et al. (2008)
Octopus bimaculoides	Non–TTX	–	Whitelaw et al. (2020)
Octopus kaurna	Phospholipase A2, chitinase	Enzymes	Almeida et al. (2020); Fry et al. (2009)
Octopus minor (= *Callistoctopus minor*?).	Non–TTX	–	Whitelaw et al. (2020)
Octopus vulgaris	α-cephalotoxin; β-cephalotoxin	Neurotoxins	Gonçalves and Costa (2021)
-Do-	OctTK-I and OctTK-II; OctTKRP-I to OctTKRP-VII	Tachykinin-related peptides	Cosmo et al. (2013)
-Do-	Serine proteases, phospholipase A2, chitinase, metalloprotease,hyaluronidase, pacifastin, protein NP2, protein Group 375, protein Group 1045	Enzymes	Almeida et al. (2020)

Produced by the bacteria, Bacillus, Pseudomonas alteromonas and Vibrio spp. (Lago et al., 2015).

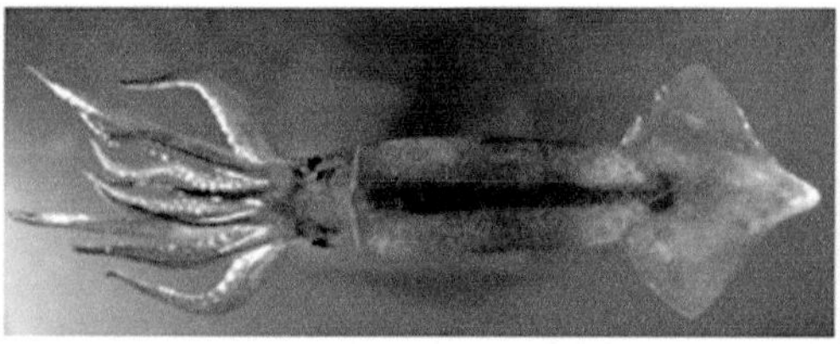

6

7

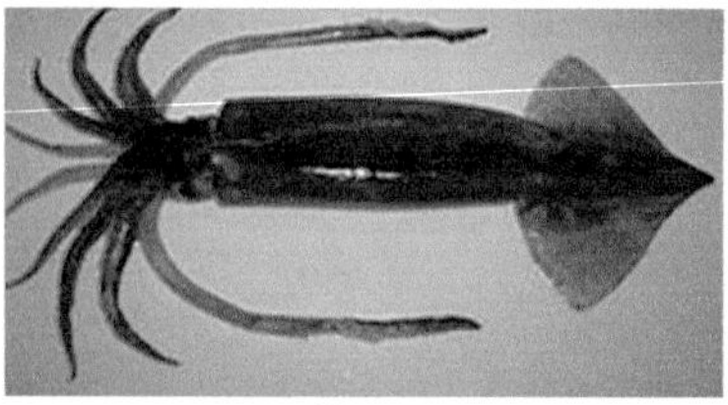

8

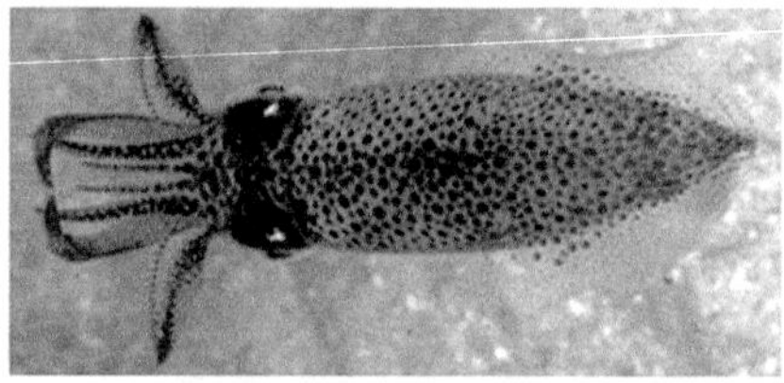

9

10

11

12

13

14

6. Illex illecebrosus.; 7. Sepioteuthis australis; 8. Todarodes pacificius; 9. Uroteuthis (Aestuariolus) noctiluca; 10. Eledone cirrhosa; 11. Eledone moschata; 12. Enteroctopus dofleini; 13. Hapalochlaena fasciata; 14. Hapalochlaena lunulata.

6. Image credit: Lisa Hendrickson. Public domain

7. Image credit: Lek Khauv. https://creativecommons.org/licenses/by/4.0/

8. Image credit: GFDL&CC. https://creativecommons.org/licenses/by-sa/3.0/
9. Image credit: Mark Norman. https://creativecommons.org/licenses/by/3.0/
10. Image credit: Comingio Merculiano. Public domain
11. Image credit: Hectonichus . https://creativecommons.org/licenses/by-sa/4.0/
12. Image credit: NOAA/R. N. Lea. Public domain
13. Image credit: A. Pollock. Public domain
14. Image credit: Reproduced with permission from Dr. Roy L. CALDWELL

15

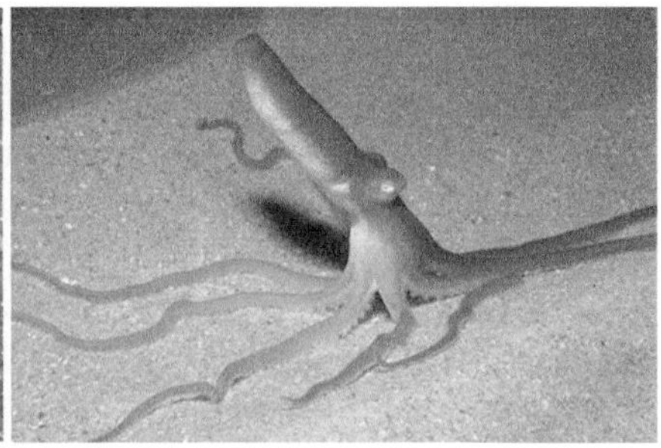

16

17

18

15. Hapalochlaena maculosa; 16. Octopus kaurna; 17. Octopus minor; 18. Octopus vulgaris

15. Image credit: Julian Finn. https://creativecommons.org/licenses/by/4.0/

16. Image credit: Julian Finn. https://creativecommons.org/licenses/by/3.0/

17. Image credit: Bo-Mi Kim et. al. https://creativecommons.org/licenses/by/4.0/

18. Image credit: Comingio Merculiano. Public domain

Octopus vulgaris: The proteinaceous toxins (complex glycoproteins) viz. cephalotoxins derived from the salivary glands of this species have been reported to block glutamate-mediated synaptic transmission, a signaling process involved in pain sensation. Further, its Tachykinin-I and II (OctTK-I an OctTK-II) peptides have shown hypotensive properties (Gonçalves and Costa, 2021).

Amphioctopus ageina (= *Octopus ageina*): The posterior salivary gland extracts containing unidentified toxins of this species displayed anticancer activity against human colon carcinoma COLO 205 cells at a concentration of 100 μg mL^{-1}. Further, this extract also showed similar activity against 2-dimethylhydrazine (DMH)-induced colon cancer in male Albino Wistar rats. Furthermore, the same extract displayed anti-β-glucuronidase activity in these rats (Karthigayan et al., 2006).

Eledone moschata **and** ***Eledone cirrhosa*** *(= Eledone aldovandi)*: The toxin eledoisin of the tachykinin family of neuropeptide of these octopuses has shown a wide range of pharmacological activities such as vasodilation, hypertension, and stimulation of extravascular smooth muscle (Gonçalves and Costa, 2021).

Acanthosepion pharaonis *(=Sepia pharaonis)* **and** ***Rhombosepion prashadi*** *(= Sepia prashadi)***:** The unidentified toxins present in the extracts of the salivary glands of this species have shown antimicrobial activity and anti-metastatic effects. Further, the extract of *Sepia prashadi*. has also shown potential antiviral effects (Gonçalves and Costa, 2021).

Acanthosepion pharaonis (= *Sepia pharaonis)***:** The posterior salivary gland (PSG) toxins of this species displayed highest cytotoxicity against MCF-7 breast cancer cells (IC50-10.64 μM) followed by KB, HeLa, and A549 cells. Further, these toxins also displayed antibacterial activity against *Escherichia coli* and *Klebsiella pneumoniae* with 10 mm as the value of zone of inhibition (Karthik et al., 2017).

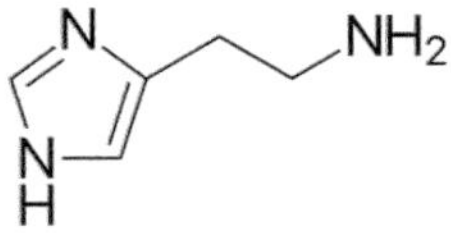

Histamine.

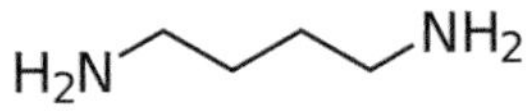

Putrescine

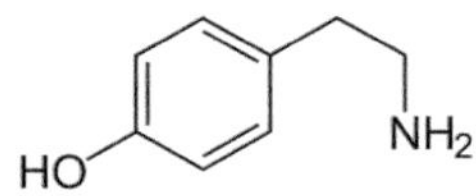

Tyramine

Cadaverine

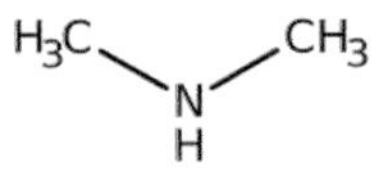

Dimethylamine

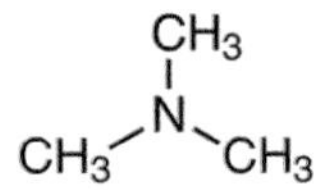

Trimethylamine

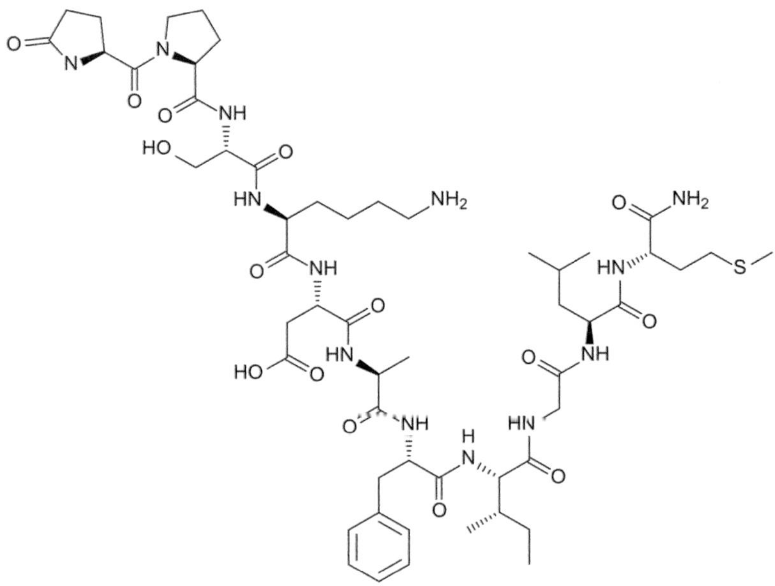

Eledoisin

7.9 TOXINS OF ECHINODERMS AND THEIR THERAPEUTIC POTENTIAL

The echinoderms (Phylum: Echinodermata) are exclusively marine invertebrates and they occur in cold and tropical seas. The habitats of these organisms vary from the intertidal zone down to the bottom of the deep sea trenches and from sand to rubble to coral reefs of this phylum. There are about 7000 described species of echinoderms worldwide and are divided into five classes including asteroids (Starfish), ophiuroids (brittle stars), echinoids (sea urchins), sea lilies (crinoids), and holothuroids (sea cucumbers). Echinoderms are characterized by a radial symmetry, and many arms (5 or more), mostly radiating from a central body (i.e., pentamerous). They have no heart, brain, or eye, but have tentacle-like structures called tube feet with suction pads located at their extremities. Among the various classes of echinoderms, the holothuroids are known for their holotoxins which are of great pharmacologic and therapeutic values (Santhanam et al., 2029a).

7.9.1 SEA STARS (CLASS: ASTEROIDEA)

Toxins of Starfish

The species of starfish have been reported to mostly accumulate the toxins such as TTX and STX obtained through the food sources of higher trophic levels and such species are given in the following table.

Therapeutic potential of starfish toxin

Acanthaster planci: The plancitoxin 1 isolated from its spine venom has been reported to possess antiproliferative activity by significantly decreasing the viability of human melanoma A375.S2 cells in a dose-dependent manner. The modes of action involved in this activity were induction of oxidative stress, mitochondrial dysfunction, and ER stress associated apoptosis (Lee et al., 2015). Further, the active component of its toxin viz. plancinin was found to be an anticoagulant (https://reefs.com/magazine/toxins-venoms-and-inhibitory-chemicals-in-marine-organisms/).

7.9.2 BRITTLE STAR (CLASS: OPHIUROIDEA)

Therapeutic potential of brittle star toxin

Brittle stars are not used as food, though they are not toxic, because of their strong skeleton. Even if some species have blunt spines, no brittle star is known to be dangerous, or venomous. I have never heard of a brittle star releasing toxin before and have had other brittle stars before with no problems.

Ophiocoma scolopendrina: This species has been reported to possess the lipopeptide toxins viz. curacin A and its congener viz. curacin E. Among these toxins, curacin A that has cyanobacterial origin (i.e., from *Lyngbya majuscula, Lyngbya polychroa,* and *Lyngbya sordida*) was found to possess antimitotic and antiproliferative activities against leukemia P388 cells (Ueoka et al., 2016; Du et al., 2019).

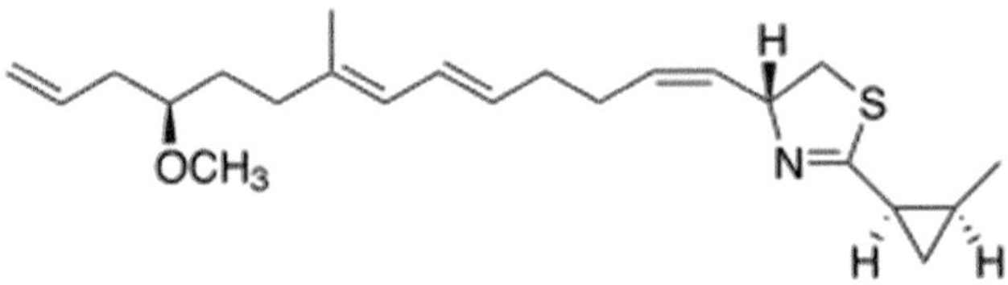

Curacin A

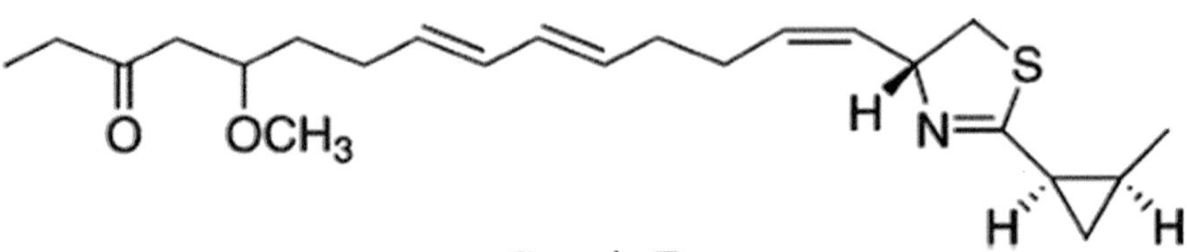

Curacin E

7.9.3 *SEA URCHINS (CLASS: ECHINOIDEA)*

Toxins of sea urchins and their therapeutic potential

Sea urchins possess venom, but are not poisonous when they sting or bite humans. When stung by their spines or bitten by their pedicellaria, the venom will only leave puncture wounds or may trigger some allergic reactions. Yet, they cannot poison humans or other animals. Gelman et al. (2022) reported that the venoms of sea urchins include toxins such as glycosides, hemolysins, and proteases. The toxins isolated from certain species of sea urchins are given in the following table.

TABLE 7.19 Toxins of Starfish.

Species	Toxin Type	Conc. Max/Range	References
Acanthaster planci.	Plancitoxin 1 protein; spine venom A, B	–	Aniya et al. (1998); Lee et al. (2015)
Asterias forbesi	Saponin–triterpene glycoside	–	Rio et al. (1965)
Astropecten latespinosus.	TTX	7800 µg/kg	Antonelli et al. (2022)
Astropecten polyacanthus	TTX*	15,080 µg/kg	Antonelli et al. (2022)
Astropecten scoparius	TTX	700–2400 µg/kg	Antonelli et al. (2022)
Astrostole scabra.	TTX	170 µg/kg	Antonelli et al. (2022)
Choriaster granulatus	Holothurin A2 (= echinoside A) (triterpene glycoside)	–	Ivanchina et al. (2019)
Crossaster papposus	STX	99–>40,000 µg/kg	Dean et al. (2021)
Ophidiaster ophidianus	TTX	42,604 µg/kg	Antonelli et al. (2022)
Patiria miniata	Saponin–triterpene glycoside	–	Rio et al. (1965)
Patiriella regularis.	TTX	140 µg/kg	Antonelli et al. (2022)
Pisaster ochraceous	Saponin–triterpene glycoside	–	Rio et al. (1965)
Pisaster brevispinus.	Saponin–triterpene glycoside	–	Rio et al. (1965)
Pycnopodia helianthoides	Saponin–triterpene glycoside	–	Rio et al. (1965)
Stellaster childreni (*= Stellaster equestris)*	Crude toxin	531 µg/mL (protein content)	Kanagarajan et al. (2008)

**Produced by the symbiotic bacterium, Vibrio alginolyticus (Lago et al., 2015).*

TABLE 7.20 Toxins Isolated from Certain Species of Sea Urchins.

Species	Toxin Type	Conc. Max/Range	References
Echinus esculentus	TTX	<5.5 μg/kg	Antonelli et al. (2022)
Fellaster zelandiae (= Arachnoides. zelandiae)	TTX	250 μg/kg	Antonelli et al. (2022)
Tripneustes gratilla	Contractin A (protein)	–	(1)
-Do-	CTXs (P-CTX-3B, 51-OH-P-CTX-3C, P-CTX-3C, P-CTX-4A, and P-CTX-4B)	–	Darius et al. (2018)
Toxopneustes pileolus	UT841 (acidic protein)	–	Zhang et al. (2001b)
-Do-	Peditoxin	–	Kuwabara (1994)

(1) https://reefkeeping.com/issues/2004-08/rhf/index.php.

***Tripneustes gratilla*:** Its toxic protein viz. Contractin A has been reported to induce smooth muscle contraction (Zhang et al., 2001b).

Toxopneustes pileolus: Its toxic, acidic protein viz. UT841 was found to influence Ca(2+)mobilization in nerve cells. Calcium ions (Ca2+) are known to play an important role in calcium signaling to communicate and drive intracellular processes (Zhang et al., 2001b).

7.9.4 FEATHER STAR (CLASS: CRINOIDEA)

Toxin of sea feather

Tropiometra carinata: This species has been reported to possess unidentified toxin. Cuttlefish exposed to this toxin was found to eject its ink thrice followed by distress and death. Further research is however required to identify and characterize its toxin (Rao et al., 1985).

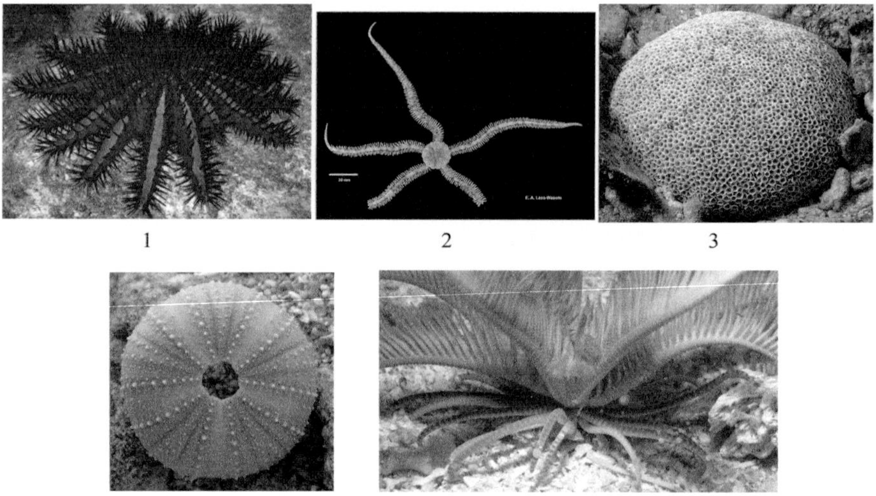

1 2 3 4 5

1: Acanthaster planci; 2: Ophiocoma scolopendrina; 3: Toxopneustes pileolus 4: Tripneustes gratilla; 5: *Tropiometra carinata.*

1. *Image credit*: jon hanson. https://creativecommons.org/licenses/by-sa/2.0/deed.en
2. *Image credit*: Eric A. Lazo-Wasem. https://creativecommons.org/publicdomain/zero/1.0/deed.en
3. *Image credit*: François Michonneau. https://creativecommons.org/licenses/by/3.0/
4. *Image credit*: Philippe Bourjon. https://creativecommons.org/licenses/by-sa/4.0/deed.en
5. *Image credit*: Seascapeza. https://creativecommons.org/licenses/by-sa/3.0/

7.9.5 SEA CUCUMBERS (CLASS: HOLOTHUROIDEA)

Among the echinoderms, the holothuroids (sea cucumbers or aquatic antagonists) are very important due to their species diversity and therapeutic importance. Approximately there are 1250 described species of sea cucumbers worldwide, of which about 20 species are edible and commercially important, and 30 species produce potent toxins collectively referred to as holothurin. Different species of holothuroids produce a variety of holothurin toxins with unique chemical structures and each toxin is nothing but a variation of a specific triterpene glycoside called saponins. Saponins are glycosides of triterpenes and steroids and these saponins comprise a large family of structurally related compounds containing a steroid or triterpenoid aglycone (sapogenin). It is worthy of mention here that holothurin was the first saponin to be isolated in animals and the only animals known to contain holothurin are the echinoderms, including sea cucumbers and sea stars. Further, holothurins A and B are the two groups of holothurin toxins produced specifically by the holothuroid sea cucumbers. These toxins consist of about 60% glycosides and pigment; 30% free amino acids (alanine, arginine, cysteine, glycine, glutamic acid, histidine, serine, and valine); 5 to 10% insoluble proteins; and 1% cholesterol, salts, and polypeptides 3 (Ellis et al., 2021). Kim and Himaya (2012) reported that more than 100 triterpenoid glycosides have been derived from the sea cucumbers in the past 20 years and based on their aglycone structure, these toxins are grouped into four main structural categories viz. 3β-hydroxyholost-9(ll)-ene aglycone skeleton, 3β-hydroxyholost-7-ene skeleton, other holostane type aglycones, and nonholostane aglycone. Though most of these triterpenoid glycosides are found to possess potential bioactivities, anticancer and antiviral activities are the most widely studied and researched areas.

Bioactivities of toxic triterpenoid glycosides of sea cucumbers

i) Cytotoxicity/antitumor activity

The species of sea cucumbers possessing toxic triterpenoid glycosides with cytotoxicity/antitumor activity are given in the following table.

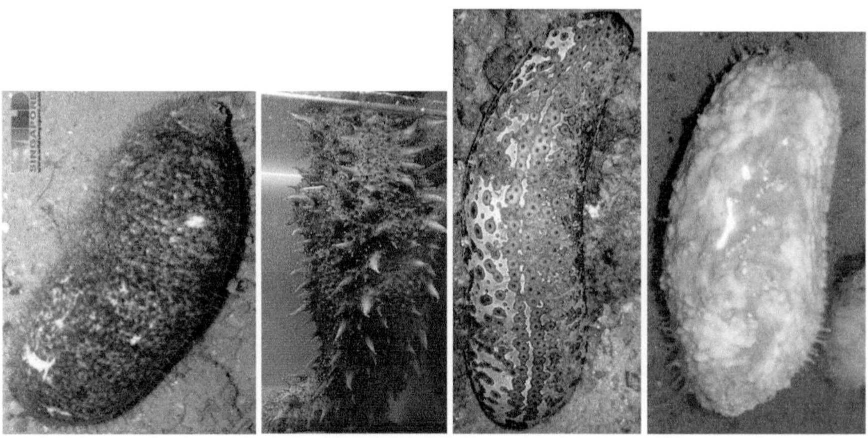

1 2 3 4

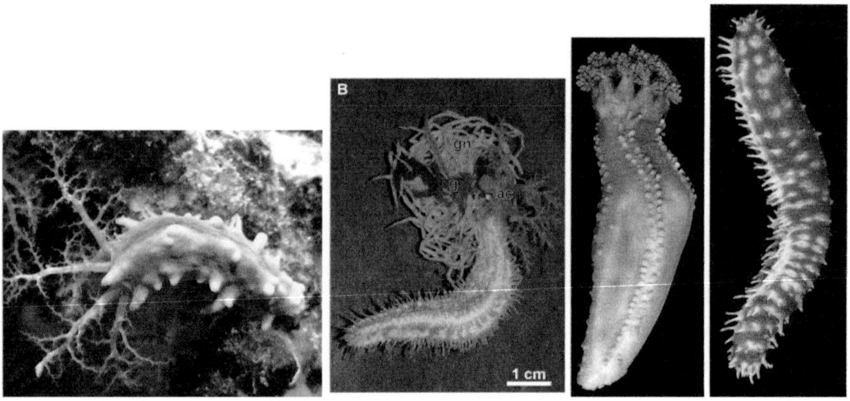

5 6 7 8

1: Actinopyga lecanora; 2: Apostichopus japonicus; 3: Bohadschia argus; 4: Cercodemas anceps; 5: Colochirus quadrangularis; 6: Eupentacta fraudatrix; 7: Cucumaria frondosa; 8: Holothuria (Mertensiothuria) hilla.

1. Image credit: Ria Tan. https://creativecommons.org/licenses/by-sa/2.0/deed.en

2. Image credit: OpenCage. https://creativecommons.org/licenses/by-sa/2.5/

3. Image credit: Anders Poulsen. https://creativecommons.org/licenses/by-sa/3.0/

4. Image credit: Ria Tan. https://creativecommons.org/licenses/by-sa/2.0/deed.en

5. Image credit: Craig D. https://creativecommons.org/licenses/by-sa/2.0/

6. Image credit: Reproduced with permission from Dr. Igor Dolmatov

7. Image credit: Creator Eric A. Lazo-Wasem Gall L. https://creativecommons.org/publicdomain/zero/1.0/deed.en

8. Image credit: François Michonneau. https://creativecommons.org/licenses/by/3.0/

TABLE 7.21 Toxic Triterpenoid Glycosides of Sea Cucumbers with Cytotoxicity/Antitumor Activity.

Species	Name of Toxin	References
Actinopyga lecanora	Lecanoroside A, B: holothurin A, B	Aminin et al. (2016)
Apostichopus japonicus	Holotoxin A1	Day et al. (2020)
Bohadschia argus	Arguside A-E	Liu et al. (2008); Aminin et al. (2016)
Colochirus anceps	Colochiroside A	Aminin et al. (2015)
Colochirus quadrangularis	Philinopside A, E; E; Pentactasides I, II, III	Zhang et al. (2006); Yi et al. (2006); Wargasetia and Widodo (2017)
Cucumaria frondosa	Frondoside A-C; Frondoside A2-7 and A2-8	Aminin et al. (2015); Wargasetia and Widodo (2017)
Cucumaria japonica	Cucumariosides A2- 2/A7; A2-2 and A4-2	Bahrami (2015); Aminin et al. (2015)
Cucumaria okhotensis	Okhotosides B1, B2, and B3, Frondoside A, Cucumarioside A2-5 and Koreoside A; Frondoside A	Bahrami (2015); Aminin et al. (2015)
Eupentacta fraudatrix.	Cucumariosides A1, A3, A4, A5, A6, A12, and A15	Silchenko et al. (2012)
Holothuria atra	Holothurin A5; holothurin A; Echinoside A and 24-dehydrowchisode A	Grauso et al. (2019)
Holothuria (Panningothuria) forskali	Holothurinoside A-D	Rodriguez et al. (1991)
Holothuria (Stauropora) fuscocinerea	Fuscocineroside A-C; pervicoside C; and holothurin A	Zhang et al. (2006); Aminin et al. (2016)
Holothuria grisea	17-dehydroxyholothurinoside A and griseaside A	Aminin et al. (2016)
Holothuria (Mertensiothuria) hilla	Hillaside A-C	Wu et al. (2006); Aminin et al. (2016)
Holothuria leucospilota	Leucospilotaside B	Bahrami (2015)
Holothuria (Thymiosycia) impatiens	Impatienside A and Bivittoside D	Bahrami (2015)
Holothuria (Microthele) nobilis	Nobiliside A-C; Echinoside A	Bahrami (2015); Wargasetia and Widodo (2017)

TABLE 7.21 *(Continued)*

Species	Name of Toxin	References
Holothuria (Roweothuria) poli	Bivittoside	Wargosetia and Widodo (2017)
Holothuria (Metriatyla) scabra	Holothurin A, B; Holothurin A1, A3, and A4; holothurin B4; scabraside A,B,D; echinoside A, 24-dehydroechinoside A, HS-1, fuscocineroside C, and Impatienside A	Aminin et al. (2016); Dang et al. (2007)
Mensamaria intercedens	Intercedenside A-C	Zou et al. (2003)
Pearsonothuria graeffei	Holothurin A, Ds-echinoside A, Echinoside A, 24-dehydro echinoside A	Aminin et al. (2015)
Pentamera calcigera	Calcigeroside B, C1, and C2; cucumarioside G2	Avilov et al. (2000)
Plesiocolochirus australis	Penaustroside A-D	Bahrami (2015)
Pseudocolochirus violaceus	Intercedenside B; Violaceusides A and B	Bahrami (2015); Wargasetia and Widodo (2017)
Pseudocnus echinatus	Cucumechinoside A-F	Bahrami (2015)
Psolus fabricii	Psolusosides B, E, F, G, H, H1, and I	Silchenko et al. (2019)
Psolus patagonicus	Patagonicoside A	Aminin et al. (2015)
Staurocucumis liouvillei	Liouvillosides A/B	Bahrami (2015)
Stichopus horrens	Stichorrenoside C and B	Wargasetia and Widodo (2017)
Synapta maculata	Synaptosides A and A1	Bahrami (2015)
Thelenota anax	Stichoposide C; D	Aminin et al. (2015); Wargasetia and Widodo (2017)

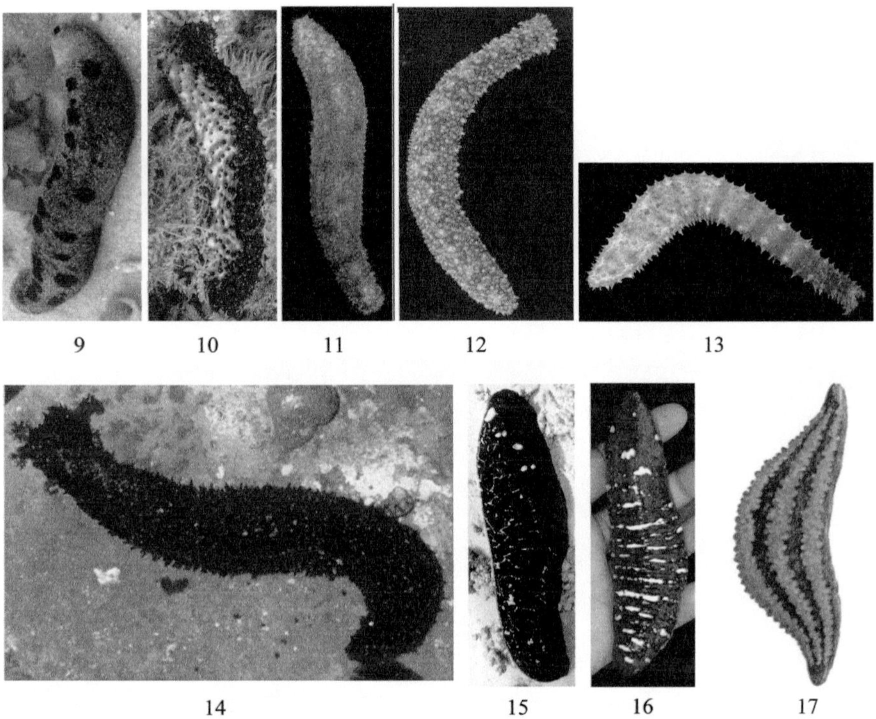

9: Holothuria atra; 10: Holothuria (Panningothuria) forskali; 11: Holothuria (Stauropora) fuscocinerea; 12: Holothuria grisea; 13: Holothuria (Thymiosycia) impatiens; 14: Holothuria leucospilota; 15: Holothuria (Microthele) nobilis; 16: Holothuria (Metriatyla) scabra; 17: Mensamaria intercedens.

9. Image credit: Chaloklum Diving- Koh Phangan. https://creativecommons.org/licenses/by-sa/4.0/

10. Image credit: Diego Delso. https://creativecommons.org/licenses/by-sa/4.0/

11. Image credit: François Michonneau. https://creativecommons.org/licenses/by/3.0/deed.en

12. Image credit: Gondim A, Dias T, Christoffersen M. https://creativecommons.org/licenses/by/4.0/

13. Image credit: François Michonneau. https://creativecommons.org/licenses/by/3.0/deed.en

14. Image credit: Keisotyo. https://creativecommons.org/licenses/by-sa/3.0/deed.en

15. Image credit: Fernando Herranz Martín. https://www.gnu.org/licenses/gpl-3.0.html

16. Image credit: Gihan Jayaweera. https://creativecommons.org/licenses/by-sa/4.0/

17. Image credit: Hubert Lyman Clark. https://creativecommons.org/publicdomain/zero/1.0/ (PD)

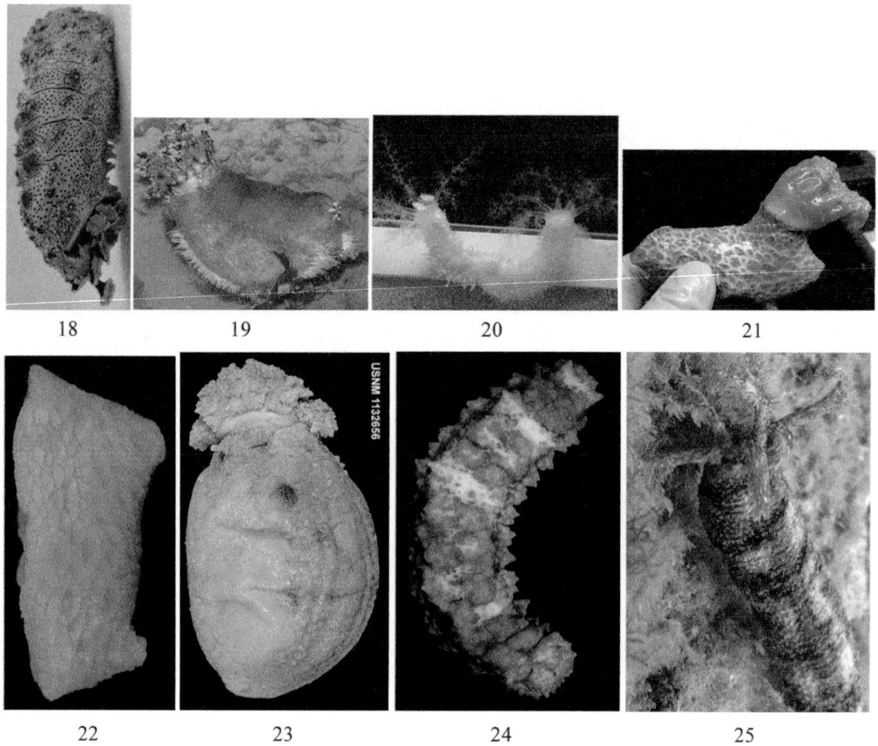

18: Pearsonothuria graeffei; 19: Pseudocolochirus violaceus; 20: Plesiocolochirus sp.; 21: Pseudocnus echinatus; 22: Psolus patagonicus; 23: Staurocucumis liouvillei; 24: Stichopus horrens; 25: Synapta maculate.

18. Image credit: Frédéric Ducarme. https://creativecommons.org/licenses/by-sa/4.0/deed.en

19. Image credit: Ria Tan. https://creativecommons.org/licenses/by-sa/2.0/deed.en

20. Image credit: OpenCage. https://creativecommons.org/licenses/by-sa/2.5/deed.en

21. Image credit: Rebecca Evans. https://creativecommons.org/licenses/by-sa/4.0/deed.en

22. Image credit: Trang Nguyen. https://creativecommons.org/publicdomain/zero/1.0/deed.en

23. Image credit: Trang Nguyen. https://creativecommons.org/publicdomain/zero/1.0/deed.en

24. Image credit: François Michonneau. https://creativecommons.org/licenses/by-sa/4.0/deed.en

25. Image credit: Philippe Bourjon. https://creativecommons.org/licenses/by-sa/3.0/deed.en

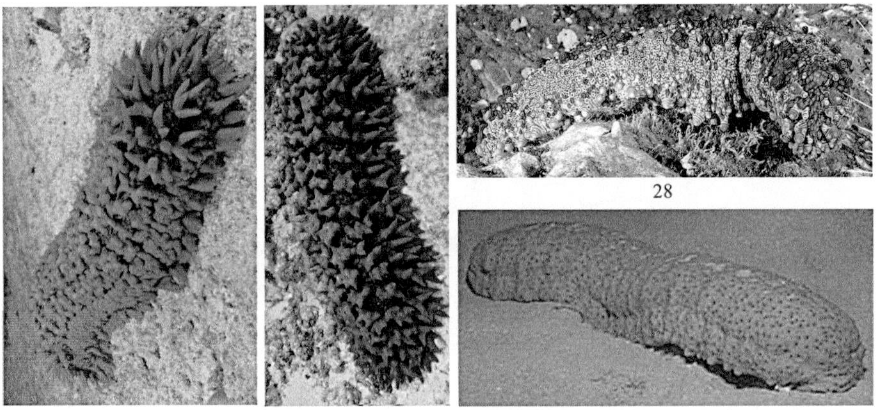

26: Thelenota anax; 27: Thelenata ananas; 28: Actinopyga agassizii; 29: Stichopus sp.

26. Image credit: EOL Learning and Education Group. https://creativecommons.org/licenses/by/2.0/deed.ru

27. Image credit: Aexander Vasenin. https://creativecommons.org/licenses/by-sa/3.0/deed.en

28. Image credit: Flower Garden Banks - Nature Marine Sanctuary- NOAA. https://creativecommons.org/publicdomain/zero/1.0/ (PD)

29. Image credit: François Michonneau . https://creativecommons.org/licenses/by/3.0/

Major sea cucumber species with cytotoxic/antitumor activity

Actinopyga lecanora: Its lecanoroside A displayed cytotoxic activity against human leukemiaHL-60 and hepatomaBEL-7402 cell lines with EC50 values of 4.5 and 17.5 μg/mL, respectively. On the other hand, its holothurin A had the EC50 values of 1.68 and.2 μg/mL, respectively (Aminin et al., 2016; Zhang et al., 2008).

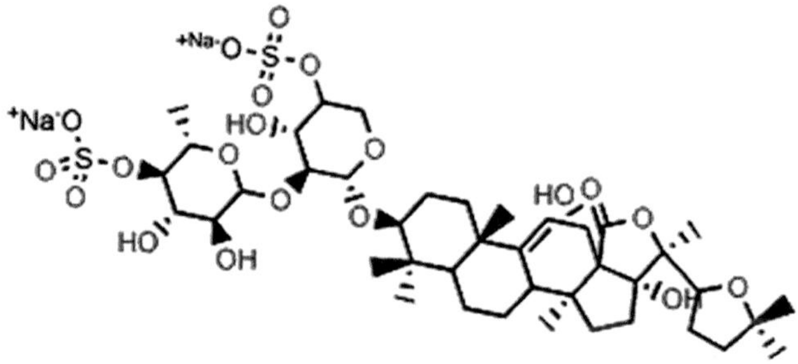

Lecanoroside A

Apostichopus japonicus *(Stichopus japonicus)*: The 70% ethanol fraction of this species containing triterpenoid glycosides had tumor-suppressing activity toward HeLa cells, A-549 lung cancer cells, SGC-7901 stomach cancer cells, and Bel-7402 liver cancer cells besides inducing apoptosis of these tumor cells (Bahrami, 2015). Day et al. (2020) reported that its Holotoxin A1 displayed anticancer activity by reducing the growth of HL-60 with an IC50 value of 23.56 μg/mL.

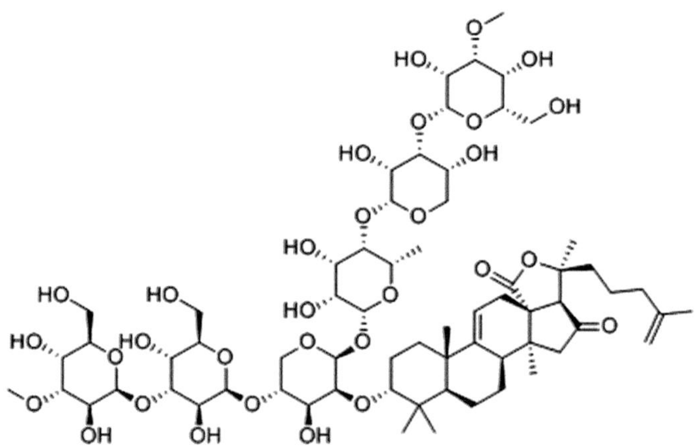

Holotoxin A1

Bohadschia argus: The triterpene glycosides of this species viz. arguside A-E displayed significant cytotoxicities against human tumor cell lines. Arguside A showed cytotoxicity against HL60, HCT116, BEI7402, and MKN45 with EC50 values of 4.4, 0.14, 0.76, and 0.24 uM, respectively (Liu et al., 2008; Aminin et al., 2016). The EC50 values of arguside

B-E reported on A549, HCT-116, HepG2, and MCF-7 cell lines are given in the following table.

TABLE 7.22 EC50 Values of Arguside B-E.

Glycoside/Cancer Cell Line	MCF7	A549	HCT116	HEpG2
Arguside B	7.8	5.4	3.5	6.5
Arguside C	7.5	5.4	3.4	6.5
Arguside D	1.6	0.5	0.5	2.6
Arguside E	1.6	0.5	0.4	2.6

Source: Aminin et al. (2016).

Cercodemas anceps *(= Colochirus anceps)***:** The toxin colochiroside A of this species has shown antitumor activity against p388, HL60, A-549, SpC-A4, and MKN28 cancer cell lines (Wargaseti and Widodo, 2017).

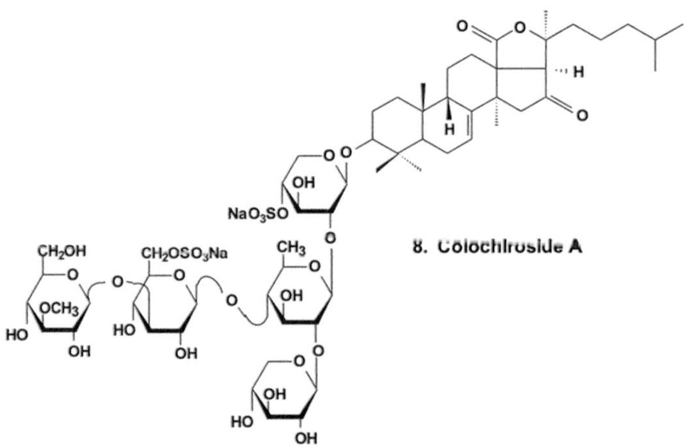

Colochiroside A

Colochirus quadrangularis *(= Pentacta quadrangularis)*: Its sulfated triterpene glycoside, philinopside E, displayed potent cytotoxicity against (IC50 = 0.75–3.50 μg/mL) various tumor cell lines such as human leukemia cells-HL60, mouse lymphocytic leukemia cells-P388, human non-small lung carcinoma cells-A549, gastric carcinoma cells-MKN28, lung adenocarcinoma cells-SPC-A4, gastric carcinoma cells-SGC7901, human epithelial carcinoma cells-A431, human hepatoma cells-BEL7402, human ovarian carcinoma-HO8901, and human fetal lung fibroblasts-W138 with

IC50 values ranging from 0.75 to 3.50 μg/mL. Further, its philinopside A has been reported to inhibit proliferation, migration, angiogenesis, RTKs in several cancer cell lines (Wargasetia and Widado, 2017). Yi et al. (2006) and Han et al. (2010) reported that its triterpene glycosides such as philinopside A and B, and pentactaside I-III displayed significant in-vitro cytotoxicity against six tumor cell lines viz. P-388, A-549, MCF-7, MKN-28, HCT-116, and U87MG with IC50 values ranging from 0.60 to 3.95 μM.

Cucumaria frondosa: This edible sea cucumber has yielded the toxic triterpene glycosides, frondoside A – C; and frondoside A2-7, and -8. Among these compounds, frondoside A and B have been reported to show cytotoxicity against MDA-MB-23, A549,MDA-MB-435, MCF-7, HepG2, HL-60, AsPC-1, AsPC-1, S2013, AsPC-1 cell lines (Wargasetia and Widodo,2017). Further, its frondoside C displayed intense cytotoxic activity against mice and human tumoral cell lines such as P-388, Schabel, A-549, HT-29, and Mel-28 with an IC50 value of 1 μg/ml (https://cdnsciencepub.com/doi/abs/10.1139/v97-218?journalCode=cjc).

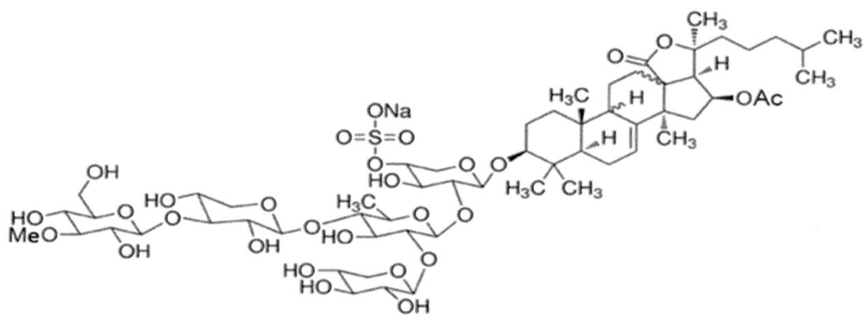

Frondoside A

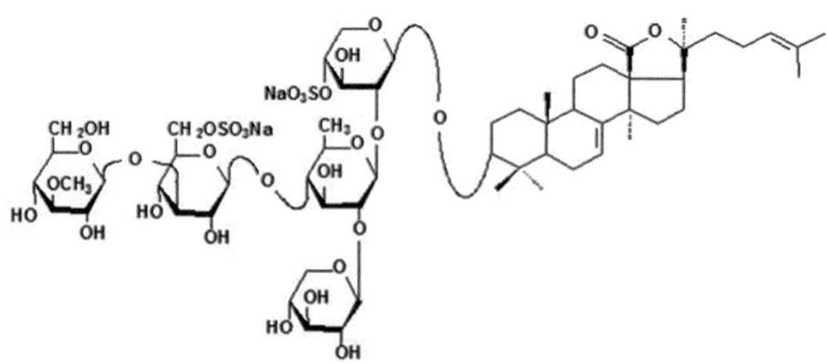

Frondoside. B

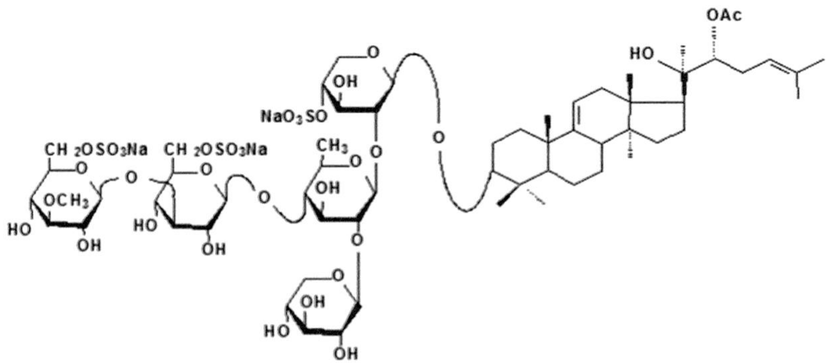

Frondoside

Cucumaria frondosa japonica *(= Cucumaria japonica):* Its triterpene glycosides viz. cucumariosides A2- 2/A7-1 have been reported to exhibit cytotoxicity against murine mammary tumor cell line (BALB/c-MC) (Bahrami, 2015). Wargasetia and Widodo (2017) reported that its cucumarioside A2-2 also showed cytotoxicity against Ehrlich ascite carcinoma, HL-60 cell lines by inhibiting the growth of multidrug-resistant tumor cells and inducing apoptosis in a caspase-dependent way.

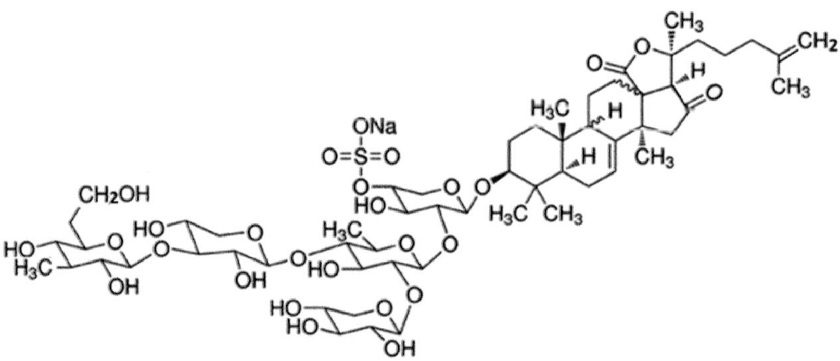

Cucumarioside A2-2

Cucumaria okhotensis: Its triterpene glycosides viz. okhotosides B1, B2, and B3, frondoside A, cucumarioside A2-5, and koreoside A have shown cytotoxicity against human cervical carcinoma (HeLa) and leukemia monocytic (THP-1) cell lines (Bahrami, 2015). Wargasetia and Widodo (2017) reported that its frondoside A was active against THP-1, HeLa, RT112, RT4, HT-1197 cell lines.

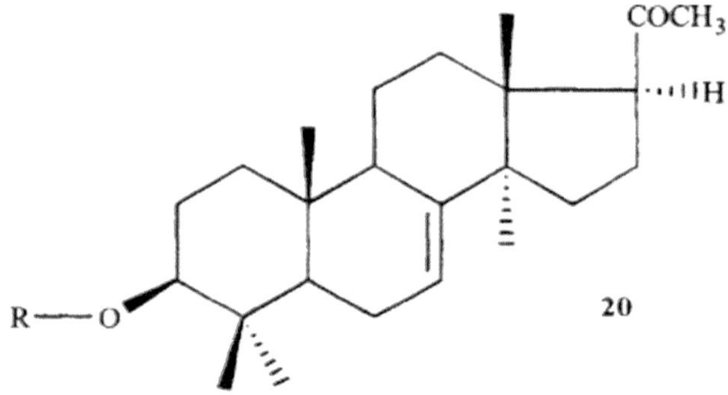

Koreoside A

Eupentacta fraudatrix. *(= Cucumaria fraudatrix.)*: Among the minor triterpene glycosides viz. cucumarioside A1, A3, A4, A5, A6, A12, and A15-produced by this species, cucumarioside A1 and A6 have been reported to show cytotoxicity against mouse spleen lymphocytes and the cells of the ascite form of mouse Ehrlich carcinoma (Silchenko et al., 2012).

Holothuria atra: The triterpene glycoside saponins, holothurin A5, hoolthurin A, echinoside A, and 24-dehydroechinoside A isolated from this species displayed cytotoxicity against HeLa cancer cells with IC50 values of 1.9, 1.4, 1.2, and 2.5 ug/mL, respectively (Grauso et al., 2019).

Holothuria (Panningothuria) forskali *(= Holothuria forskalii):* Five triterpene glycosides viz. holothurinoside A, B, C, and D; and desholothurin A (nobiliside 2a), have been derived from this species. Among these toxins, holothurinoside A and B are non-sulfated pentasaccharide saponins; holothurinoside C and D are di and tetrasaccharides; and desholothurin A is a tetrasaccharide and a synthetic intermediate. All these compounds have shown antitumor and antiviral activities and holothurinoside A and desholothurin A were found to be most active against P388 cells with IC50 values of 0.46 and 0.38 mg/mL, respectively (Rodriguez et al., 1991).

Holothuria (Stauropora) fuscocinerea *(= Holothuria fuscocinerea):* The triterpene glycosides, fuscocineroside A-C; and pervicoside C and holothurin A have been isolated from this species. Among these compounds, fuscocineroside C, pervicoside C, and holothurin A have been reported to show more significant cytotoxicity against human promyelocytic leukemia cell line (HL-60) and human hepatoma cell line (BEL-7402) (Zhang et al., 2006; Aminin et al., 2016).

Holothuria grisea: The glycosides (holostane nonsulfated pentaosides), 17-dehydroxyholothurinoside A, and griseaside A displayed significant in-vitro cytotoxic activity against four human tumor cell lines viz. A549, Bel-7402, HL-60, and Molt-4 with IC50 values of 0.89, 0.97, 0.25, and 0.34 uM, respectively (17-dehydroxyholothurinoside A); and 1.07, 1.11, 0.43, and 0.52 uM, respectively (griseaside A) (Aminin et al., 2016).

Holothuria (Mertensiothuria) hilla *(= Holothuria hilla)*: lanostane-type triterpene glycosides, hillaside A, B, and C of this species showed significant cytotoxicity against eight human tumor cell viz. A-549, MCF-7, IA9, CAKI-1, PC-3, KB, KB-VIN, and HCT-8 with IC50 values ranging between 0.10 and 3.80; 0.26 and 3.78; and 0.15 and 3.20; µg/mL, respectively (Wu et al., 2006).

Holothuria (Thymiosycia) impatiens *(= Holothuria impatiens):* The triterpene glycosides, impatienside A and bivittoside D of this species, had anticancer activity against HCT-116, HT-29, A549, HepG2, DU145, MCF-7, and KB cell lines (Bahrami, 2015).

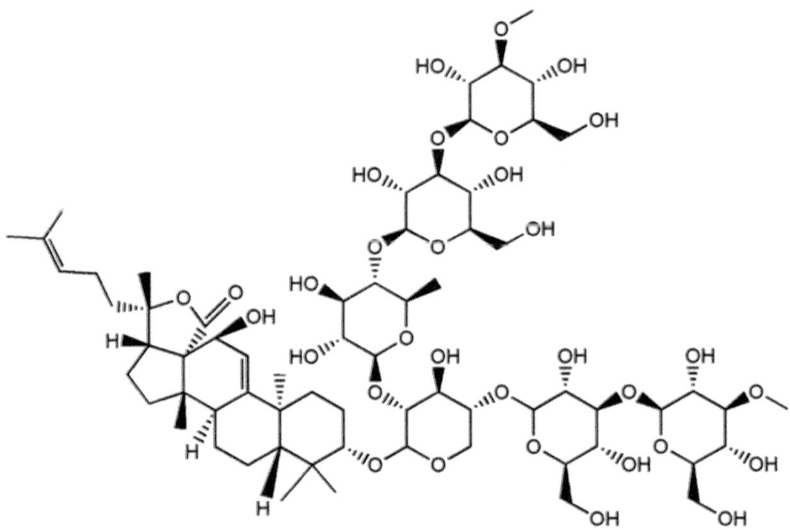

Impatienside A

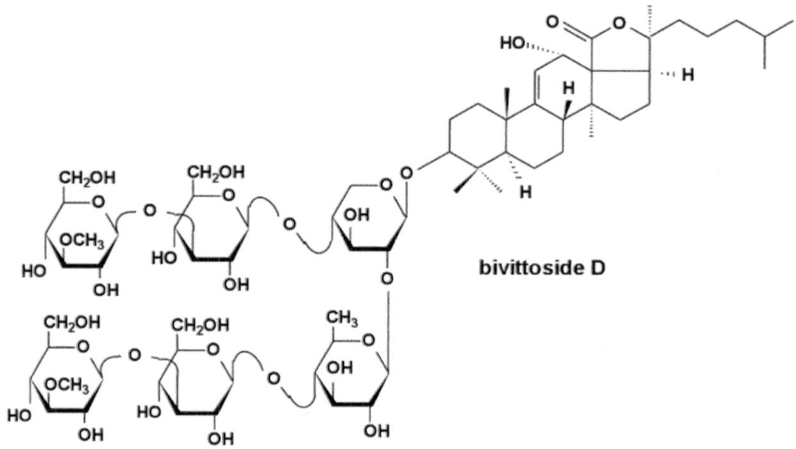

Bivittoside D

Holothuria leucospilota: The triterpene glycoside, leucospilotaside B of this species, showed strong cytotoxicity against A549, BEL-7402, HL-60, and Molt-4 cell lines with EC50 values of 1.06, 2.62, 0.44, and 0.45 μM, respectively (Bahrami, 2015). Wargasetia and Widodo (201) reported that its crude saponin had cytotoxicity against B16F10 cell line by inducing apoptosis and upregulating caspase-3 and 9.

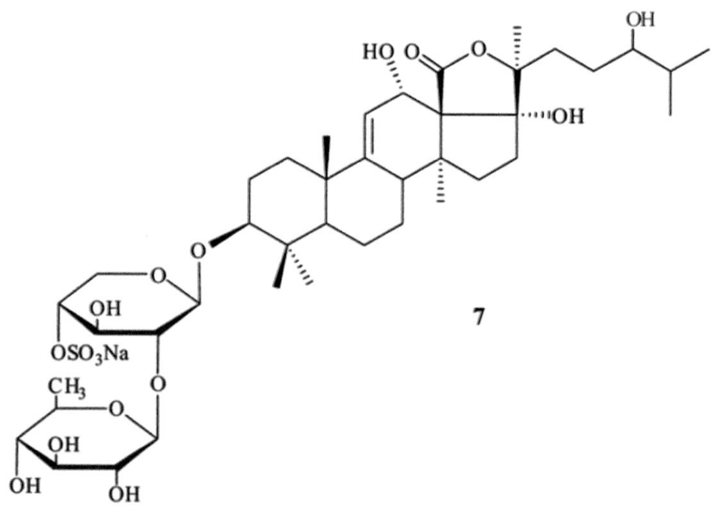

7

Leucospilotaside B

Holothuria (Microthele) nobilis *(= Holthuria nobilis):* The triterpene glycosides, nobilisides A, B, and C of this species showed cytotoxicity against A-549, MCF-7, IA9, CAKI-1, PC-3, KB, KBVIN, and HCT-8 cell lines (Bahrami, 2015). Wargasetia and Widodo (2017) reported that its echinoside A had cytotoxicity against several cancer cell lines by inducing apoptosis and cell cycle arrest.

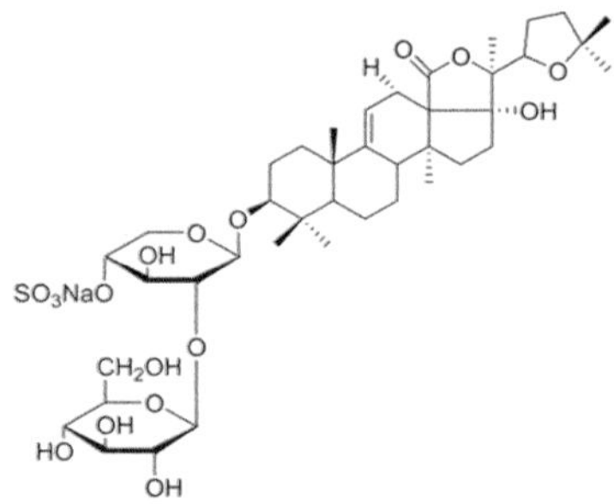

Nobiliside B.

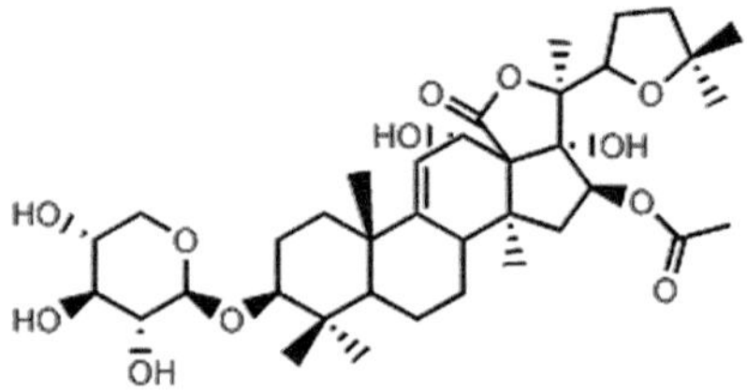

Nobiliside C

Holothuria (Roweothuria) poli *(= Holothuria poli):* The triterpene glycoside bivittoside of this species possessed cytotoxicity against HCT116, and MCF7 cell lines (Wargosetia and Widodo, 2017).

Holothuria (Metriatyla) scabra (= Holothuria scabra): This species has been reported to produce several triterpene glycosides that possess cytotoxicity against several cancer cell lines. The glycosides scabrasides A and B exhibited cytotoxicity against HL-60 cells and human acute lymphoblastic leukemia cell line (MOLT-4) with EC50 values of 0.05 and 0.25 μM, respectively; and 0.09 and 0.08 μM, respectively. Further, its holothurins A3 and A4 exhibited significant cytotoxic activity against human epidermoid carcinoma cells (KB) with EC50 values of 0.87 and 1.12 μg/mL, respectively; and human hepatocellular carcinoma (HepG2) cell lines with IC50 values of 0.32 and 0.57 μg/mL, respectively. Furthermore, its dehydroechinoside A and scabraside D have been reported to act against P388, A549, MKN28, HCT116, and MCF7 with EC50 values of 1.2, 1.4, 2.2, 1.1, and 1.8 uM respectively; and 1.0, 1.7, 1.3, 1.6, and 1.8 uM respectively. The other toxins of this species and their cytotoxic activities are given in the following tables.

TABLE 7.23 Cytotoxicity (Ec50uM) of Triterpene Glycosides of *Holothuria (Metriatyla) scabra.*

Compound/Cell Line	HeLa	K562	HepG2	HL7702
Holothurin A	3.8	8.9	3.5	3.9
Scabraside B	4.4	11.9	7.3	11.0
Echinoside A	1.3	1.6	1.5	3.1
24-dehydroechinoside A	2.1	6.2	2.0	3.8
Holothurin B	2.1	3.6	1.8	2.7
Holothurin B4	2.7	3.6	2.7	4.6
HS-1	3.2	13.0	6.1	7.2
Holothurin A1	2.8	6.5	2.9	3.6
Scabraside D	3.8	11.1	3.3	4.4

Source: Aminin et al. (2016).

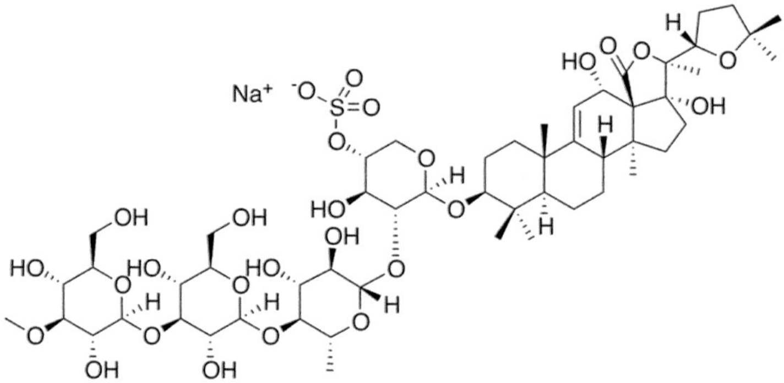

Holothurin A

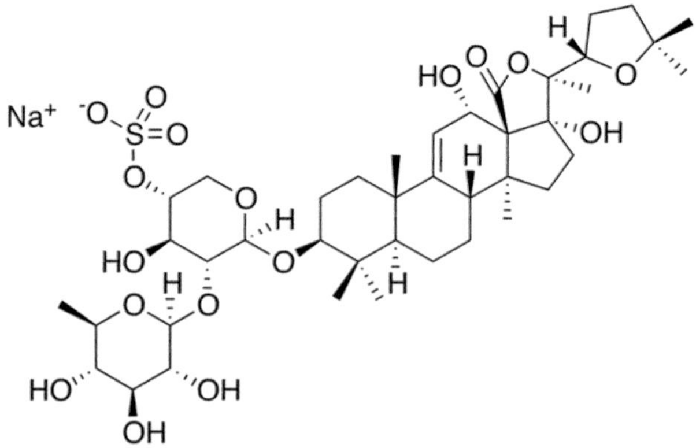

Holothurin B

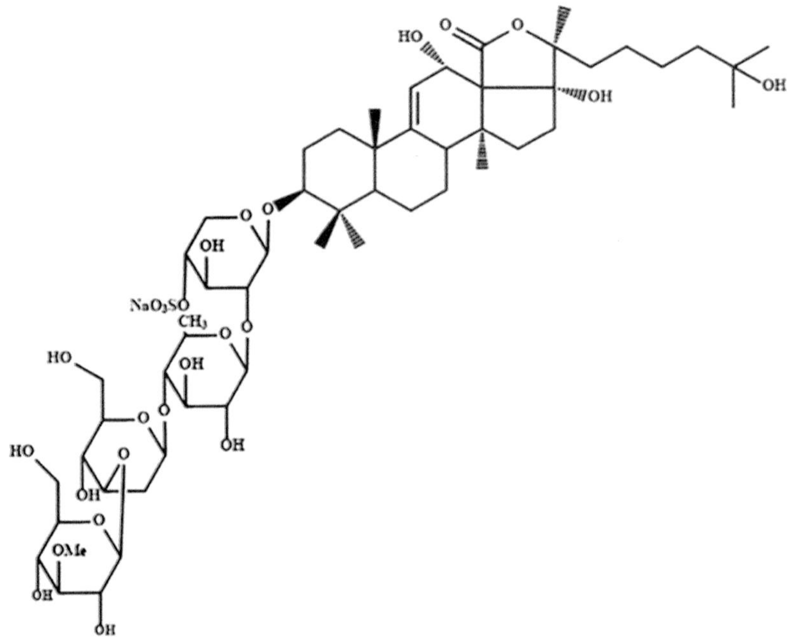

Scabraside D.

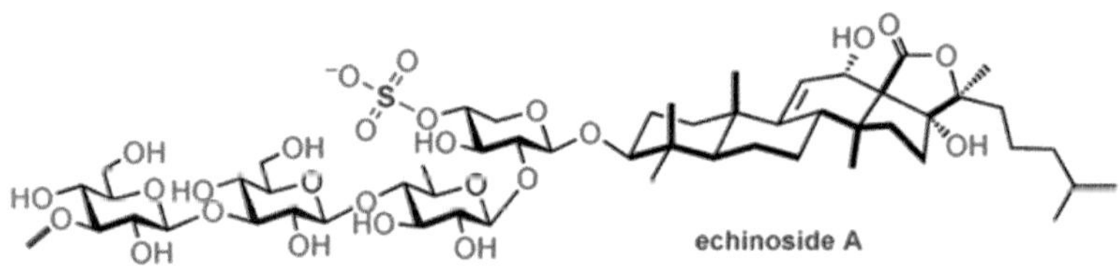

Echinoside A

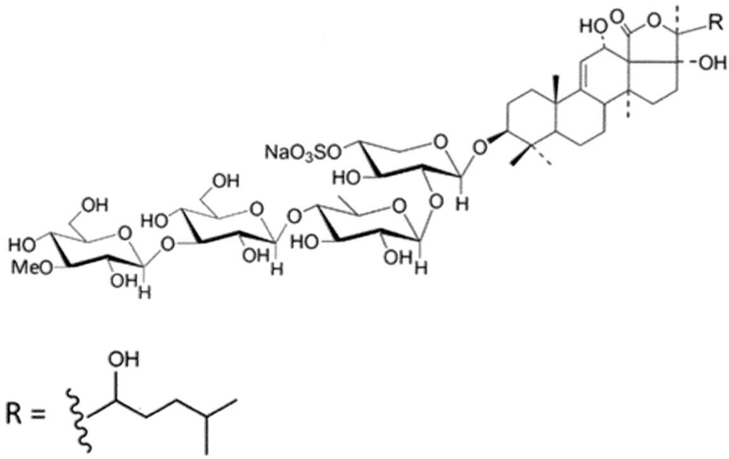

Holothurin A1

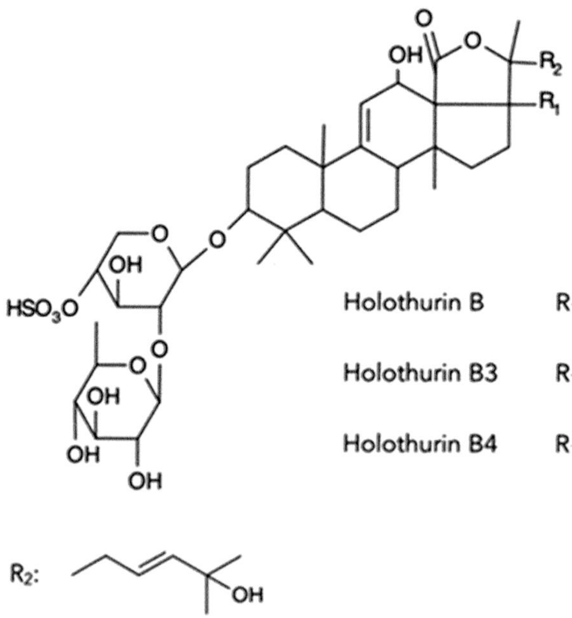

Holothurin B4

Mensamaria intercedens: The triterpene glycosides, intercedenside A, B, and C of this species, exhibited significant cytotoxicity and anti-neoplastic activity against 10 human tumor cell lines with ED 50 values ranging from 0.6 to 4.0 μg/mL. Further, its intercedenside A also showed significant in-vivo antineoplastic activity against mouse Lewis lung cancer and mouse S180 sarcoma (Zou et al., 2003; Wargasetia and Widdodo, 2015).

Pentamera calcigera: This species has yielded monosulfated triterpene glycosides (i.e., sulfated holothurins) viz. calcigerosides B, C1, C2, D1, D2, and E; and cucumarioside G2. Among these compounds, calcigerosides B, C1, and C2 showed moderate cytotoxicity against human and mouse tumoral cell lines such as P-388, A-549, HT-29, and Mel-98 with an IC50 value of 5 μg/mL (Avilov et al., 2000a).

Pearsonothuria graeffei: Among the triterpene glycosides produced by this species viz. holothurin A, Ds-echinoside A, echinoside A, and 24-dehydroechinoside A, the latter three compounds have been reported to exhibit cytotoxicity against Hep G2 by inhibiting the cell migration and invasion (Wargasetia and Widodo, 2017; Aminin et al., 2015).

Plesiocolochirus australis *(= Pentacta australis):* The triterpene glycosides, DS-penaustroside A-D of this species showed cytotoxicity against L 1210 and KB cells (Bahrami, 2015).

Pseudocolochirus violaceus: The triterpene glycoside, intercedenside B of this species showed cytotoxicity against MKN-45 and HCT-116 cell lines (Bahrami, 2015). Further, its violaceusides A and B have also been reported to possess similar activity against HL-60, BEL-7402 cell lines (Wargasetia and Widodo, 2017).

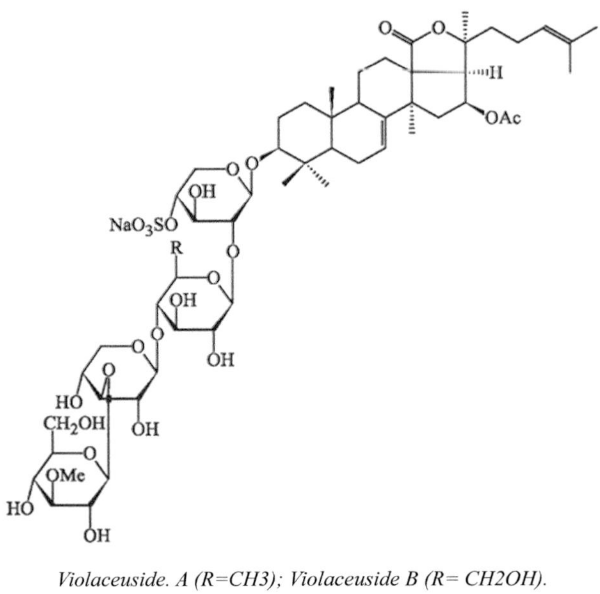

Violaceuside. A (R=CH3); Violaceuside B (R= CH2OH).

Pseudocnus echinatus (= *Cucumaria echinata*): The triterpene glycosides, cucumechinoside A-F of this species have shown cytotoxicity against L 1210 and KB cells (Bahrami, 2015).

***Psolus fabricii*.:** A total of eight sulfated triterpene glycosides viz. psolusoside A, B, E, F, G, H, H1, and I have been derived from this species. All these compounds have shown cytotoxicity against human colorectal adenocarcinoma HT-29 cells with IC50 values of >20 μM. Among these compounds, the highest inhibitory activities were exhibited by psolusosides E and F (Silchenko et al., 2019).

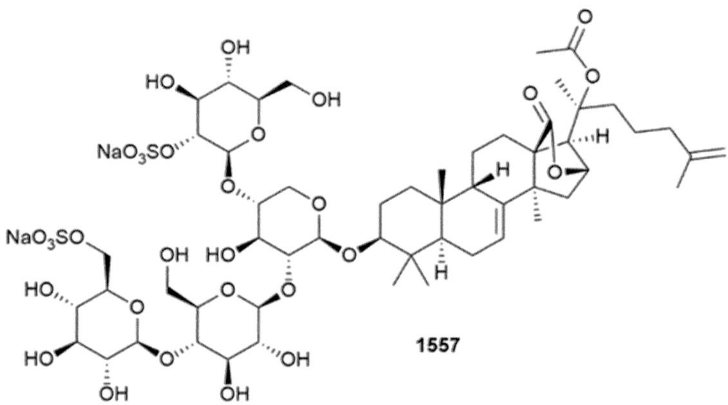

1557

Psolusoside E

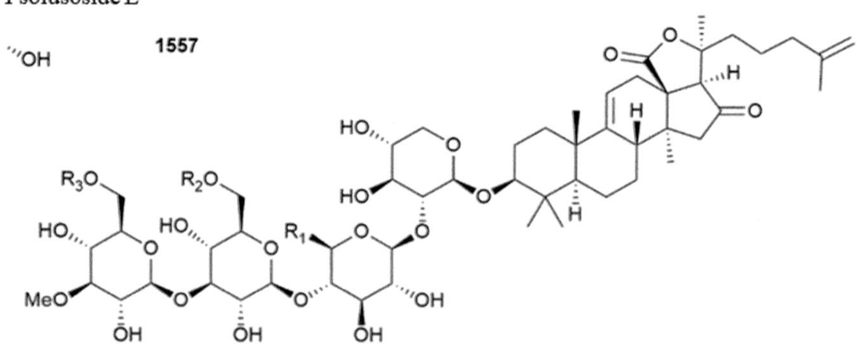

1558 R_1 = Me, R_2 = SO_3Na, R_3 = H
1559 R_1 = Me, R_2 = H, R_3 = SO_3Na
1560 R_1 = CH_2OH, R_2 = H, R_3 = SO_3Na

Psolusoside F; Ppsolusoside G; Ppsolusoside H

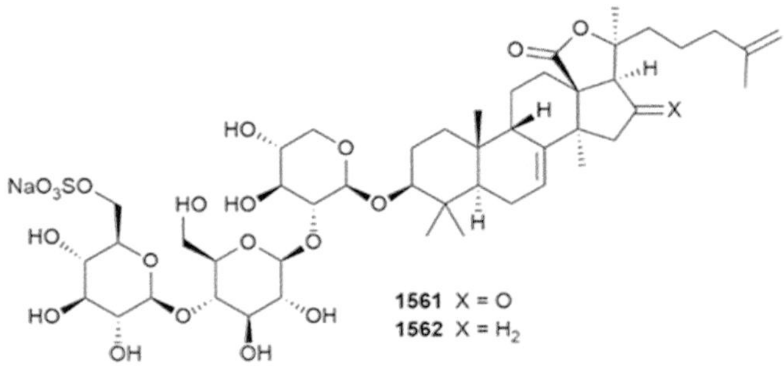

Psolusoside H1(X=O); Psolusoside I (X=H2).

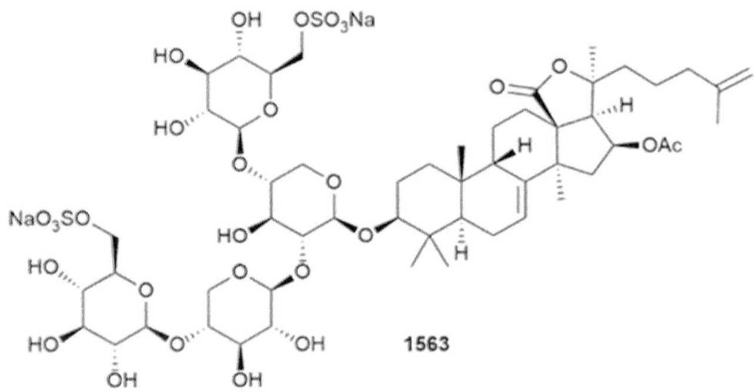

Psolusoside B

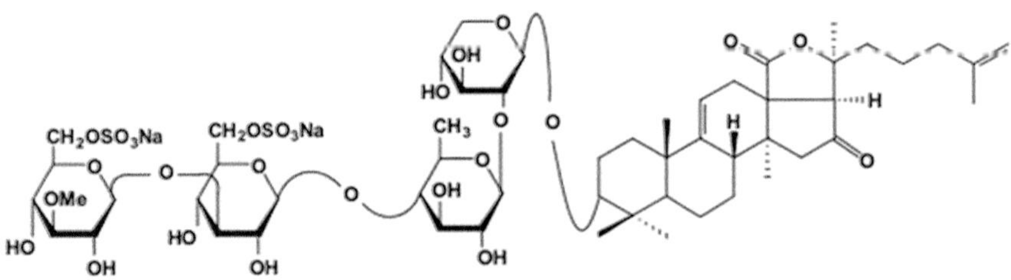

Psolusoside A

Psolus patagonicus: Two unidentified glycosides of this species have been reported to exhibit cytotoxicity against Hep3B, MDA-MB231, and A549 cell lines by suppressing the growth of (Hep3B, MDA-MB231, and A549) cell lines and inducing the activation of NF-κB which is a key player linking chronic inflammation and cancer (Careaga et al., 2009).

Staurocucumis liouvillei: The triterpene glycosides, liouvillosides A/B- of this species have been reported to exhibit cytotoxicity against vero cells (Bahrami, 2015).

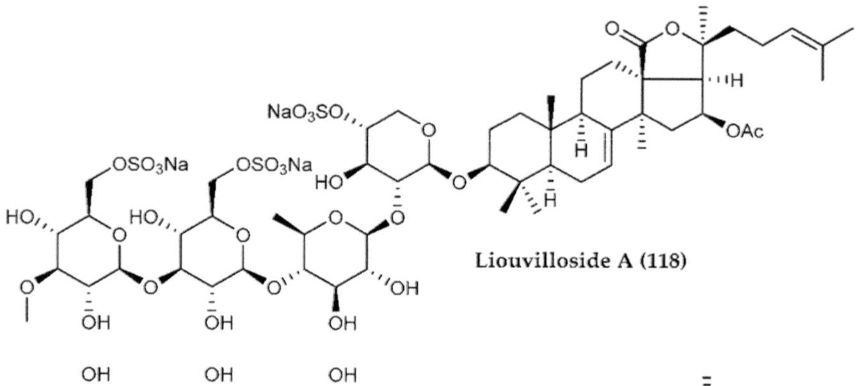

Liouvilloside A (118)

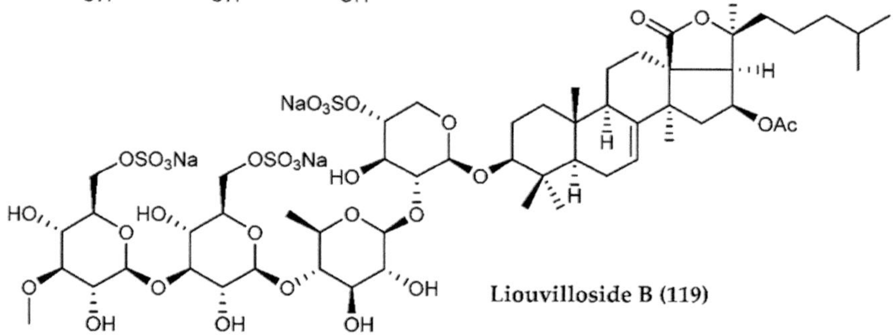

Liouvilloside B (119)

Stichopus horrens: The triterpene glycosides, stichorrenoside B and C exhibited cytotoxic activity against Hep-G2, KB, LNCaP, MCF7, and SK-Mel2 cell lines (Wargasetia and Widodo, 2017).

Synapta maculata: The triterpene glycosides, synaptosides A and A1 of this species showed cytotoxicity against Hela cells (Bahrami, 2015).

TABLE 7.24 Sea Cucumbers Possessing Toxic Triterpenoid Glycosides with Other Bioactivities.

Species	Name of Toxin	Bioactivity	References
Staurocucumis liouvillei	Liouvilloside A, B	Antiviral activity	Maier et al. (2001)
Holothuria forskali	Holothurinoside A, C and D	-Do-	Bahrami (2015)
Telenata ananas	Bivittoside D	-Do-	Bahrami (2015)
Actinopyga agassizii	Holothurin A, A1	Antiobesity; and anti-melanogenic, and antiwrinkling	Kitagawa et al. (1978); (1)
Cucumaria frondosa	Echinoside A	Antiobesity	Hu et al. (2016)
Pearsonothuria graeffei	Echinoside A and holothurin A	-Do-	Wang et al. (2014)
Cucumaria frondosa	Frondoside A	Immunomodulatory	Aminin et al. (2008)
Cucumaria japonica	Cucumarioside A2-2	-Do-	Aminin et al. (2008)
Pearsonothuria graeffei	Saponins	-Do-	Bahrami (2015)
Stichopus hermanni and Stichopus horrens	Saponins	Antidiabetic	Bahrami (2015)
Actinopyga lecanora	Holothurin A,B	Leishmanicidal activity	Singh et al. (2008)
Apostichopus japonicus	Holothurin	Antiprotozoal activity	(2)
Cucumaria echinata	Glycosides	-Do-	(2)

(1): https://en.wikipedia.org/wiki/Holothurin; (2): https://www.sciencedirect.com/topics/chemistry/antiprotozoal-activity.

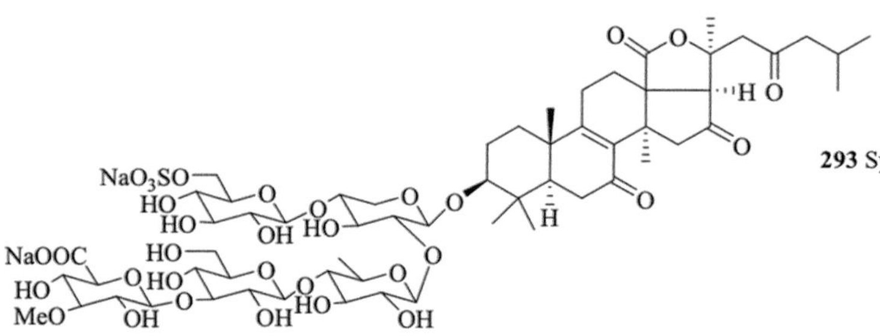

293 Synaptoside A_1

Synaptoside A1

Thelenota anax: Among its triterpene glycosides, stichoposide C and D, the former compound showed cytoxicity against HL-60 and K562 cell lines, and the latter compound induced apoptosis of leukemia and colorectal cancer cells (Wargasetia and Widodo, 2017).

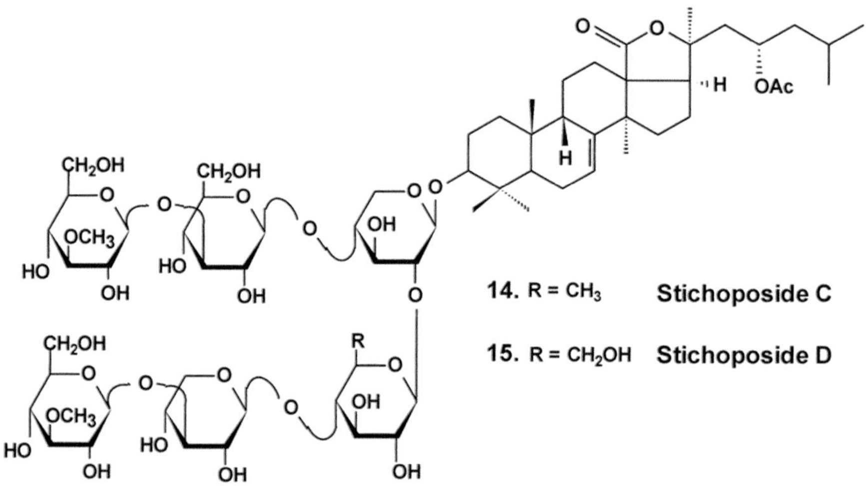

14. R = CH_3 **Stichoposide C**

15. R = CH_2OH **Stichoposide D**

Stichoposide C (R=CH3); Stichoposide D (R=CH2OH).

ii) Other bioactivities

The species of sea cucumbers possessing toxic triterpenoid glycosides with other bioactivities are given in the following table.

a) Antiviral activity

Staurocucumis liouvillei: The trisulfated triterpene glycosides viz. liouvillosides A and B of this species have been reported to be virucidal against herpes simplex virus type 1 (HSV-1) at concentrations <10 μg/mL (Maier et al., 2001). Bahrami (2015) reported that these compounds also interfered with an early step of the viral replicative cycle.

Holothuria (Panningothuria) forskali *(= Holothuria forskali):* The triterpene glycosides, holothurinoside A, C, and D of this species have been reported to show activity against HSV-1 by inhibiting cytopathic effect (Bahrami, 2015).

Thelenata ananas: The triterpene glycoside, bivittoside D of this species exhibited HIV-1-inhibitory activity in chemokine receptor subtype 5 (CRC 5) (Bahrami, 2015).

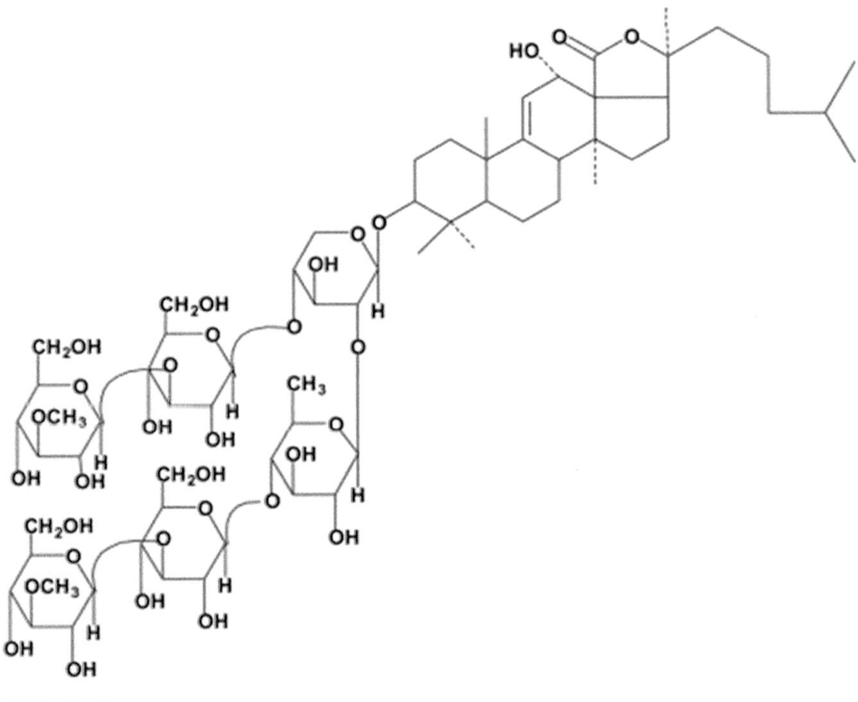

Bivittoside D

b) Antiobesity

Actinopyga agassizii: The holothurin A and holothurin A1 of this species have been reported to show antiobesity effects by reducing weight gain in mice. They were also found to inhibit the absorption of lipids in the intestine, reduce levels of lipids in blood and liver, and improve glucose tolerance, Further, they also inhibit the activity of pancreatic lipase and decrease the growth of white adipocytes, a factor that contributes to obesity (Kitagawa et al.1978).

Cucumaria frondosa: The saponin of this species has been reported to show antiobesity effects by exerting a dose-dependent inhibition against pancreatic lipase with an IC50 value of 1.7 mg/mL (Hu et al., 2016).

Pearsonothuria graeffei: Its echinoside A strongly suppressed pancreatic lipase activity with an IC50 value of 9.6 µg/mL (7.9 µM). Further, its holothurin A was found to show much weaker inhibitory effect with IC50 values of > 35 µM (Wang et al., 2014).

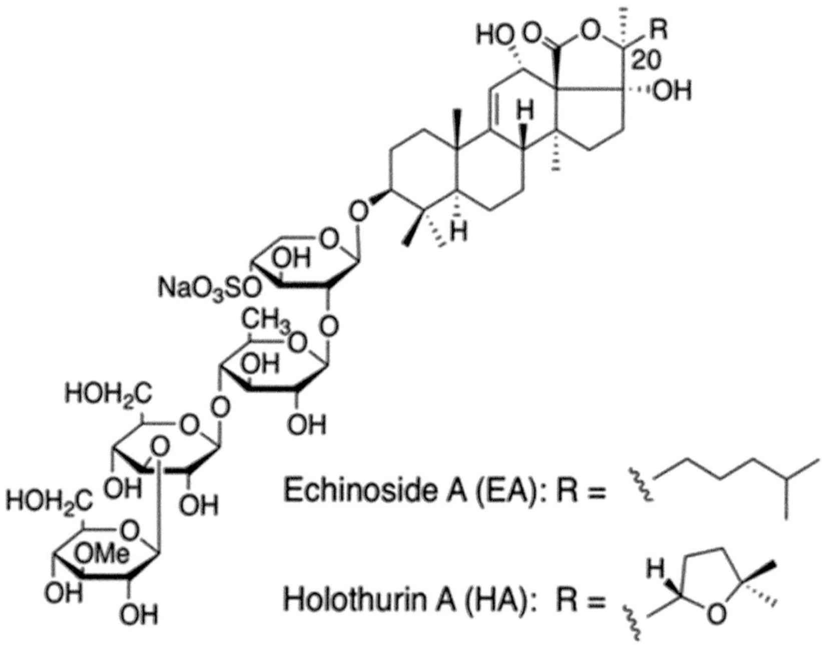

Top: Echinoside A; Bottom: Holothurin A

c) Immunomodulatory effects

Cucumaria frondosa: The triterpene glycoside, frondoside A, possessed strong immunomodulatory properties in subtoxic doses. Frondoside A stimulated lysosomal activity of mouse macrophages in vivo at a maximal dose of 0.2 µg per mouse. Frondoside A was found to enhance macrophage phagocytosis of the bacterium *Staphylococcus aureus* in vitro at a maximal concentration of 0.001 µg/mL. Further, this compound has been reported

TABLE 7.25 Antifungal Activity of Toxic Triterpene Glycosides of Sea Cucumbers.

Toxic Compound(s)	**Source Specie of Holothurian**	**Against Pathogenic Fungi**
Typicosides A1, A2, B1, C1 and C2	*Actinocucumis typica*	*Aspergillus niger, Candida albicans*
Holotoxin A1, Holotoxin B1, Stichloroside C2, Bivittosides A, B, C, and D, Echinosides A and B	*Actinopyga echinites*	*Aspergillus niger, Candida albicans, Penicillium citricum*
Holotoxin A1	*Apostichopus japonica*	*Candida albicans, Aspergillus niger, Penicillum niger*
Neothyonidioside and Mollisoside A	*Australostichopus mollis*	*Saccharomyces cerevisiae*
Marmoratoside A, 17α-hydroxy impatienside A, Marmoratoside B, 25-acetoxy bivittoside D, Impatienside A and Bivittoside D	*Bohadschia marmorata*	*Candida albicans, Cryptococcus neoformans, Aspergillus fumigatus*
Holotoxin A1, Holotoxin B1, Stichloroside C2, Bivittosides A, B, C, and D, Echinosides A and B	*Bohadschia bivittata*	*Aspergillus niger, Candida albicans, Penicillium citricum*
17-hydroxy fuscocineroside B, 25-hydroxy fuscocineroside B, Fuscocineroside B	*Bohadschia marmorata*	*Candida albicans, Candida neoformans, Aspergillus fumigatus*
Cucumechinosides A, B, C, D, E, and F	*Cucumaria echinata*	*Candida albicans, Trichomonas mentagrophytes, Aspergillus niger, Penicillum chrysogenum*
Cucumariosides I and II	*Cucumaria japonica*	*Candida albicans, Candida tropicalis*
Cucumariosides I-1, I-3, and I-4	*Eupentacta fraudatrix*	*Candida albicans, Aspergillus niger*
Cucumariosides A1, A3, A4, A5, A6, A12, and A15	*Eupentacta fraudatrix*	*Candida albicans, Aspergillus niger*
A2, A7, A8, A9, A10, A11, A13, and A14	*Eupentacta fraudatrix*	*Candida albicans. Aspergillus niger*
Scabraside A, Echinoside A, and Holothurin A	*Holothuria scabra*	*Candida albicans, Candida pseudotropicalis, Candida neoformans, Aspergillus fumigates, Fonsecaea compacta, Microsporum gypseum*

TABLE 7.25 *(Continued)*

Toxic Compound(s)	Source Specie of Holothurian	Against Pathogenic Fungi
Axilogoside A and Holothurin B	*Holothuria. (Microthele) axiloga*	*Candida. albicans, C. andidaneoformans, Aspergillus fumigatus*
Arguside F, Impatienside B, and Pervicoside D	*Holothuria. (Microthele) axiloga*	*Candida albicans, Candida neoformu, Trichophyton rubrum, Candida tropicalls, Candida krusei*
Patagonicosides A,	*Psolus patagonicus*	*Cladosporium cladosporoides*
Stichoposides A, C, D, E	*Stichopus chloronotus*	*Candida albicans, Aspergillus niger, Penicillum niger, Saccharomyces pastorianus (=Saccharomyces carlsbergensis)*
Thelenotosides A and B	*Thelenota ananas*	*Candida albicans, Aspergillus niger, Penicillum niger, Hormodendro pedrosoi, Saccharomyces pastorianus (=Saccharomyces carlsbergensis)*

Source: Bahrami (2015).

to stimulate an increase in the number of antibody plaque-forming cells (normally B-cells in spleen) at a concentration of 0.2 μg per mouse. Based on these findings, frondoside A may provide curative and/or preventive treatment options against diseases with a depleted immune status (Aminin et al., 2008; Bahrami, 2015).

Cucumaria frondosa japonica *(= Cucumaria japonica):* A monosulphated triterpene glycoside, cucumarioside A2-2 (derived from this species) and cholesterol in a molar ratio of 1:2 (termed Cumaside) exhibited immunomodulatory properties (Aminin et al. 2006; Bahrami, 2015).

Pearsonothuria graeffei: The dietary saponins of this sea cucumber have been reported to alleviate orotic acid-induced fatty liver in rats via PPARα and SREBP-1c signaling (Bahrami, 2015).

d) Antidiabetic activity

Stichopus hermanni* and *Stichopus horrens: The saponins of these sea cucumbers have been reported to possess antidiabetic effects by inhibiting the absorption of glucose (hypoglycemia). These sea cucumber species are being consumed in traditional medicines to decrease/control blood glucose level in patients suffering from diabetes mellitus (Bahrami, 2015).

e) Antiprotozoal activity

Actinopyga lecanora: At a concentration of 86.08 μM, the triterpene glycoside, holothurin A of this species exhibited antiprotozoal activity with 73 and 66% inhibition of *Leishmania donovani* promastigotes and amastigotes, respectively. Further, its holothurin B showed similar activity with 83 and 47% inhibition (Tiwari et al., 2019).

Apostichopus japonicus: The triterpene glycoside, holothurin of this species has been reported to act against the protozoan parasites viz. *Trichomonas vaginalis* and *Amoeba proteus*. Further, its crude holothurin showed inhibitory activity against *Tripanosoma lewisi* in rats that were intraperitoneally inoculated with the holothurin (Kalinin, et al., 2008).

Pseudocnus echinatus *(= Cucumaria echinata):* The glycosides of this species have shown antiprotozoal activity (Kalinin, et al., 2008).

Others: Jangoux (2020) reported that holothurin of sea cucumber is active in vitro against the protozoan parasite, *Amoeba proteus,* and nemathelminths. Further, he stated that the holotoxins A, B, and C of sea cucumbers are active against *Trichomonas vaginalis*.

f) Antifungal activity

The antifungal activity of toxic triterpene glycosides of the different species of sea cucumbers is given in the following table.

g) Anti-melanogenic and anti-wrinkling effects

Actinopyga agassizii: the holothurins of this species have shown anti-melanogenic and anti-wrinkling effects on human skin by inhibiting melanin production in Melan-A cells and promoting collagen production in human dermal fibroblasts (https://en.wikipedia.org/wiki/Holothurin).

KEYWORDS

- **marine zootoxins**
- **sponge toxins**
- **cnidarian toxins**
- **TTX of flatworms**
- **TTX of ribbon worms**
- **molluskan toxins**
- **echinoderm toxins**

CHAPTER 8

Toxins of Ascidians and Their Therapeutic Potential

ABSTRACT

Aspects relating to the toxic ascidian diversity; toxic compounds, and chemical classes of ascidian toxins; cytotoxic ascidians; and the therapeutic potential of ascidian toxins are presented in this chapter.

Ascidians or sea-squirts (Phylum, Chordata; Sub-phylum; Tunicata; Class, Ascidiacea) which are commonly called urochordates/tunicates are exclusively marine. There are about 3000 species of ascidians. They are solitary or colonial and are found distributed in all depths of the seas of the world. Ascidians, along with invertebrates such as sponges, cnidarians, and bryozoans, are known to produce a rich variety of secondary metabolites including cyclic peptides and depsipeptides and aromatic alkaloids that are produced by the endosymbiotic micro-organisms and not produced by the ascidian themselves. Most of the metabolites of these ascidians are considered to be toxins with diverse activities, and they have the potential for development as therapeutic drugs for a wide variety of diseases. It is worth mentioning here that two ascidian toxins have been marketed for cancer treatment. Among them, Ecteinascidin (ET-743, trabectedin) which has been derived from *Ecteinascidia turbinata* is FDA approved and marketed under the trade name Yondelis®; and Aplidin® (dehyrodidemnin B, plitidepsin) sourced from *Aplidium albicans* has attained orphan drug status. Both are marketed by PharmaMar (Madrid, Spain) (Watters, 2018; Santhanam et al., 2019b). The species of ascidians possessing therapeutic properties are shown in the following tables.

Therapeutic Potential of Marine Biotoxins: From Poison to Pharmacy. Santhanam Ramesh, Ramasamy Santhanam, & Subbiah Balasundari (Authors)

DOI: 10.1201/9781003610991-8

TABLE 8.1 Cytotoxic/Anticancer Potential of Toxins of Ascidians.

Species	Toxic compound	Chemical class
Stolonica australis (= Amphicarpa meridiana)	Meridine	Pyridoacridine alkaloid
Aplidia albicans	Plitidepsin (dehydrodidemnin B) (Aplidin®)	Cyclic depsipeptide (anticancer drug)
Aplidium conicum	Thiaplidiaquinones	Thiazinoquinones
Botryllus sp.	Botryllamides A-J	Brominated tyrosine derivatives
Diazona angulata	Diazonamide A	Cyclic peptide
Didemnum candidum	Euseynstelamide B	Bis-indole alkaloid
Didemnum sp.	Lamellarins	DOPA/TOPA derived pyrrole alkaloids
Didemnum sp.	Ascididemin	Pyridoacridine alkaloid
Ecteinascidia turbinata	Trabectidin (ET-473) Yondelis®	Tetrahydroisoquinoline alkaloid (anticancer drug)
Eudistoma sp.	Eudistomin C	β-Carboline alkaloid
Lissoclinum badium	Lissoclibadins	Polysulfur aromatic alkaloids
Lissoclinum bistratum	Bistratamides	Cyanobactins
Lissoclinum bistratum	Bistramide A(bisA)	Spiroketal
Lissoclinum mandelai	Mandelalides A & B	Polyketides
Lissoclinum patella	Lissoclinamides	Cyanobactins
Lissoclinum patella	Patellamides	Cyanobactins
Lissoclinum patella	Patellazoles A–C	Polyketides
Ritterella tokiaka	Ritterazines	Dimeric steroidal pyrazine alkaloids
Tridemnum solidum	Didemnin B	Cyclic depsipeptide
Unidentified Brazilian species	Tamandarins	Cyclic depsipeptides, closely related to didemnin B

Source: Watters (2018).

TABLE 8.2 Other Ascidian Toxins and Their Therapeutic Potential.

Species	Toxic compound	Chemical class	Bioactivity
Stolonica australis (= *Amphicarpa meridiana)*	Meridine	Pyridoacridine alkaloid	Not known
Aplidium conicum	Thiaplidiaquinones	Thiazinoquinones	Anti-malarial
Aplidium meridianum	Meridianins	Indole alkaloid	Kinase inhibitor
Ciona intestinalis(1)	Citron kinase protein	Tachykinin	Neuroprotective
Didemnum granulatum	Granulatimides	Alkaloids	Kinase inhibitor
Didemnum molle	Antollamides A and B	Dimeric hexapeptides	Not known
Didemnum molle	Mollecarbamates A–D	Polyketides	Not known
Didemnum sp.	Ningalins	DOPA/TOPA derived pyrrole alkaloids	Kinase inhibitor
Eudistoma sp.	Eudistidines	Alkaloids	Antimalarial
Eudistoma sp	Eudistomin C	β-Carboline alkaloid	Antiviral
Halocynthia roretzi(2)	GTXs; STXs		Healing; anesthetic
Lissoclinum vareau	Varamine A	Pyridoacridine Alkaloid	Not known
Microcosmus spp.(3)	Domoic acid	Excitatory amino acid (Neurotoxin)	Not known
Phallusia nigra(2)	GTX 1,3 and 2, STX, neo-STX and TTX	Neurotoxic alkaloids; neurotoxin	Potential therapeutic agent
Polyandrocarpa sp.	Polyandrocarpamines A and B	2-aminoimidazolone alkaloids	Kinase inhibitor
Aplidium elegans (= *Sidnyum elegans*)	Phosphoeleganin	Polyketide	Phosphatase inhibitor
Styela clava(4)	TTX	Neurotoxin	Potential therapeutic agent for paint
Tridemnum solidum	Didemnin B	Cyclic depsipeptide	Antiviral

Source: Watters (2018); (1): Satake (2016); (2): Freitast et al. (1996); (3): Kvrgić et al. (2022); (4): Antonelli et al. (2022)

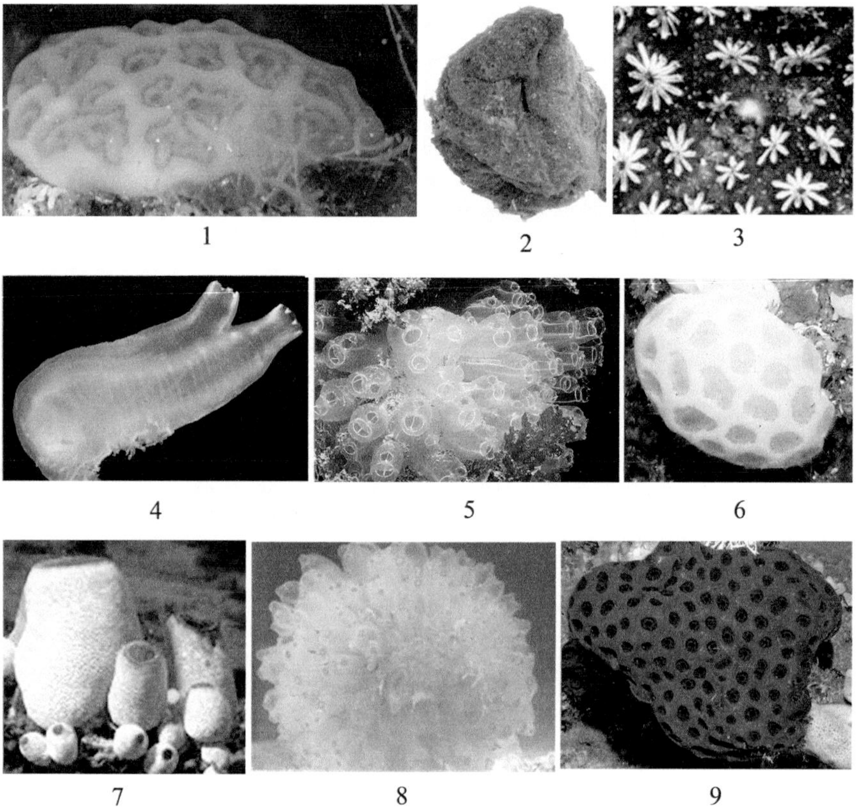

1: Aplidium conicum; 2: Aplidium meridianum; 3: Botryllus sp.; 4: Ciona intestinalis*; 5: Diazona angulata;.6: Didemnum candidum; 7: Didemnum molle; 8:Ecteinascidia turbinate; 9: Eudistoma* sp.

1. Image credit: Diego Delso. https://creativecommons.org/licenses/by-sa/4.0/deed.en

2. Image credit: Reproduced with permission from Françoise Monniot.

3. Image credit: Massimiliano De Martino. https://creativecommons.org/licenses/by-sa/3.0/

4. Image credit: Eric A. Lazo-Wasem. https://creativecommons.org/publicdomain/zero/1.0/deed.en

5. Image credit: Bernard Picton. https://creativecommons.org/licenses/by-sa/4.0/

6. Image credit: Anna Barnett. https://creativecommons.org/licenses/by/2.0/deed.en

7. Image credit: Silke Baron. https://creativecommons.org/licenses/by/2.0/

8. Image credit: Davis, Alison. Public domain

9. Image credit: Bernard Dupont. https://creativecommons.org/licenses/by-sa/2.0/

10

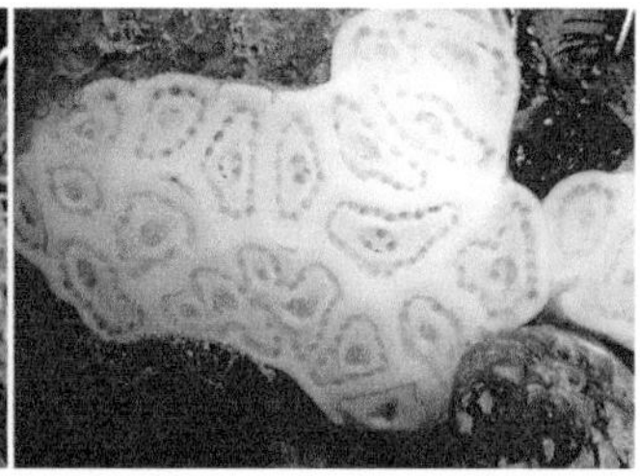

11

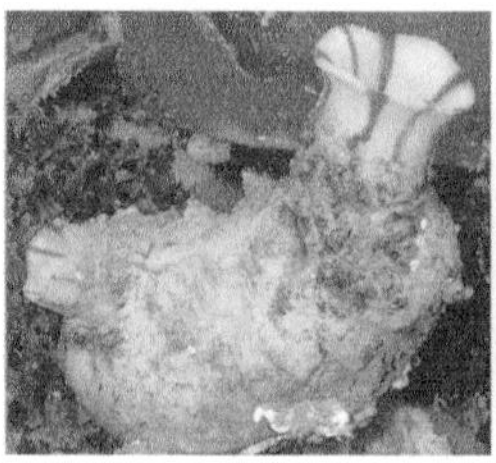

12

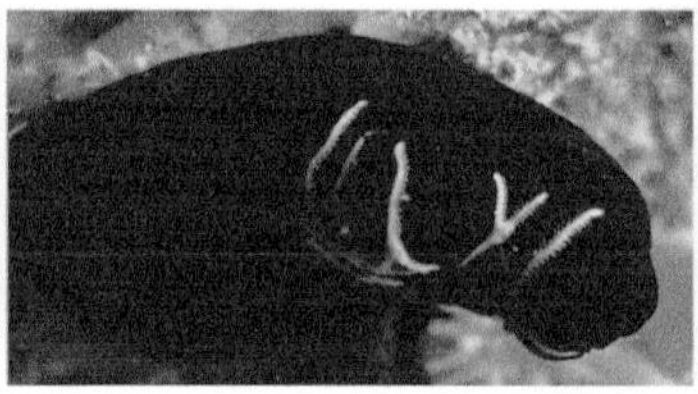

13

14

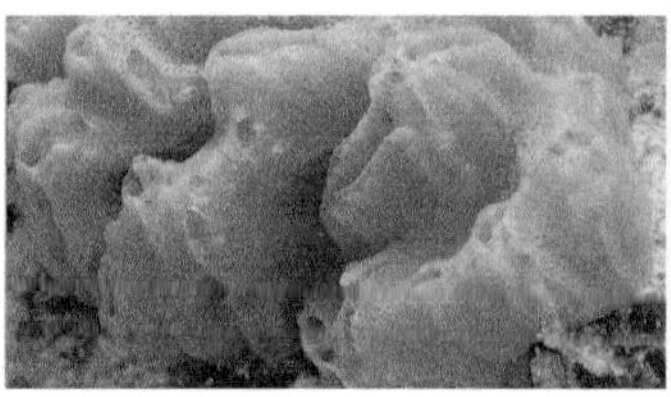

15

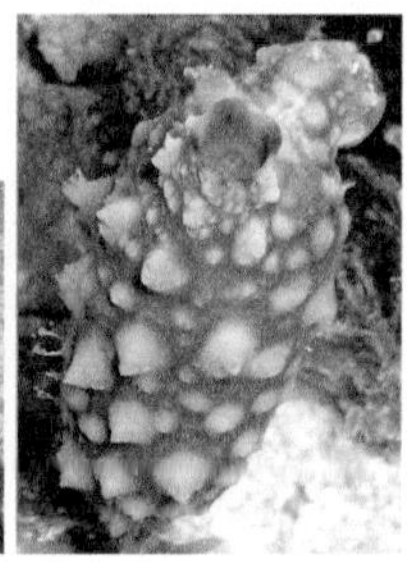

16

10: Halocynthia roretzi,; 11:Lissoclinum patella; 12: Microcosmus sp.; 13: Phallusia nigra; 14: Ritterella tokiaka; 15: Aplidium elegans*; 16: Styela clava*

10. Image credit: bifyu. https://creativecommons.org/licenses/by-sa/2.0/

11. Image credit: Nick Hobgood. https://creativecommons.org/licenses/by-sa/3.0/

12. Image credit: Sylvain Le Bris. https://creativecommons.org/licenses/by-sa/4.0/

13. Image credit: Marcelo Checoli Mantelatto et al. https://creativecommons.org/licenses/by/4.0/

14. Image credit: Alex Heyman. https://creativecommons.org/publicdomain/zero/1.0/deed.en

15. Image credit: Matthieu Sontag. https://creativecommons.org/licenses/by/3.0/deed.en

16. Image credit: Matthieu Sontag. https://creativecommons.org/licenses/by-sa/3.0/

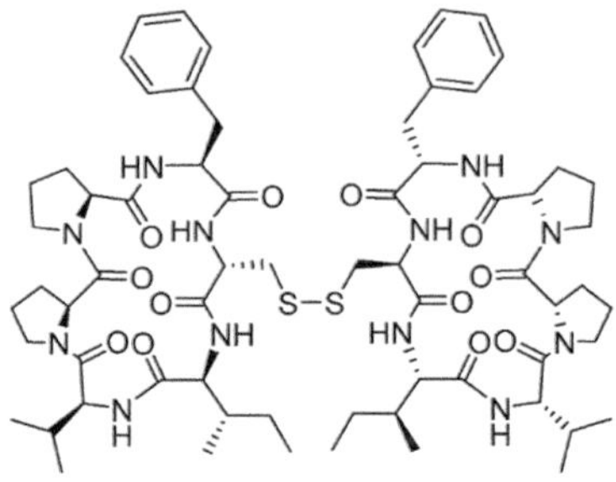

s-trans-Pro-1, s-cis-Pro-2
s-cis-Pro-1, s-trans-Pro-2

Antollamides A and B

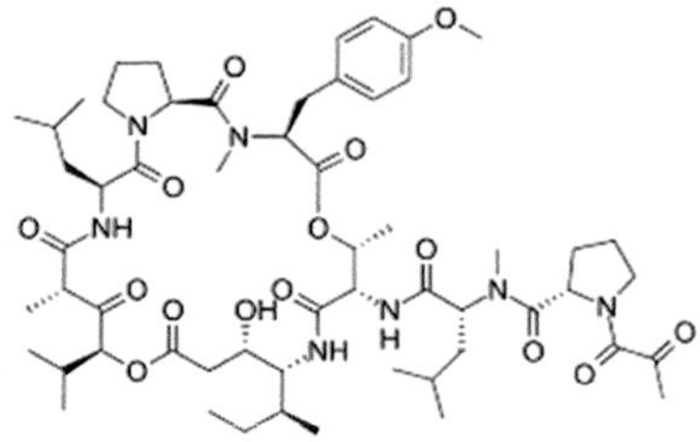

Plitidepsin (=dehydrodidemnin B,) (Aplidin)

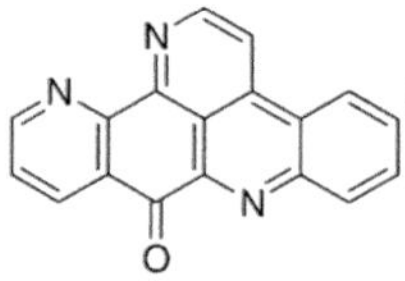

Ascididemin.

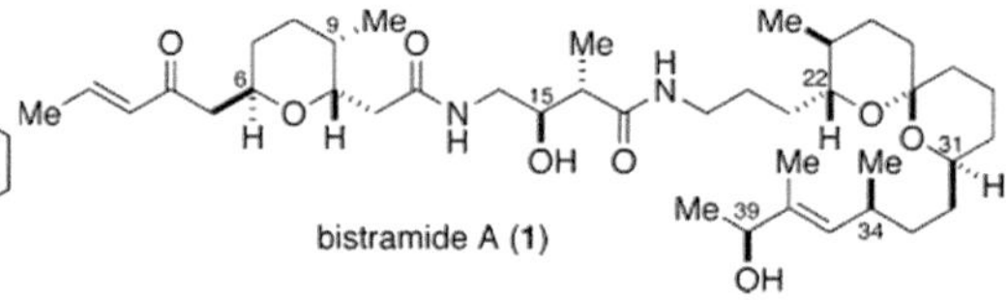

Bistramide A(bisA)

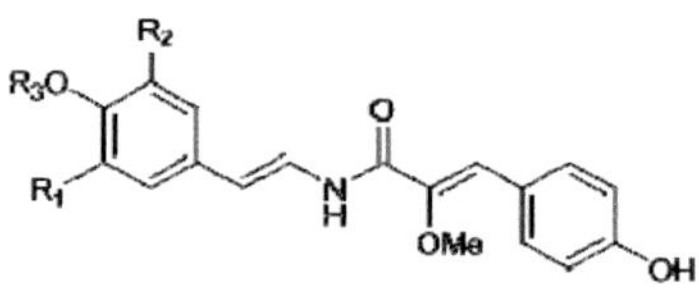

178 R_1=R_2=Br; R_3=Me
180 R_1=Br; R_2=H; R_3=Me
182 R_1=R_2=H; R_3=Me
183 R_1=R_2=H; R_3=H
184 R_1=R_2=Br; R_3=H

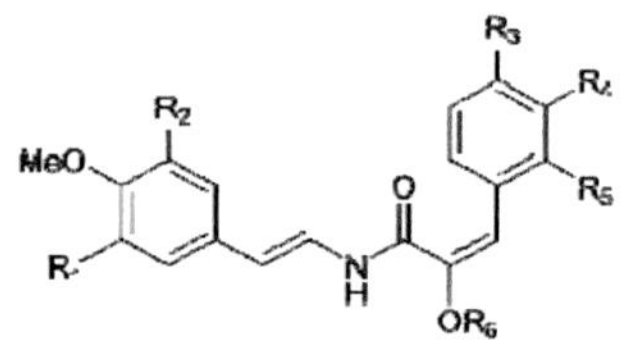

179 R_1=R_2=Br; R_3=OH; R_4=R_5=H; R_6=Me
181 R_1=Br; R_2=H; R_3=OH; R_4=R_5=H; R_6=Me
185 R_1=R_2=R_3=H; R_4=OH; R_5=NC; R_6=H
186 R_1=R_2=H; R_3=OH; R_4=R_5=H; R_6=Me

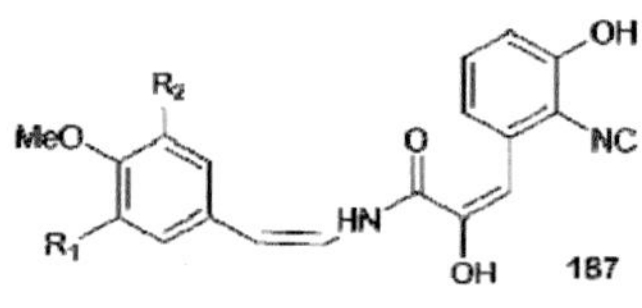

Botryllamides A-J

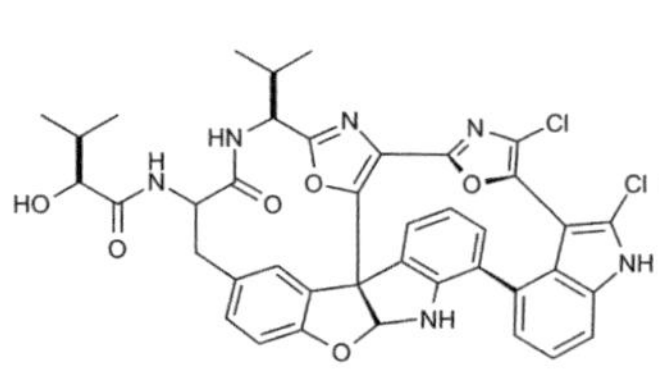

Diazonamide A.

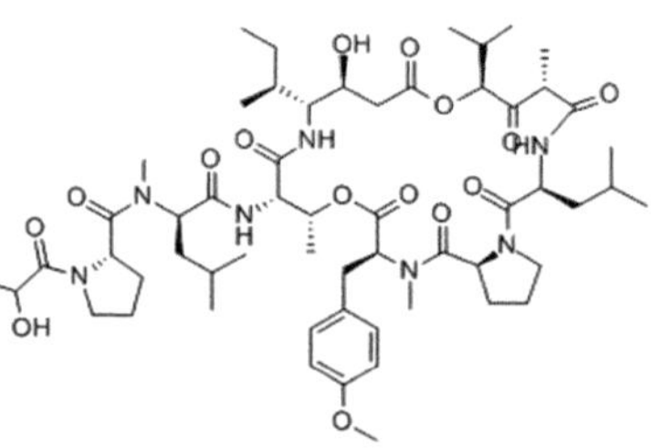

Didemnin B

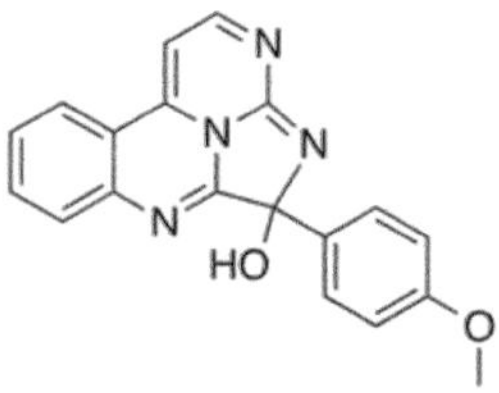

Eudistidine.

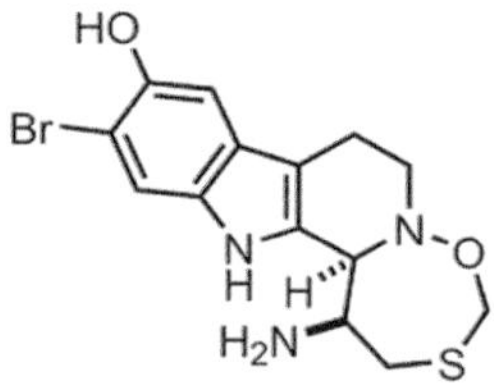

Eudistomin C.

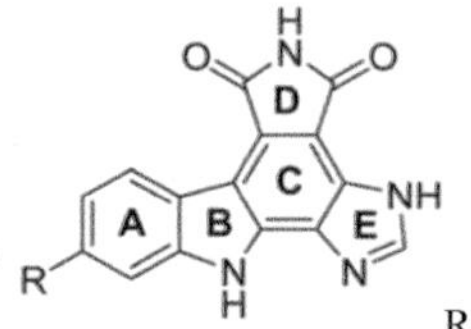

R= H

Granulatimide

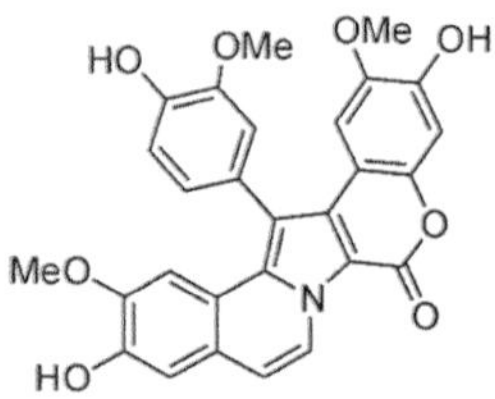

Lamellarin.

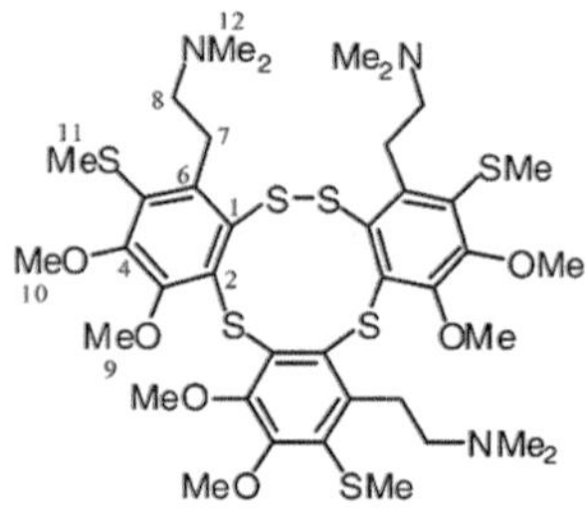

Lissoclibadin.

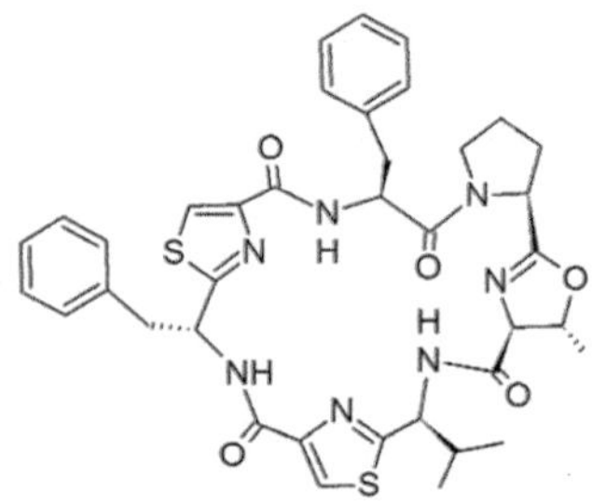

Lissoclinamide

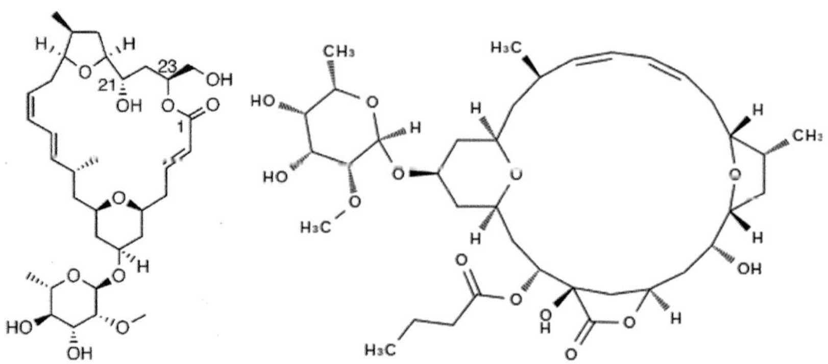

Mandelalides A & B

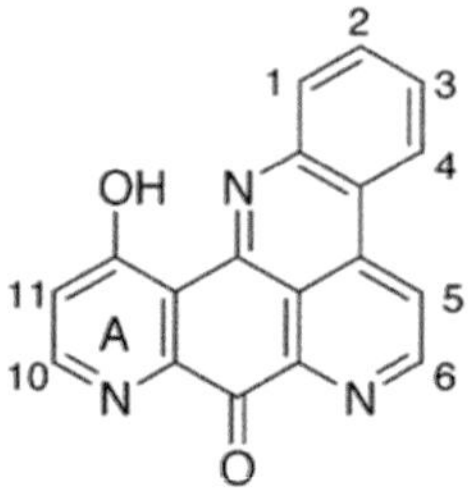

Meridine.

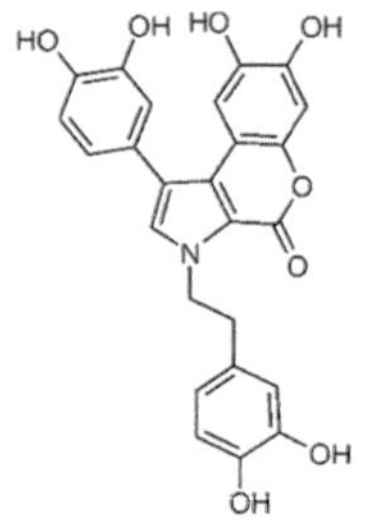

Ningalin.

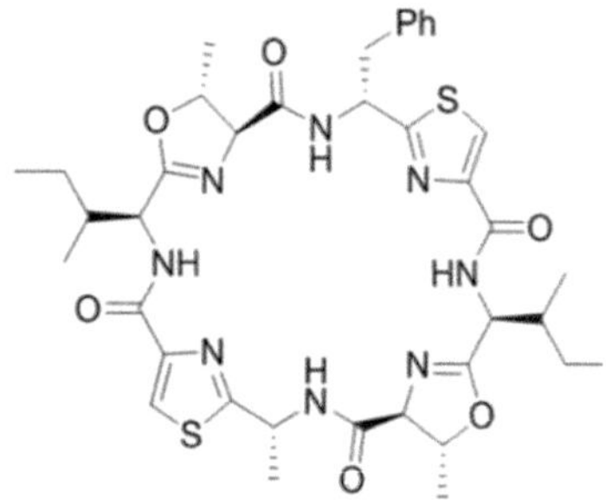

Patellamide

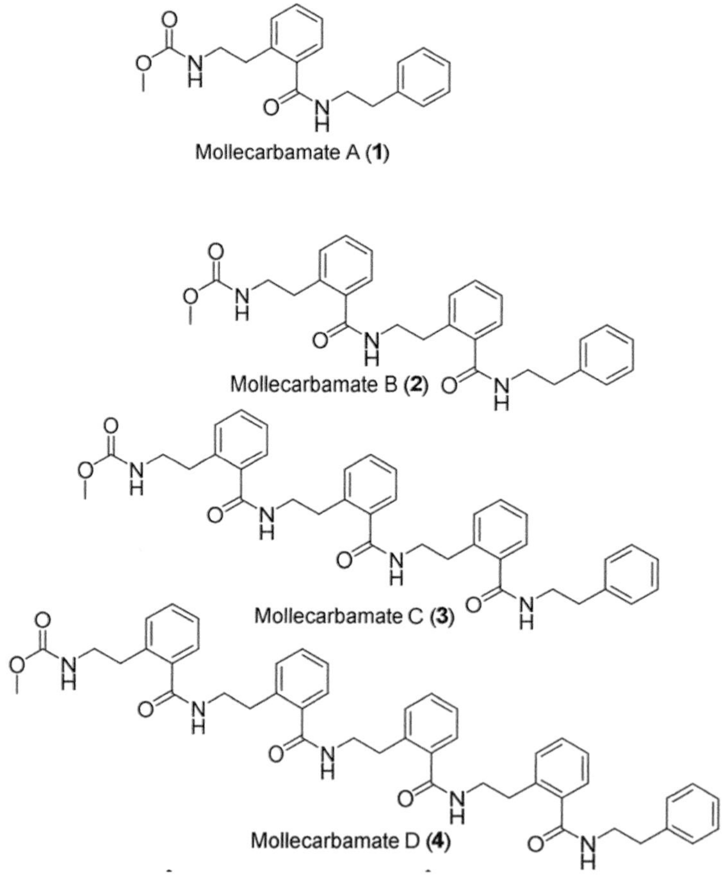

(1)–(4): Mollecarbamates A–D

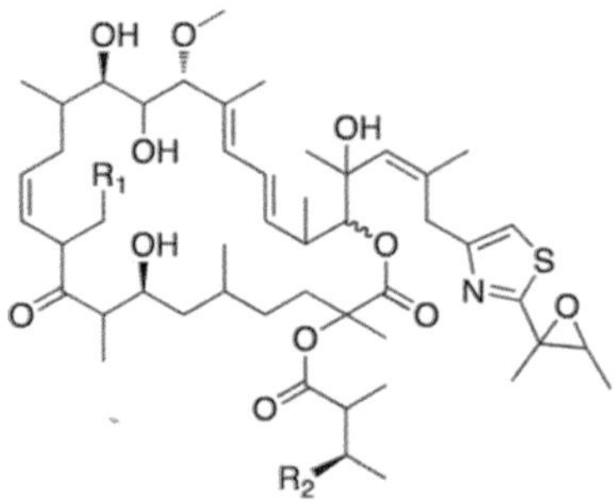

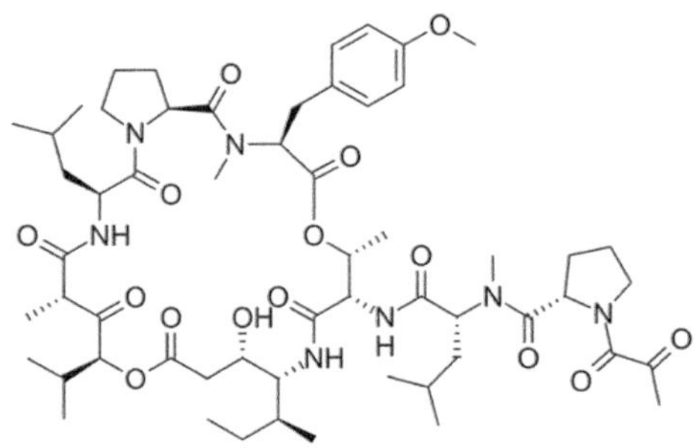

Patellazole A: R1=H; R2= H;

Patellazole B: R1=H; R2= OH;

Patellazole C: R1=OH; R2= OH,

Plitidepsin (dehydrodidemnin B)

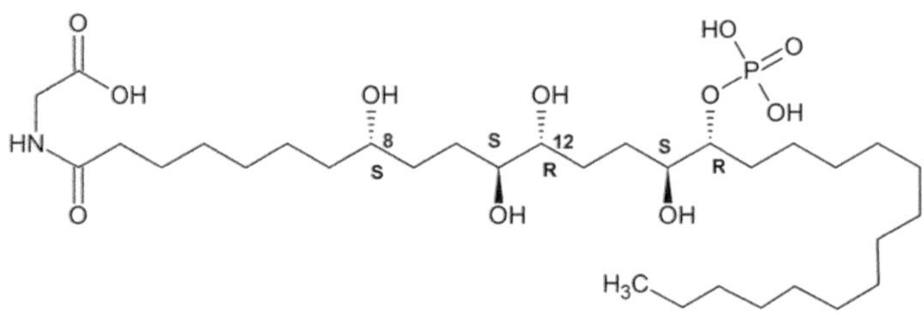

Phosphoeleganin

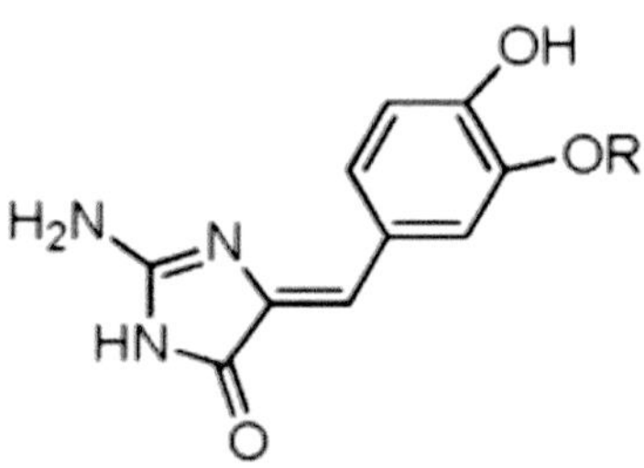

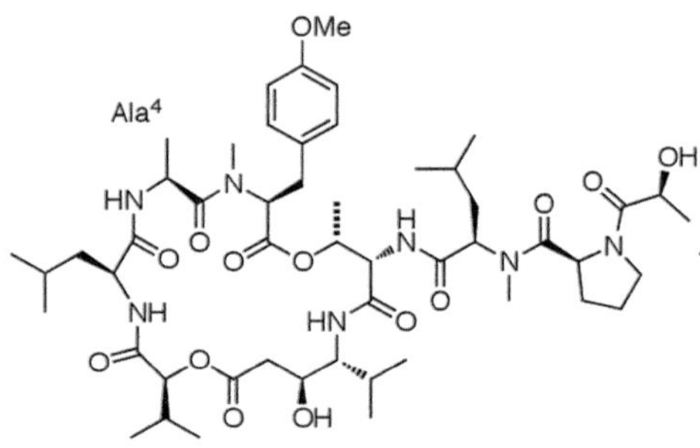

Polyandrocarpamine A (R=Me)

Polyandrocarpamine B (R=H).

Tamandarin

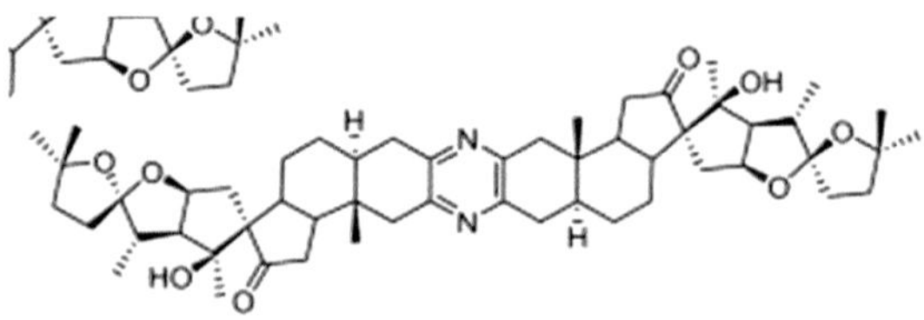

Ritterazine.

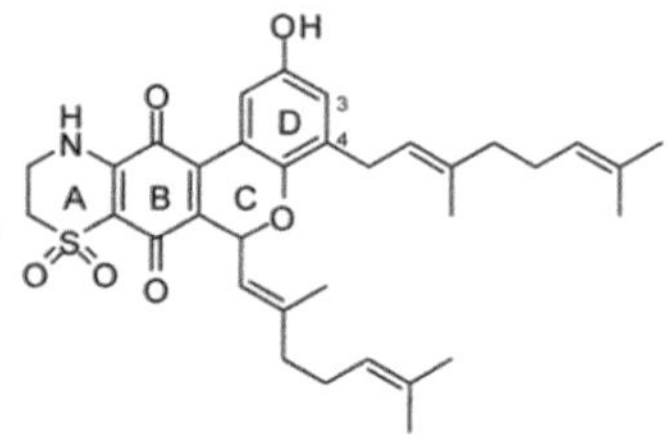

Thiaplidiaquinone

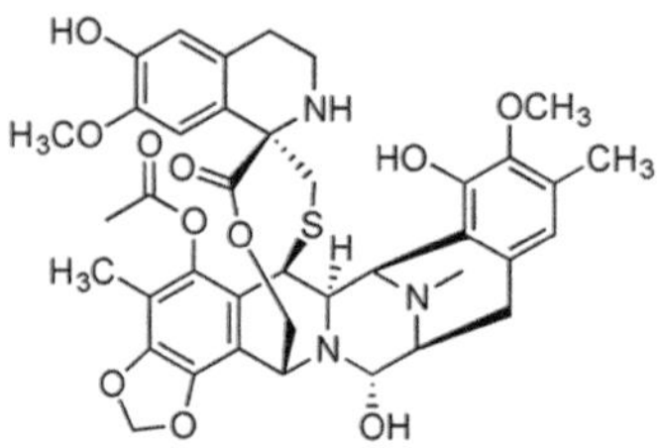

Trabectidin (ET-473) Yondelis®

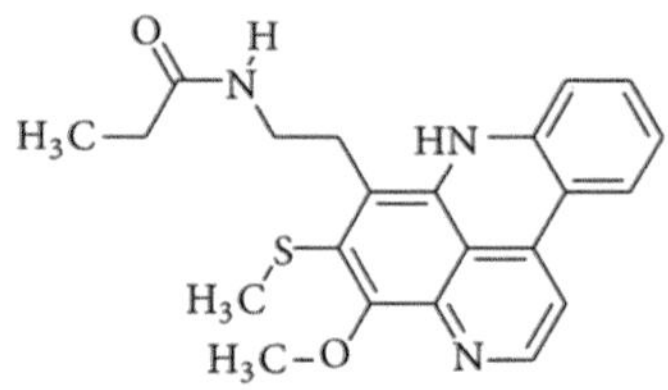

Varamine A

KEYWORDS

- **toxic ascidian diversity**
- **ascidian toxins**
- **toxic compounds**
- **cytotoxic ascidian**

CHAPTER 9

Toxins of Marine Fishes and Their Therapeutic Potential

ABSTRACT

This chapter deals with the toxins of marine fishes and their therapeutic potentials. Aspects such as toxins of marine elasmobranch and teleost fishes and their therapeutic potentials; and the biomedical applications of TTX derived from the bacteria associated with the marine pufferfish are given in detail.

The bioactive compounds of marine fish are well known to support human body growth and function with their therapeutic potentials. These compounds help alleviate as well as manage disease conditions such as cancer, cardiovascular diseases, diabetes, obesity, brain function disorders, immune system disorder, and oxidative stress (Awuchi et al., 2022; Santhanam et al., 2022). However, the toxins present in the marine elasmobranch and teleost fishes and their therapeutic potentials are receiving greater attention only now.

9.1 MARINE ELASMOBRANCHS

Elasmobranch fish (cartilaginous fish) comprise sharks, skates, and rays, and are mostly marine. Freshwater elasmobranchs constitute only 5% about 1200 described species. Many of these species have been reported to possess advantages as animal models for biomedical research. The crude venom of shark tissues has exhibited antitumoral activity. These fish lack both bone marrow and a true lymphatic system, but they have evolved specific lymphomyeloid tissues that are known to play an important role in immune system development and function.

Therapeutic Potential of Marine Biotoxins: From Poison to Pharmacy. Santhanam Ramesh, Ramasamy Santhanam, & Subbiah Balasundari (Authors)

DOI: 10.1201/9781003610991-9

TABLE 9.1 Therapeutic Values of Crude Venoms of Elasmobranchs.

Species	Toxin	Bioactivity	References
Aetobatus narinari	Fibrinogenolytic enzymes	Anticoagulant and gelatinolytic activity	Kumar et al. (2011)
Carcharhinus leucas	CTX; I-CTX analogues (I-CTX-1&2 and I-CTX-3&4); Gambieric acid D	Cytotoxicity	Diogene et al. (2017)
Dasyatis pastinaca	Crude venom	Cytotoxicity	Kirchhoff et al. (2022)
Pastinachus sephen (= Dasyatis sephen)	Fibrinogenolytic enzymes	Anticoagulant and gelatinolytic activity; cytotoxicity	Kumar et al. (2011), Rajeshkumar et al. (2015)
Pteroplatytrygon violacea	Crude venom	Cytotoxicity	Kirchhoff et al. (2022)
Somniosus microcephalus	TMAO	Cardiovascular	Huc et al. (2018)

Therapeutic values of crude venoms of elasmobranchs

Aetobatus narinari* and *Pastinachus sephen *(= Dasyatis sephen):* The toxic fibrin(ogen)olytic enzyme containing venom extracts of these species have been reported to possess anticoagulant property, and it is suggested that this extract may be of great use for antithrombosis therapy (Kumar et al., 2011).

Dasyatis pastinaca* and *Pteroplatytrygon violacea: The crude venom of these species was found to possess cytotoxicity of HeLa cells (Kirchhoff et al., 2022).

Dasyatis sephen: The crude venom containing fibrinogenolytic enzymes has shown cytotoxicity against HeLa cells at concentrations of 4, 8, 12, and 16 μg/mL. Further, these venom concentrations increased apoptotic cell levels to 30, 50, 74, and 78 %, respectively (Rajeshkumar et al., 2015).

Carcharhinus leucas: This species is known to possess Gambieric acid D which is produced by the dinoflagellate, *Gambierdiscus* spp., and is transferred to this fish species through food chain. A mixture of Gambieric acid C and Gambieric acid D was found to be moderately toxic to mouse lymphoma cells L5178Y1 (Diogene et al., 2017).

Somniosus microcephalus: This species possesses TMAO (trimethylamine oxide). Its gut bacteria product, trimethylamine (TMA), is oxidized by the liver to TMAO. Treatment with low-dose TMAO of this fish species helped reduce hypertension-related heart disease symptoms.

Further, increased dietary TMAO seemed to reduce diastolic dysfunction in pressure-overloaded hearts in rats (Huc et al. 2018).

Sphyrna tiburo: The extracts of the epigonal cells of this species have been reported to possess antitumor activity by inhibiting the growth of mammalian tumor cell lines such as breast carcinoma cell lines (MCF7, HCC38, and Hs578T), T-cell leukemia (Jurkat), pancreatic cancer (PANC-1), ovarian cancer (NIH:OVCAR-3), fibrosarcoma (WEHI-164), melanoma (A375.S2), and B-cell lymphoma (Daudi). Among the cancer cell lines, the fibrosarcoma (WEHI 164) cell line was found to be the most sensitive (93% inhibition) followed by the B-cell lymphoma (Daudi; 92%), malignant melanoma (A375.S2; 66%), T-cell leukemia (Jurkat; 58%), MCF-7 (52%), pancreatic carcinoma (PANC-1; 49%), Hs 578T (27%), HCC38 (26%), Hs 578Bst (16%), HCC38 BL (15%), and ovarian carcinoma (NIH:OVCAR-3) cell line (5%) (Walsh et al., 2006).

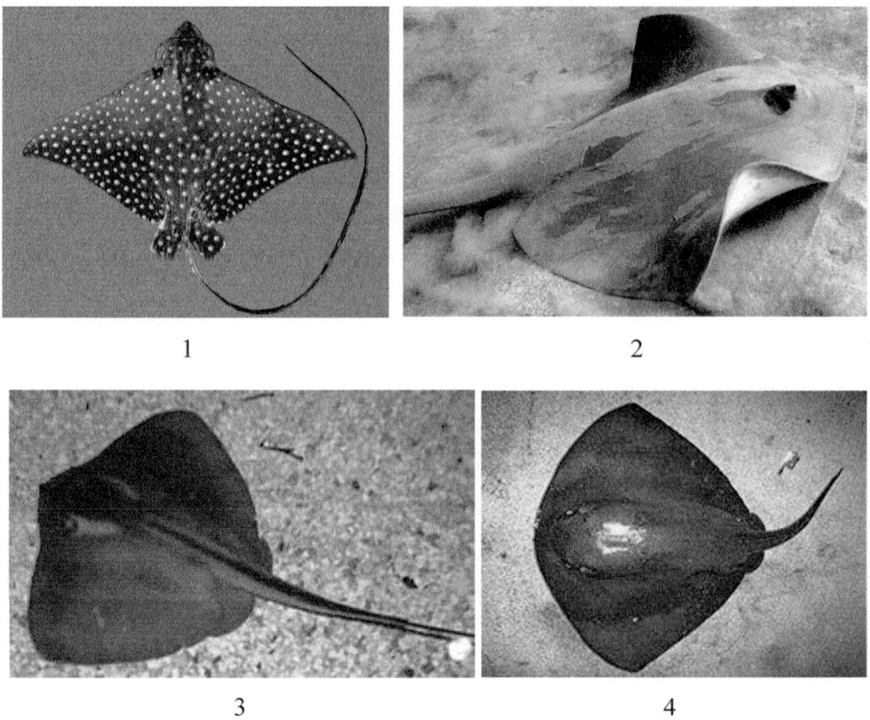

1: Aetobatus narinari; 2: Pastinachus sephen; 3: Dasyatis Pastinaca.; 4: Pteroplatytrygon violacea.

1. *Image credit*: D Ross Robertson. https://creativecommons.org/publicdomain/zero/1.0/

2. *Image credit*: Julien Willem. https://creativecommons.org/licenses/by-sa/3.0/

3. *Image credit*: Glen Bowman. https://creativecommons.org/licenses/by/2.0/

4. *Image credit*: NOAA Observer Program. https://creativecommons.org/publicdomain/zero/1.0/ (PD)

5

6

7

5: Carcharhinus leucas; 6: Somniosus microcephalus; 7: Sphyrna tiburo.

5. *Image credit*: Reproduced with permission from Brenda Bowling

6. *Image credit*: Gervais et Boulart. https://creativecommons.org/publicdomain/zero/1.0/ (PD)

7. *Image credit*: D Ross Robertson. https://creativecommons.org/publicdomain/zero/1.0/ (PD)

9.2 MARINE TELEOSTS

The therapeutic compounds of marine teleost fish (bony fish) such as sardine, mackerel, tuna, pollock, and yellowfin sole have been reported to be of great use in treating diseases relating to cancer, cardiovascular, oxygen stress, brain function, immune system oxidative stress, etc.

(Awuchi et al., 2022). The teleost species are also known to possess toxins including TTX and are rarely harming humans. The therapeutic potential of such toxins is receiving attention nowadays.

9.2.1 MARINE TEOLESTS POSSESSING TTX AND THEIR THERAPEUTIC POTENTIAL

The tetrodotoxin (TTX) was first derived from the pufferfish (Family: Tetraodontidae). TTX of pufferfish is not endogenous and is obtained either from the food sources of the different trophic levels of the food chain or from the symbiotic bacteria hosted by the pufferfish inside their body. However, the amount of TTX produced by bacteria is too little to account for the TTX accumulation in pufferfish and bioconcentration plays a larger role in the accumulation of TTX in pufferfish (Bucciarelli et al., 2021). The TTX-producing bacteria isolated from the species of marine pufferfish are given in the following table.

TABLE 9.2 TTX-Producing Bacteria Isolated from the Species of Marine Pufferfish.

Species of pufferfish	Bacteria
Chelonodon patoca	*Microbacterium arabinogalactanolyticum, Serratia marcescens, Vibrio alginolyticus*
Takifugu flavipterus (= Takifugu poecilonotus;Fugu poecilonotus)	*Vibrio, Alteromonas*, and *Shewanella*
Takifugu niphobles	*Raoultella terrigena*
Takifugu obscurus(= Fugu obscurus)	*Aeromonas, Bacillus* spp.
Takifugu rubripes (=Fugu rubripes)	*Bacillus* and *Actinomycetes*
Takifugu snyderi	*Vibrio* strain
Takifugu vermicularis, (=Fugu vermicularis radiatus)	*Vibrio* strain LM-*1*

Source: Lago et al. (2015).

Hapalochlaena maculosa *(= Octopus maculosus)*

Apart from the family Tetraodontidae, TTX has also been reported from the pufferfish families of Diodontidae and Ostraciidae, and from gobiid fishes (Family: Gobiidae). Among the TTX-containing fish families, the Tetraodontidae has the maximum number of species (about 30). In pufferfish, highest concentrations of TTX have been reported from the gonads (mainly ovaries) and liver, followed by muscles and skin. The measured toxicity of these organs always exceeds the safety level. While a dose of 1–2 mg of purified toxin may be lethal, reported cases from the Centers for Disease Control and Prevention (CDC) documented toxicity with ingestion of as little as 1.4 ounces of puffer fish (Wikipedia). To humans, tetrodotoxin has been reported to be deadly, up to 1200 times more poisonous than cyanide. There is enough toxin in one pufferfish to kill 30 adult humans, and there is no known antidote till date. Further, it is worthy of mention here that the meat of certain species of edible pufferfish is considered a delicacy in Japan and is called fugu. It is extremely expensive and should only be prepared by trained, licensed chefs who only know that bad cut may lead to almost death for a customer. Despite this many such deaths do occur annually in Japan (https://twofishdivers.com/blog/pufferfish-are-the-second-most-poisonous-vertebrate-in-the-world/). Above safety level, TTX is known to cause paralysis by blocking voltage-gated sodium channel. On the other hand, TTX has got its own therapeutic benefits. For example, it can be applied as a potential pain relief. Further, the analgesic activity of TTX may be of great value in treating various types of pains such as severe cancer pain or in reducing cue-induced increases in heroin craving and associated anxiety (Lago et al., 2015). The TTX-containing species of marine fishes are given in the following table.

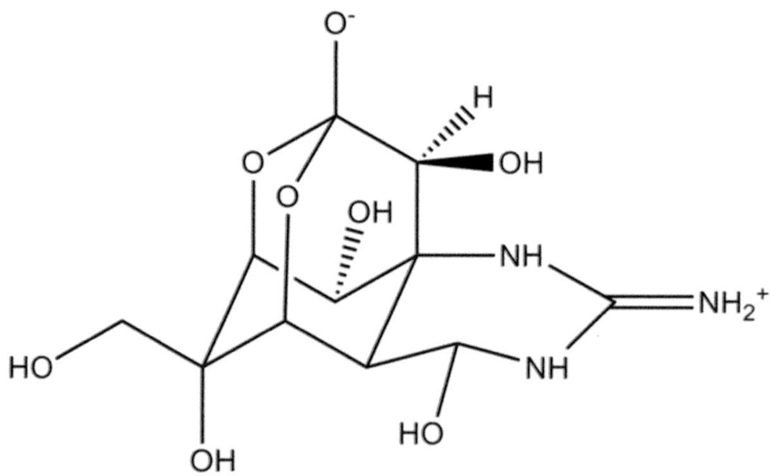

TTX

Therapeutic potential of TTX

The neurotoxin, tetrodotoxin (TTX), was first isolated from the pufferfish and subsequently from a variety of species which host TTX-producing bacteria. Though it is harmful to humans, levels far below its LD50 have

TABLE 9.3 TTX-Containing Species of Marine Fishes.

Family	Species	References
Gobiidae	*Acentrogobius caninus (= Amoya caninus; Radigobius caninus*	You et al. (2015), Arakawa et al. (2010)
-Do-	*Acentrogobius nebulosus (= Yongeichthys nebulosus).*	You et al. (2015)
-Do-	*Parachaeturichthys polynema (= Prachaeturichtys palynena),*	Arakawa et al. (2010)
-Do-	*Yongeichthys criniger (= Gobius criniger)*	Noguch and Arakawa (2008)
-Do-	*Yongeichthys nebulosus*	Arakawa et al. (2010)
Diodontidae.	*Chilomyceterus reticulatus (= Chilomycterus affinis)*	Noguch and Arakawa (2008)
-Do-	*Diodon holocanthus*	(1)
-do	*Diodon hystrix*	Azman et al. (2014)
Ostraciidae	*Lactoria diaphana*	Noguch and Arakawa (2008)
-Do-	*Ostracion cunicum*	Noguch and Arakawa (2008)
-Do-	*Ostracion immaculatum*	Noguch and Arakawa, 2008
-Do-	*Rhynchostracion nasus (= Ostracion nasus)*	Azman et al. (2014)
Tetraodontidae	*Arothron firmamentum*	Arakawa et al. (2010)
-Do-	*Arothron hispidus*	Mana et al. (2010)
-Do-	*Arothron immaculatus*	Azman et al. (2014)
-Do-	*Arothron manilensis*	Azman et al. (2014)
	Arothron stellatus (= Arothron alboreticulatus; Tetraodon alboreticulatus	Arakawa et al. (2010)
-Do-	*Canthigaster rivulata*	Arakawa et al. (2010)
-Do-	*Chelonodon patoca*	Yotsu-Yamashita et al. (2018)
-Do-	*Dichotomyctere nigroviridis (= Tetraodon nigroviridis)*	Arakawa et al. (2010)

TABLE 9.3 *(Continued)*

Family	Species	References
-Do-	*Lagocephalus gloveri*	Arakawa et al. (2010)
-Do-	*Lagocephalus inermis*	Arakawa et al. (2010)
-Do-	*Lagocephalus lunaris*	Arakawa et al. (2010)
-Do-	*Lagocephalus sceleratus,- (= Pleuranacanthus sceleratus)*	Hassoun et al. (2022)
-Do-	*Lagocephalus spadicus*	Azman et al. (2014)
-Do-	*Lagocephalus wheeleri*	Arakawa et al. (2010)
-Do-	*Sphoeroides marmoratus*	(2)
-Do-	*Sphoeroides pachygaster*	Arakawa et al. (2010)
-Do-	*Takifugu chinensis, Takifugu chrysops, Takifugu exascurus, Takifugu flavidus, Takifugu niphobles, Takifugu oblongus, Takifugu obscurus, Takifugu pardalis, Takifugu porphyreus, Takifugu pseudommus, Takifugu stictonotus, Takifugu snyderi, Takifugu vermicularis, Takifugu xanthopterus*	Arakawa et al. (2010)
-Do-	*Takifugu flavipterus (= Takifugu poecilonotus)*	Oshiro et al. (2021)
-Do-	*Takifugu rubripes*	Gao et al. (2020)
-Do-	*Tetractenos hamiltoni*	(3)
-Do-	*Torquigener pallimaculatus*	Azman et al. (2014)
-Do-	*Xenopterus naritus*	Azman et al. (2016)

(1): Food Safety (1993); (2): https://hero.epa.gov/hero/index.cfm/reference/details/reference_id/7425459; (3): Anon., https://en.wikipedia.org/wiki/Common_toadfish

TABLE 9.4 TTX-Containing Species of Marine Fishes and Their Toxicity Level.

Family	Species	TTX level (Mean); ug/g	References
Gobiidae	*Acentrogobius caninus (= Amoya caninus)*	2.0	You et al. (2015)
-Do-	*Yongeichthys criniger.(= Gobius criniger)*	226.5. (max)	Ito et al. (2022)
-Do-	*Acentrogobius nebulosus (= Yongeichthys nebulosus)*	3.0	You et al. (2015)
Diodontidae	*Diodon holocanthus*	21.1	Azman et al. (2014)
Ostraciidae	*Rhynchostracion nasus (= Ostracion nasus)*	9.94	Azman et al. (2014)
Tetraodontidae	*Arothron hispidus*	4.11	Mana et al. (2010)
-Do-	*Arothron immaculatus*	107	Azman et al. (2014)
	Arothron manilensis	133	Azman et al. (2014)
-Do-	*Chelonodontops patoca*	277	Azman et al. (2014)
-Do-	*Lagocephalus inermis*	75.8	Azman et al. (2014)
-Do-	*Lagocephalus sceleratus,- (= Pleuranacanthus sceleratus)*	30.3 (max)	Alabssawy (2017)
-Do-	*Lagocephalus sceleratus,- (= Pleuranacanthus sceleratus)*	1.71	Mana et al. (2010)
-Do-	*Lagocephalus lunaris*	4.96	Azman et al. (2014)
-Do-	*Lagocephalus spadiceus*	8.71	Azman et al. (2014)
-Do-	*Dichotomyctere nigroviridis (= Tetraodon nigroviridis)*	33.5	Azman et al. (2014)
-Do-	*Dichotomyctere nigroviridis (= Tetraodon nigroviridis)*	1.34	Mana et al. (2010)
-Do-	*Sphoeroides marmoratus*	15	(1)
-Do-	*Takifugu flavipterus (= Takifugu poecilonotus)-*	2.2	Oshiro et al. (2021)
-Do-	*Takifugu oblongus*	1.38	Mana et al. (2010)

(1): https://hero.epa.gov/hero/index.cfm/reference/details/reference_id/7425459

displayed therapeutic functions, especially in the treatment of cancer-related pain, visceral pain, neuropathic pain, spinal cord injuries, brain trauma, and few types of tumors (Bucciarelli et al., 2021) as detailed below.

Neuropathic pain: TTX may also be used in the treatment of neuropathic pain, and it does not see to produce deleterious side effects such as addiction, inflammation, sedation, and respiratory depression which are the characteristics of opioid use. Though the results are encouraging in rodent models, further intensive research is needed in human models to corroborate the effectiveness of TTX in the treatment of neuropathic pain.

Cancer-related pain: As an analgesic, TTX has shown promising effects in the treatment of cancer-related pain. Preclinical and clinical trials testing has confirmed the efficacy of TTX injections in the treatment of mild to severe cancer pain. Injection of TTX in doses of 1.0, 3.0, and 6.0 μg/kg was found to inhibit neuropathic pain induced by chemotherapy agents. Further, TTX was also found to provide clinically important analgesia for patients with moderate to severe cancer pain. As a result of this finding, WEX Pharmaceuticals developed TTX as a treatment for moderate to severe cancer pain and it is now in Phase III clinical trials.

Inflammatory pain: Subcutaneous TTX injection produced antinociceptive effects in a rat carrageenan model. At a concentration of 0.03–1.0 μg/20 μL, TTX was found to inhibit mechanical hyperalgesia in the carrageenan-affected muscle. These findings suggest that TTX may serve as a potential alternative to current anti-inflammatory drugs, including opioid-based treatments such as morphine.

Chronic pain: It is a long persisting pain after the healing of any underlying injury and is relating to the hyperexcitable state of these neurons. TTX could help control this sort of pain through modulating NaVs and blocking action potentials in nociceptors. Local injection of TTX was found to provide a potential antinociceptive effect on exercise-induced mechanical hyperalgesia in gastrocnemius muscle. Preclinical and clinical studies relating to TTX injections in humans yielded long-term improvement of pain with minimal adverse effects.

Allodynia and hyperalgesia: In rat models, low-dose TTX applications were found to improve the mechanical threshold in an affected hind paw. Further, TTX was found to attenuate both allodynia (pain due to a stimulus) and hyperalgesia (extreme response to pain) for up to three hours in sciatic-nerve and infraorbital-nerve ligated rats.

Postherpetic neuralgia: TTX may also be used for the treatment of another type of neuropathic pain viz. postherpetic neuralgia which is caused by the reactivation of the varicella-zoster virus. Further, TTX pellets were found to inhibit PHN with an ED50 value of 5.9 μg/kg and with a much higher therapeutic index of 88.5.

Visceral Pain: Visceral pain comprises a wide variety of both acute and chronic pain conditions and it originates in the internal organs. Subcutaneous injection of TTX at 1–6 μg/kg reduced visceral pain in Swiss Webster mice without any side effects. TTX was also found to be effective in treating interstitial cystitis/bladder pain syndrome in mice.

TTX as a therapeutic: for brain trauma and spinal cord injury: TTX treatment after a traumatic brain injury has been reported to prevent the development of posttraumatic epilepsy. TTX may also enhance pruning through suppression of neuronal activity in the injured cortex. In rat models, TTX was also found to prevent the tissue damage that normally follows hypoxia and ischemia. Further, TTX is also known to block the abnormal increase in intra-axonal calcium ion levels in an in vitro axonal stretch injury paradigm. Furthermore, microinjection of TTX into an injury site 15 min after spinal cord injuries in a rat model significantly reduced white matter loss.

TTX as a tumor-suppressing therapeutic: the overexpression of TTX-sensitive NaV1.7 is known to promote the invasion of cancerous cells. In human H460 non-small cell lung carcinoma, TTX has been reported to reduce such cell invasion by 50% without affecting noninvasive wild-type cells. Further, very high dose (30 uM) of TTX treatment significantly reduced migration and invasion of two ovarian cancer cell lines viz. Caov-3 and SKOV-3. Further, TTX has also been reported to serve as an anticancer agent. Two groups of Swiss albino mice were injected with TTX in addition to Ehrlich ascites carcinoma (EAC) cells. Mice administered 1/10 LD50 TTX displayed the highest increase in lifespan, (days), that is, 46.6% and those administered 1/20 LD50 TTX showed an increase of only 26.7%. TTX also exhibited cytotoxic effect on the cancer cells, causing apoptosis.

TTX as an anesthetic: Single application of TTX to rabbit cornea was found to produce a long-lasting anesthetic effect up to 8 h with no symptoms of systemic toxicity, ocular irritation, and corneal thickening. Similar effects prevailed even when the TTX is applied repeatedly over a period of 24 h. Further, this application did not prevent epithelial healing in the affected cornea. These findings suggest that TTX may serve as a potential anesthetic for corneal procedures. Further, use of TTX in combination with conventional local anesthetics proved useful in providing prolonged anesthesia with minimal side effects.

TTX as a therapeutic for heroin and cocaine addiction: Intramuscular injection of 5 or 10 μg TTX three times a day was found to significantly reduce withdrawal symptoms by the third day in newly abstinent addicts however, without adverse effects on respiration or blood pressure. These findings suggested that TTX may be effective in reducing the symptoms of opiate withdrawal and preventing relapse. TTX may also provide a potential alternative to opioid treatments because it is not addictive and recognized effective doses do not pose a cardiovascular or neuronal risk. Further, bilateral TTX injections were found to reduce cocaine drug-craving behavior in rats.

TTX- as an antibacterial and antifungal agent: The crude extracts of TTX isolated from liver, skin, and muscles of *Lagocephalus sceleratus* exhibited antibacterial activity against *Escherichia coli, Bacillus subtilis,* and *Vibiro cholerae*; and antifungal activity against *Aspergillus fumigatus* and *Trichophyton rubrum*. The TTX concentrations against these three bacterial species and two fungal species were reported as 22.24, 17.51, and 9.22 ug/g, respectively, and 19.21 and 5.33 ug/g respectively (Alabssawy, 2017).

1

2

3

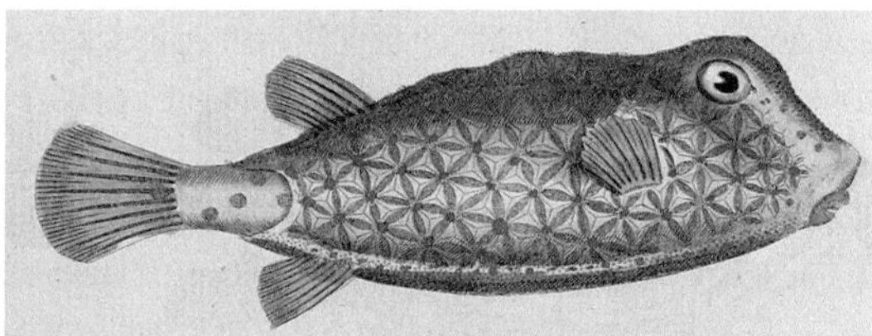

4

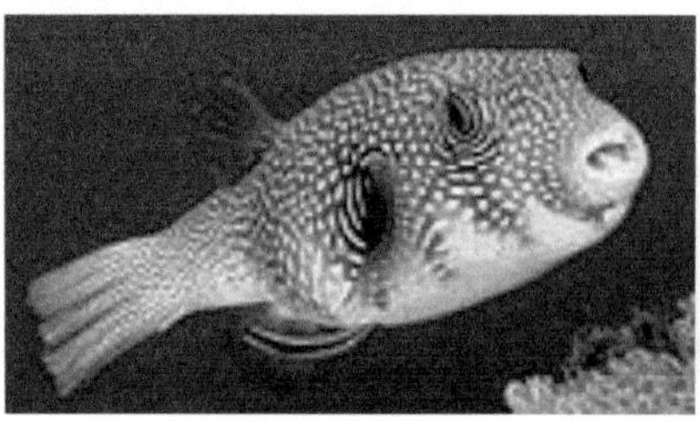

5

6

7

8

1. Acentrogobius caninus; 2. Yongeichthys criniger; 3. Diodon holocanthus; 4. Rhynchostracion nasus; 5. Arothron hispidus; 6. Chelonodon patoca; 7. Lagocephalus sceleratus; 8. Dichotomyctere nigroviridis;

1. *Image credit*: Iconographia Zoologica. https://creativecommons.org/publicdomain/zero/1.0/
2. *Image credit*: Totti. https://creativecommons.org/licenses/by-sa/4.0/deed.en
3. *Image credit*: LASZLO ILYES. https://creativecommons.org/licenses/by/2.0/deed.en
4. *Image credit*: Biodiversity Heritage Library. https://creativecommons.org/licenses/by/2.0/deed.en
5. *Image credit*: Albert kok. https://creativecommons.org/licenses/by-sa/3.0/
6. *Image credit*: Totti. https://creativecommons.org/licenses/by-sa/4.0/
7. *Image credit*: Reproduced with permission from Lew Reynolds and Claire Vince
8. *Image credit*: Steven G. Johnson. https://creativecommons.org/licenses/by-sa/3.0/

9

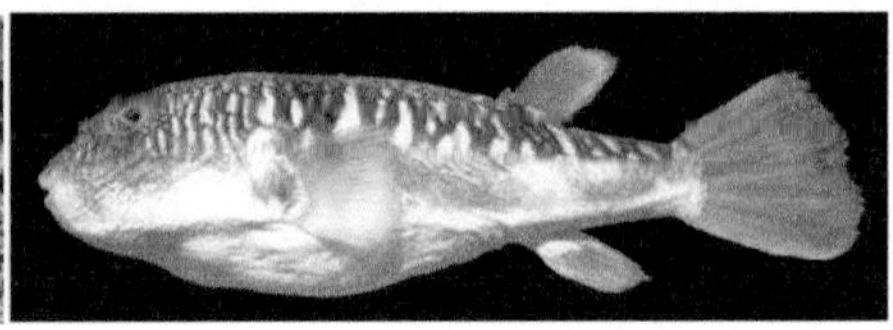

10

9. Sphoeroides marmoratus; 10. Takifugu oblongus

9. *Image credit*: Fernando Losada Rodríguez. https://creativecommons.org/licenses/by-sa/4.0/deed.en

10.Takifugu oblongus- Image credit: Reproduced with permission from Seishi Kimura

9.2.2 MARINE TEOLESTS POSSESSING OTHER TOXINS AND THEIR THERAPEUTIC POTENTIAL

The species of marine teleosts possessing toxins other than TTX are listed in the following table:

TABLE 9.5 The Species of Marine Teleosts Possessing Toxins other Than TTX.

Family	Species	Toxin	References
Ariidae	*Cathrops spixii*	Wap65	Ziegman and Alewood (2015)
Batrachoididae	*Thalassophryne maculosa*	TmC4-47.2	Ziegman and Alewood (2015)
-Do-	*Thalassophryne nattereri*	Nattectin	Ziegman and Alewood (2015)
Lutjanidae	*Lutjanus sebae*	I-CTX	Diogene et al. (2017)
-Do-	*Lutjanus bohar*	CTX and its congeners, CTX3C-group (2,3,51-trihydroxyCTX3C, 2,3,-dihydroxyCTX3C, 2-hydroxy CTX3C, M-seco-CTX3C, 51-hydroxyCTX3C, CTX3C	Spielmeyer et al. (2022)
Plotosidae	*Plotosus canius*	PC	Ziegman and Alewood (2015)
Scatophagidae	*Scatophagus argus*	SA-HT	Karmakar et al. (2004)
Scorpaenidae	*Dendrochirus zebra*	Karatoxin	Ziegman and Alewood (2015)
-Do-	*Pterois antennata*	Nocitoxin	Ziegman and Alewood (2015)
-Do-	*Pterois lunulata*	Nocitoxin	Ziegman and Alewood (2015)
-Do-	*Pterois volitans*	Nocitoxin	Ziegman and Alewood (2015)
-Do-	*Scorpaena plumieri*	Sp-CTx, Plumieribetin and SP-CL 1-5	Ziegman and Alewood (2015)
-Do-	*Scorpaenopsis oxycephala*	Karatoxin	Ziegman and Alewood (2015)
-Do-	*Sebastapistes stongia*	Karatoxin	Ziegman and Alewood (2015)
-Do-	*Sebasticus marmoratus*	Karatoxin	Ziegman and Alewood (2015)
Serranidae	*Epinephelus coioides*	Epinecidin-1	Chee et al. (2019)
-Do-	*Grammistes sexlineatus*	Grammistins	Sugiyama et al. (2006)
-Do-	*Pogonoperca punctata*	Grammistins	Hashimoto and Oshima (1972)
Soleidae	*Pardachirus marmoratus*	Pardaxin	Huang et al. (2011); Wu et al. (2012)

TABLE 9.5 *(Continued)*

Family	Species	Toxin	References
Synanceiidae	*Inimicus japonicus*	Nocitoxin	Ziegman and Alewood (2015)
-Do-	*Synanceia trachynis* (= *Synanceia horrida)*	Trachynilysin (TLY); Stonustoxin (SNTX)	Ziegman and Alewood (2015)
-Do-	*Synanceia verrucosa*	Verrucotoxin (VTX), Neoverrucotoxin (neoVTX), Cardioleputin	Ziegman and Alewood (2015)
Tetrarogidae	*Notesthes robusta*	Nocitoxin	Ziegman and Alewood (2015)
-Do-	*Paracentropogon rubripinnis* (= *Hypodytes rubripinnis)*	Nocitoxin, Karatoxin	Ziegman and Alewood (2015), Nagasaka et al. (2009)
Trachinidae	*Trachinus draco*	Dracotoxin	Ziegman and Alewood (2015)
-Do-	*Trachinus vipera*	Trachinine	Ziegman and Alewood (2015)

Therapeutic potential of Teleost toxins other than TTX

***Thalassophryne nattereri* (Family: Batrachoididae):** A C-type lectin toxin viz. nattectin derived from this species has displayed anti-inflammatory and antiallergic potential (Ziegman and Alewood, 2015; https://www.eurekalert.org/news-releases/569329)

***Plotosus lineatus* (Family: Plotosidae):** The spine venom containing unidentified toxin has been reported to possess potential cytotoxicity activity against Human Hep2 (HeLa derivative) cell line. At an extract concentration of 12ug/100ul/well, the viability of this cancer cell line was 18% (Pachaiyappan et al., 2015).

***Pterois volitans* (Family: scorpaenidae):** The venom of this lionfish (with unidentified toxin) has been reported to activate Caspase-3, causing apoptosis in Ehrlichs ascites carcinoma xenografted mice (Ziegman and Alewood, 2015). Further, a peptide termed 7.6 kDa from this lionfish venom caused apoptosis on cultivated HEp2 and HeLa cells, while sparing normal cells (Ndandala et al., 2023).

***Scorpaena plumieri* (Family: scorpaenidae):** The crude venom (SPB) and a gelatinolytic protease purified from this venom (SPGP) from this species displayed significant antitumor activity against p53-wild-type glioblastoma cells with LD50 values of 3.9 μg/mL and 8.00×10^{-12} M, respectively; and against Ehrlich ascites carcinoma cells with 14.05 μg/mL and 1.22×10^{-11} M respectively (Juliana, https://inis.iaea.org/search/search.aspx?orig_q=RN:39107589). Further, two toxins viz. Sp-CTx, Plumieribetin and SP-CL 1-5 have also been reported from this species (Ziegman and Alewood, 2015).

***Epinephelus coioides* (Family: Serranidae):** The peptide toxin, epinecidin-1, derived from this species has shown antibacterial activity against Gram-negative bacterial strains such as *Vibrio parahaemolyticus, Vibrio alginolyticus, Pasturella multocida, Morganella morganii*, and *Aeromonas sobrio* with a minimum concentration (MBC) of <2 μM. However, this toxin was found to be slightly less effective against Vibrio vulnificus and Pseudomonas fluorescens with the MBC values of 4.19 and 67.04 μM, respectively (Chee et al., 2019)

***Grammistes sexlineatus* and *Pogonoperca punctata* (Family: Serranidae):** The peptide toxins viz. grammistins, Gs 1 and Gs 2 of Grammistes sexlineatus; and grammistins Pp 1 and Pp 3 of Pogonoperca punctata have shown antibacterial activity against *Bacillus subtilis, Staphylococcus aureus, Escherichia coli, Solmonella typhimurium, Vibrio anguillarum* (= Listonella anguillarum), and *Vibrio parahaemlyticus* (Sugiyama et al., 2006) and the MIC values are given in the following table.

TABLE 9.6 Antibacterial Activity of Grammistins (MIC, ug/mL).

Toxin/Bacterium	B. s	S.a	E.c	S.t	V. a	V. p
Gs 1	25	6.3	50	>50	50	25
Gs 2	12.5	6.3	50	>50	25	25
Pp 1	3.1	3.1	>50	>50	12.5	25
Pp 3	50	50	>50	>50	50	50

Gs and Pp: Grammistins; Bs: *Bacillus subtilis;* Sa: *Staphylococcus aureus*; Es: *Escherichia coli;* St: *Solmonella typhimurium*; Va: *Vibrio anguillarum (= Listella anguillarum*); Vp: *Vibrio parahaemolyticus.*

Source: Sugiyama et al., 2006

Pardachirus marmoratus **(Family: Soleidae):** The peptide toxin, pardaxin, derived from this species has shown antitumor activity against human fibrosarcoma HT-1080 cells. This toxin was found to inhibit cell proliferation by inducing apoptosis. Further, pardaxin-treated cells exhibited elevation of caspase-3/7 activities, accumulation of reactive oxygen species production. These reports suggest that pardaxin may serve as a potential anticancer agent for selectively inducing apoptosis in cancer cells (Huang et al., 2011).

Pseudosynanceia melanostigm **(Family: Synanceiidae):** The crude venom of this stonefish has been reported to cause apoptosis in U87 glioblastoma cell mitochondria by mediating increased ROS. Based on these findings, this venom may serve as a potential candidate for the treatment of glioblastoma (brain tumors) (Ramezani et al., 2021).

Synanceia trachynis (= *Synanceia horrida*) **(Family: Synanceiidae):** The Stonustoxin (SNTX) has been reported to display cytolytic properties against MCF-7 breast cancer cells with an IC50 value of 8.28 μg/ml (Wijanarko et al., 2022). Ziegman and Alewood (2015) reported on the presence of another toxin viz. Trachynilysin (TLY) from this species.

Synanceia verrucosa **(Family: Synanceiidae):** The toxins,-Verrucotoxin (VTX), Neoverrucotoxin (neoVTX) and Cardioleputin have been reported from this species (Ziegman and Alewood, 2015). Among these toxins, Verrucotoxin acted as β1-adrenoceptor agonist. The β1-adrenoceptor (β1AR) is the site of action of beta blockers used in the treatment of cardiac-related illnesses (Kirchhoff et al., 2022). Further, cardioleputin has been reported to possess cardioactive properties (https://en.wikipedia.org/wiki/Synanceia_verrucosa). Ndandala et al. (2023) stated that the venom of this species has shown cytotoxicity.

Paracentropogon rubripinnis (= *Hypodytes rubripinnis)* **(Family: Tetrarogidae):** The glycoprotein, Karatoxin derived from this species has shown cytotoxic activity on murine P388 leukemic cells (Nagasaka et al., 2009).

1: Plotosus lineatus; 2: Pterois volitans; 3: Epinephelus coioides; 4: Grammistes sexlineatus; 5: Pogonoperca punctata; 6: Pardachirus marmoratus; 7: Pseudosynanceia melanostigma; 8: Synanceia verrucose; 9: Paracentropogon rubripinnis

1. *Image credit*: Jens Petersen. https://creativecommons.org/licenses/by/2.5/
2. *Image credit*: Michael Gäbler. https://creativecommons.org/licenses/by/3.0/deed.en
3. *Image credit*: Sahat Ratmuangkhwang. https://creativecommons.org/licenses/by/3.0/
4. *Image credit*: Izuzuki. https://creativecommons.org/licenses/by-sa/3.0/
5. *Image credit*: Hectonichus. https://creativecommons.org/licenses/by-sa/4.0/
6. *Image credit*: Serge Planes. https://creativecommons.org/publicdomain/zero/1.0/
7. *Image credit*: George Henry Ford. https://creativecommons.org/licenses/by-sa/4.0/deed.nl
8. *Image credit*: Karelj. https://creativecommons.org/licenses/by-sa/3.0/
9. *Image credit*: harum.koh. https://creativecommons.org/licenses/by-sa/2.0/deed.en

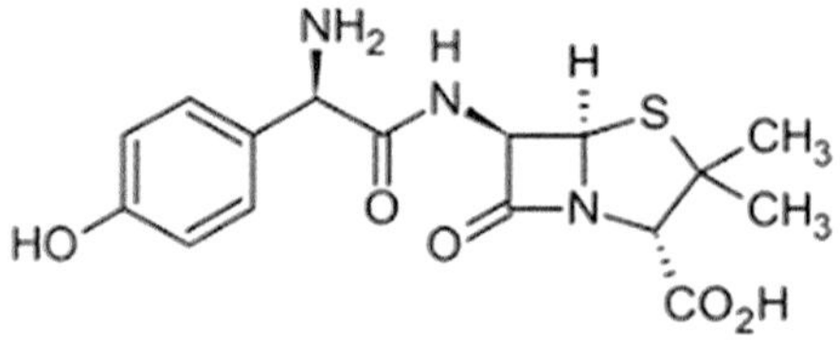

Pardaxin

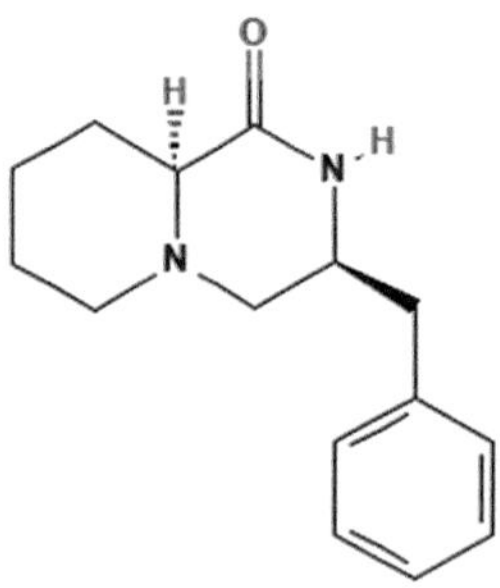

Verrucotoxin

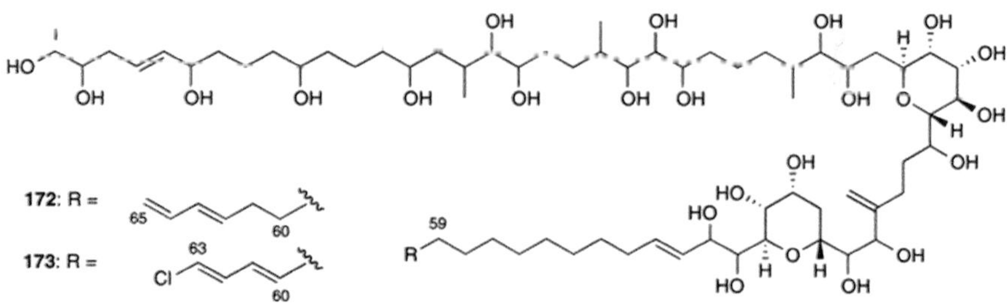

Karatoxin

KEYWORDS

- **marine fish toxins**
- **elasmobranch toxins**
- **teleost toxins**
- **therapeutic fish toxins**
- **TTX-producing bacteria**
- **therapeutic TTX**

References

Accoroni, S. et al. *Ostreopsis fattorussoi* sp. nov. (Dinophyceae), a New Benthic Toxic *Ostreopsis* Species from the Eastern Mediterranean Sea. *J. Phycol.* **2016,** *52*, 1064–1084.

Akakabe, M.; Kumagai, K.; Tsuda, M.; Konishi, Y. ChemInform Abstract: Amphirionin-5, a Novel Linear Polyketide from a Cultured Marine Dinoflagellate *Amphidinium* Species with a Potent Cell Proliferation-Promoting Activity. *Tetrahedron Lett.* **2014,** *55*, 3491–3494.

Alabssawy, A. N. Antimicrobial Activity of Tetrodotoxin Extracted from Liver, Skin and Muscles of Puffer Fish, *Lagocephalus sceleratus* Inhabiting Mediterranean Sea, Egypt. *Int. J. Biomed. Cancer Res.* **2017,** *1*, 2–10.

Alberghina, M.; Gould, R. Characterization of Phospholipase A2 and Acyltransferase Activities in Squid (*Loligo pealei*) Axoplasm: Comparison with Enzyme Activities in Other Neural Tissues, Axolemma and Axoplasmic Subfractions. *Neurochem. Int.* **1993,** *21*, 563–571.

Almeida, D. et al. Putative Antimicrobial Peptides of the Posterior Salivary Glands from the Cephalopod *Octopus vulgaris* Revealed by Exploring a Composite Protein Database. *Antibiotics* **2020,** *9*, 757.

Álvarez, G. et al. Rapid Domoic Acid Depuration in the Scallop *Argopecten purpuratus* and Its Transfer from the Digestive Gland to Other Organs. *Toxins* **2020,** *12*, 698.

Aminin, D. L.; Agafonova, I. G.; Kalinin, V. I.; Silchenko, A. S.; Avilov, S. A.; Stonik, V. A.; Collin, P. D.; Woodward, C. Immunomodulatory Properties of Frondoside A, a Major Triterpene Glycoside from the North Atlantic Commercially Harvested Sea Cucumber *Cucumaria frondosa*. *J. Med. Food* **2008,** *11*, 443–453.

Aminin, D. L.; Menchinskaya, E. S.; Pislyagin, E. A.; Silchenko, A. S.; Avilov, S. A.; Kalinin, V. I. Sea Cucumber Triterpene Glycosides as Anticancer Agents. *Stud. Nat. Prod. Chem.*, **2016,** *49*, 55–105.

Aminin, D. L.; Menchinskaya, E. S.; Pisliagin, E. A.; Silchenko, A. S.; Avilov, S. A.; Kalinin, V. I. Anticancer Activity of Sea Cucumber Triterpene Glycosides *Mar. Drugs* **2015,** *13*, 1202–1223.

Ansdell. https://wwwnc.cdc.gov/travel/yellowbook/2024/environmental-hazards-risks/food-poisoning-from-marine-toxins

Aniya, Y.; Terukina, R.; Minamitake, Y.; Shiohira, S. Effect of the Spine Venom from the Crown-of-Thorns Starfish, *Acanthaster planci*, on Drug-Metabolizing Enzyme in Rat Liver. *J. Toxicol. Sci.* **1998,** *23*, 419–423.

Anon. Amnesic Shellfish Poisoning: Domoic Acid Production by *Pseudo-nitzschia* Diatoms. Aqua Info Aquaculture Notes. http://www.gov.pe.ca/photos/original/af_domoic_acid.pdf

Anon. Report No. 3476. Pectenotoxins and Okadaic Acid Group Toxins in New Zealand Bivalve Molluskan Shellfish, 2009–2019: Risk Assessment. New Zealand Food Safety Tech. Paper **2020,** *12*, 2020.

Anon. Identification of Anti-Tumoral Effect of a Polypeptide Isolated from Scorpionfish *Scorpaena plumieri* Venom and Assessment of Its Potential Use for Tumor Diagnosis. Soprani, Juliana. https://inis.iaea.org/search/search.aspx?orig_q=RN:39107589

Anon. Marine Biotoxins in Shellfish—Okadaic Acid and Analogues—Scientific Opinion of the Panel on Contaminants in the Food Chain. European Food Safety Authority (EFSA). *EFSA J.* **2008,** *589*, 1–62.

Anon. *Synanceia verrucosa*. https://en.wikipedia.org/wiki/Synanceia_verrucosa.

Anon. Common Toad Fish. https://en.wikipedia.org/wiki/Common_toadfish.

Antonelli, P.; Salerno, B.; Bordin, P.; Peruzzo, A.;Orsini, M.; Arcangeli, G.; Barco, L.; Losasso, C. Tetrodotoxin in Live Bivalve Mollusks from Europe: Is It to Be Considered an Emerging Concern for Food Safety? *Compr. Rev. Food Sci. Food Saf.* **2022,** *21*, 719–737.

Arakawa, O.; Hwang, D.; Taniyama, S.; Takatani, T. *Toxins of Pufferfish That Cause Human Intoxications Coastal Environmental and Ecosystem Issues of the East China Sea*; Ishimatsu, A., Lie, H.-J., Ed.; 2010; pp 227–244.

Arruda, G. L. et al. Box Jellyfish (Cnidaria, Cubozoa) Extract Increases Neuron's Connection: A Possible Neuroprotector Effect. *Biomed. Res. Int.* **2021,** *2021*, ID 8855248.

Assunção, J.; Guedes, A. C.; Malcata, F. X. Biotechnological and Pharmacological Applications of Biotoxins and Other Bioactive Molecules from Dinoflagellates. *Mar. Drugs* **2017,** *15*, 393.

Avilov, S. A. et al. Triterpene Glycosides from the Far-Eastern Sea Cucumber *Pentamera calcigera.* 1. Monosulfated Glycosides and Cytotoxicity of Their Unsulfated Derivatives. *J. Nat. Prod.* **2000,** *63*, 65–71.

Awuchi, C. G. et al. Bioactive Compounds and Therapeutics from Fish: Revisiting Their Suitability in Functional Foods to Enhance Human Wellbeing. *Biomed. Res. Int.* **2022,** *2022*, 3661866.

Ayed, Y.; Dellai, A.; Mansour, H. B.; Bacha, H.; Abid, S. Analgesic and Antibutyrylcholinestrasic Activities of the Venom Prepared from the Mediterranean Jellyfish *Pelagia noctiluca* (Forsskal, 1775). *Ann. Clin. Microbiol. Antimicrob.* **2012,** *11*, 15.

Azman, A. M. N.; Samsur, M.; Othman, M. Distribution of Tetrodotoxin Among Tissues of Pufferfish from Sabah and Sarawak Waters. *Sains Malaysiana* **2014,** *43*, 1003–1011.

Bahrami, Y. *Discovery of Novel Saponins as Potential Future Drugs from Sea Cucumber Viscera*. Ph.D. Thesis; Flinders University, Australia, 2015.

Bajpai, V. K.; Shukla, S.; Kang, S.; Hwang, S. K.; Song, X.; Huh, Y. S.; Young-Kyu Han, Y. Developments of Cyanobacteria for Nano-Marine Drugs: Relevance of Nanoformulations in Cancer Therapies. *Mar. Drugs* **2018,** *16*, 179.

Bali, D. K. Chemical and Biochemical Studies on Metabolites from Australian Marine Sponges; University of Wollongong Thesis Collections, 1992. https://ro.uow.edu.au/cgi/viewcontent.cgi?article=3779&context=theses

Barbera-Sanchez, A. L.; Soler, J. F.; de Astudillo, L. R.; Cheng-Yen, I. Paralytic Shellfish Poisoning (PSP) in Margarita Island, Venezuela. *Rev. Biol. Trop.* **2004,** *52*, ISSN 0034–7744.

Bargu, S.; Powell, C. L.; Coale, S. L.; Busman, M.; Doucette, G. J.; Silver, M. W. Krill: A Potential Vector for Domoic Acid in Marine Food Webs. *Mar. Ecol. Prog. Ser.* **2002,** *237*, 209–216.

Bates, S. S. Domoic Acid-Producing Diatoms. Another Genus Added. *J. Phycol.* **2000,** *36*, 978–985.

Bauer, I.; Maranda, L.; Young, K. A.; Shimuzu; Fairchild, C.; Cornell, L.; MacBeth, J.; Huang, S. Isolation and Structure of Caribenolide I, a Highly Potent Antitumor Macrolide from a Cultured Free-Swimming Caribbean Dinoflagellate, *Amphidinium* sp. S1–36–5. *J. Org. Chem.* **1995,** *60*, 1084–1086.

Berdalet, E. et al. Marine Harmful Algal Blooms, Human Health and Wellbeing: Challenges and Opportunities in the 21st Century. *J. Mar. Biol. Assoc. UK* **2015,** *2015*, 10.

Berti, M.; Milandri, A. Le biotossine Marine. In: *Igiene Degli Alimenti;* Schirone, M., Visciano, P., Eds.; Edagricole: Bologna, 2014.

Bhattacharya, A.; Wickenden, A. D.; Chaplan, S. R. (Eds.) Sodium Channel Blockers for the Treatment of Neuropathic Pain. *Neurotherapeutics* **2009,** *6*, 663–678.

Biessy, L.; Boundy, M.j.; Smith, K. F.; Harwood, D. T.; Hawes, I.; Sood, S. A. Tetrodotoxin in Marine Bivalves and Edible Gastropods: A Mini-Review. *Chemosphere* **2019,** *236*, 124404.

Blanco, J. Accumulation of Dinophysis Toxins in Bivalve Mollusks. *Toxins* **2018,** *10*, 453.

Bogan, Y. *Factors Affecting the Concentration of Domoic Acid in Scallop, Pecten maximus*; Ph.D. Thesis, Higher Education and Training Awards Council (HETAC), 2006.

Bordon, K. C. F. et al. From Animal Poisons and Venoms to Medicines: Achievements, Challenges and Perspectives in Drug Discovery. *Front. Pharmacol.* **2020,** *11*, 1132.

Botana, L. M.; Louzao, M. C.; Vilarino.N.; de Gruyter, W. *Climate Change and Marine and Freshwater Toxins*; GmbH & Co KG, 2020.

Botelho, M. J.; Milinovic, J.; Bandarra, N. M. Alzheimer's Disease and Toxins Produced by Marine Dinoflagellates: An Issue to Explore. *Mar. Drugs* **2022,** *20*, 253.

Boundy, M. J.; Harwood, D. T.; Kiermeier, A.; McLeod, C.; Nicolas, J.; Finch, S. Risk Assessment of Pectenotoxins in New Zealand Bivalve Molluskan Shellfish, 2009–2019. *Toxins* **2020,** *12*, 776.

Bousquet, M. S. et al. Seaweed Natural Products Modify the Host Inflammatory Response via Nrf2 Signaling and Alter Colon Microbiota Composition and Gene Expression. *Free Radic. Biol. Med.* **2020,** *146*, 306–323.

Bruce, K. L.; Leterme, S. C.; Ellis A. V.; Lenehan, C. E. Approaches for the Detection of Harmful Algal Blooms Using Oligonucleotide Interactions. *Anal. Bioanal. Chem.* **2015,** *407*, 95–116.

Bruhn, T.; Schaller, C.; Schultze, C.; Sánchez-Rodríguez, J. Isolation and Characterisation of Five Neurotoxic and Cardiotoxic Polypeptides from the Sea Anemone *Anthopleura elegantissima*. *Toxicon* **2001,** *39*, 693–702.

Bucciarelli, G. M.; Lechner, M.; Fontes, A.;Kats, L. B.; Eisthen, H. L.; Shaffer, H. B. From Poison to Promise: The Evolution of Tetrodotoxin and Its Potential as a Therapeutic. *Toxins* **2021,** *13*, 517.

Cabral, I. M. F. F. *A Computational Approach to Identify Target Receptors of Marine Toxins in the Human Proteome: Potential Biotechnological Applications*; Master Thesis, Nova University, Lisbon, 2021.

Careaga, V. P.; Bueno, C.; Muniain, C.; Alché, L.; Maier, M. S. Antiproliferative, Cytotoxic and Hemolytic Activities of a Triterpene Glycoside from *Psolus patagonicus* and Its Desulfated Analog. *Chemotherapy* **2009,** *55*, 60–68.

Carli, A.; Bussotti, S.; Mariottini, G. L.; Robbiano, L.Toxicity of Jellyfish and Sea-Anemone Venoms on Cultured V79 Cells. *Toxicon* **1996,** *34*, 496–500.

Carvalho, P.; Catia, R.; Moukha, S.; Matias, W. g.; Creppy, E. E. Comparative Study of Domoic Acid and Okadaic Acid Induced—Chromosomal Abnormalities in the CACO-2 Cell Line. *Int. J. Environ. Res. Public Health* **2006,** *3*, 4–10.

Cuevas, N.; Martins, M.; Rodrigo, A. P.; Martins, C.; Costa, P. M. Explorations on the Ecological Role of Toxin Secretion and Delivery in Jawless Predatory Polychaeta. *Sci. Rep.* **2018,** *8*, 7635.

Chand, P. Seafood Neurotoxins I: Shellfish Poisoning and the Nervous System. *Clin. Neurotoxicol.* **2009**. DOI: 10.1016/B978-032305260-3.50046-0.

Chee, P. Y. et al. Epinecidin-1, an Antimicrobial Peptide Derived From Grouper (*Epinephelus coioides*): Pharmacological Activities and Applications. *Front. Microbiol.* **2019,** *10*, 2631.

Chen, T. et al. Food-Borne Disease Outbreak of Diarrhetic Shellfish Poisoning Due to Toxic Mussel Consumption: The First Recorded Outbreak in China. *PLoS One,* **2013,** *8,* e65049.

Chen, C.; Chou, H.; Chen, Y.; Lee, T. Detection of Tetrodotoxin by HPLC in Shellfishes and Goby from South Taiwan. *J. Nat. Toxins* **2002,** *11*, 63–68.

Cheney, D. Ch 13: Toxic and Harmful Seaweeds. In: *Seaweed in Health and Disease Prevention*; 2016; pp 407–421.

Cherry, P.; O'Hara, C.; Magee, P. J.; Mcsorley, E. M.; Allson, P. J. Risks and Benefits of Consuming Edible Seaweeds. *Nutr. Rev.* **2019,** *77*, 307–329.

Choi, K. D.; Lee, J. S.; Jong, O. K.; Lee, J. O.; Oh, K. S.; Shin, S. Investigation of Domoic Acid in Shellfish Collected from Korean Fish Retail Outlets *Food Sci. Biotechnol.* **2009,** *18*, 842–848.

Clinton, G. L. Veale, C. G. L.; Davies-Coleman, M. T. Chemistry and Biology. *Alkaloids: Chem. Biol.* **2014,** *73*, 1–347.

Cooke, I. R.; Whitelaw, B.; Norman, M.; Caruana, N.; Strugnel, J. M. Toxicity in Cephalopods. In: Malhotra, A., Ed.; *Evolution of Venomous Animals and Their Toxins: Toxinology*; Springer: Dordrecht, 2017.

Cosmo, A. D.; Polese, G. Ch 39: Molluskan Bioactive Peptides. In: Kastin, A. J., Ed.; *Handbook of Biologically Active Peptides*; Elsevier Onc, 2013.

Costa, M.; Rodrigues, J. C.; Fernandes, M. H.; Barros, P.; Vasconcelos, V.; Martins, R. Marine Cyanobacteria Compounds with Anticancer Properties: A Review on the Implication of Apoptosis. *Mar. Drugs* **2012,** *10*, 2181–2207.

Coulter-Parkhill, A.; McClean, S.; Gault, V. A.; Irwin, N. Therapeutic Potential of Peptides Derived from Animal Venoms: Current Views and Emerging Drugs for Diabetes. *Clin. Med. Insights: Endocrinol. Diabetes* **2021,** *14*, 11795514211006071.

Cousseau, A.; Siano, R.; Probert, I.; Bach, S.; Mehiri, M. Marine Dinoflagellates as a Source of New Bioactive Structures. *Stud. Nat. Prod. Chem.* **2020,** *65*, 125–171.

Crinelli, R. et al. Palytoxin and an Ostreopsis Toxin Extract Increase the Levels of mRNAs Encoding Inflammation-Related Proteins in Human Macrophages via p38 MAPK and NF-κB. *PLoS One* **2012,** *7*, e38139.

D'Ambra, I.; Lauritano, C. A Review of Toxins from Cnidaria. *Mar. Drugs* **2020,** *18*, 507.

Dang, N. H.; Thanh, N. V.; Kiem, P. V.; Huong, M.; Minh, C. V.; Kim, Y. H. Two New Triterpene Glycosides from the Vietnamese Sea Cucumber *Holothuria scabra. Arch. Pharm. Res.* **2007,** *30*, 1387–1391.

Dao, H. V.; Takata, Y.; Sato, S.; Fukuyo, Y.; Kodama, M. Frequent Occurrence of the Tetrodotoxin-Bearing Horseshoe Crab *Carcinoscorpius rotundicauda* in Vietnam. *Fish. Sci.* **2009,** *75*, 435–438.

Dao, F. et al. Recent Advances in Conotoxin Classification by Using Machine Learning Methods *Molecules.* **2017,** *22*, 1057.

Darius, H. T. et al. Toxicological Investigations on the Sea Urchin *Tripneustes gratilla* (Toxopneustidae, Echinoid) from Anaho Bay (Nuku Hiva, French Polynesia): Evidence for the Presence of Pacific Ciguatoxins. *Mar. Drugs* **2018,** *16*, 122.

Davidson, K.; Baker, C.; Higgins, C.; Higman, W.; Swan, S.; Veszelovszki, A.; Turner, A. M. Potential Threats Posed by New or Emerging Marine Biotoxins in UK Waters and Examination of Detection Methodologies Used for Their Control: Cyclic Imines. Mar. *Drugs* **2015,** *13*, 7087–7112.

Day, Y. et al. Characterization and Anti-Tumor Activity of Saponin-Rich Fractions of South Korean Sea Cucumbers (*Apostichopus japonicus*). *J. Food. Sci. Technol.* **2020,** *57*, 2283–2292.

Dean, K. J. et al. The Common Sunstar *Crossaster papposus*—A Neurotoxic Starfish. *Mar. Drugs* **2021,** *19*, 695.

Deeds, J. R.; Schwartz, M. D. Human Risk Associated with Palytoxin Exposure. *Toxicon* **2010,** *56*, 150–162.

Delcourt, N. et al. Pinnatoxins' Deleterious Effects on Cholinergic Networks: From Experimental Models to Human Health. *Mar. Drugs* **2019,** *17*, 425.

Diaz-Garcia, C. M. et al.Toxins from *Physalia physalis* (Cnidaria) Raise the Intracellular Ca2+ of Beta-Cells and Promote Insulin Secretion. *Curr. Med. Chem.* **2012,** *19*, 5414–5423.

Diogene, J. et al. Identification of Ciguatoxins in a Shark Involved in a Fatal Food Poisoning in the Indian Ocean. *Sci. Rep.* **2017,** *7*, 8240.

Du, X. et al. The Diversity of Cyanobacterial Toxins on Structural Characterization, Distribution and Identification: A Systematic Review. *Toxins* **2019,** *11*, 530.

Durán-Riveroll, L. M.; Weber, J.; Krock, B. First Identification of Amphidinols from Mexican Strains and New Analogs. *Toxins* **2023,** *15*, 163.

Edwards, I. A.; Henriques, S. T.; Blaskovich, M. A.; Elliott, A. G.; Cooper, M. A. Investigations into the Membrane Activity of Arenicin Antimicrobial Peptide AA139. *Biochim. Biophys. Acta* **2022,** *1866*, 130156.

EFSA. Scientific Opinion of the Panel on Contaminants in the Food Chain on a Request from the European Commission on Marine Biotoxins in Shellfish—Saxitoxin Group. *EFSA J.* **2009,** *7*, 1393.

Ellis, C. R.; Elston, D. M.; Joustra, J. P. L.; Haddad Jr, V. Aquatic Antagonists: Sea Cucumbers (Holothuroidea). *Cutis* **2021,** *108*, 68–70.

Erol, E.; Orhan, M. D.; Avsar, T.; Akdemir, A.; Okudan, E. S.; Toraman, G. O. A. Topcu, G.Anti-SARS-CoV-2 and Cytotoxic Activity of Two Marine Alkaloids from Green Alga *Caulerpa cylindracea* Sonder in the Dardanelles. *RSC Adv.* **2022,** *12*, 29983–29990.

Espiritu, R. A.; Tan, M. C. S.; Oyong, G. G. Evaluation of the Anti-Cancer Potential of Amphidinol 2, a Polyketide Metabolite from the Marine Dinoflagellate *Amphidinium klebsii Jordan*. *J. Biol. Sci.* **2017,** *10*, 297–302.

Estevez, P. et al. Toxicity Screening of a *Gambierdiscus australes* Strain from the Western Mediterranean Sea and Identification of a Novel Maitotoxin Analogue. *Mar. Drugs* **2021,** *19*, 460.

FAO and WHO. Toxicity Equivalence Factors for Marine Biotoxins Associated with Bivalve Mollusks; Technical Paper, 2016. https://apps.who.int/iris/bitstream/handle/10665/250663/9789241511483-eng.pdf;jsessionid=45903DFA3D1CB977BDFD77572ED2303B?sequence=1

FAO and WHO. Report of the Expert Meeting on Food Safety for Seaweed—Current Status and Future Perspectives. *Food Saf. Qual. Ser*. **2022**. https://doi.org/10.4060/cc0846en

Frazão, B.; Vasconcelos, V. Antunes A Sea Anemone (Cnidaria, Anthozoa, Actiniaria) Toxins: An Overview. *Mar. Drugs* **2012,** *10*, 1812–1851.

Freile-Pelegrín, Y.; Tasdemir, D. Seaweeds to the Rescue of Forgotten Diseases: A Review. *Bot. Mar.* **2019,** *62*, 211–226.

Freitast, J. C.; Ogata, T.; Veit, C. H.; Kodama, M. Occurrence of Teradotoxin and Paralytic Shellfish Toxins in *Phallusia nigra* (Tunicate: Ascidiacea) from the Brazilian Coast. *J. Venom. Anim. Toxins* **1996,** *2*, 86427481.

Fry, B. G.; Roelants, K.; Norman, J. A. Tentacles of Venom: Toxic Protein Convergence in the Kingdom Animalia. *J. Mol. Evol.* **2009,** *68*, 311–321.

Furstner, A. et al. Total Syntheses of Amphidinolides B1, B4, G1, H1 and Structure Revision of Amphidinolide H2. *Chemistry* **2009,** *15*, 3983–4010.

Gao, B.; Peng, C.; Yang, J.; Yi, Y.; Zhang, J.; Shi, Q. Cone Snails: A Big Store of Conotoxins for Novel Drug Discovery. *Toxins* **2017,** *9*, 397.

Gao, W. et al. Evaluation of the Tetrodotoxin Uptake Ability of Pufferfish *Takifugu rubripes* Tissues According to Age Using an In Vitro Tissue Slice Incubation Method. *Toxicon* **2020,** *174*, 8–12.

Gemma, S.; Molteni, M.; Rossetti, C. Lipopolysaccharides in Cyanobacteria: A Brief Overview. *Adv. Microbiol.* **2016,** *6*, 391–397.

Gideon, T. P.; Rengasamy, R. Toxicological Evaluation of Fucoidan from *Cladosiphon okamuranus*. *J. Med. Food* **2008,** *11*, 638–642.

Gonçalves, C.; Costa, P. M. Cephalotoxins: A Hotspot for Marine Bioprospecting? *Front. Mar. Sci.* **2021,** *8*, 647344.

González-Cano, R.; Ruiz-Cantero, M. C.; Santos-Caballero, M.; Gómez-Navas, C.; Tejada, M. A.; Nieto, F. R. Tetrodotoxin, a Potential Drug for Neuropathic and Cancer Pain Relief? *Toxins* **2021,** *13*, 483.

Göransson, U.; Jacobsson, E.; Strand, M.; Andersson, H. S. The Toxins of Nemertean Worms. *Toxins* **2019,** *11*, 120.

Gou, J.; Choi, K.; He, X.; Ahn, J. Dimethylamine, Trimethylamine, and Biogenic Amine Formation in High-Pressure Processed Semidried Squid (*Todarodes pacificius*) During Refrigerated Storage. *Food Sci.* **2010,** *75*, M489–M495.

Grauso, L.; Afsaneh Yegdaneh, A.; Mohsen Sharifi, M.; Alfonso Mangoni, A.; Behzad Zolfaghari, B.; and Virginia Lanzotti, V. Molecular Networking-Based Analysis of Cytotoxic Saponins from Sea Cucumber *Holothuria atra* Laura. *Mar. Drugs* **2019,** *17*, 86.

Gupta, D. K.; Kaur, P.; Leong, S. T.; Tan, L. T.; Prinsep, M. R.; Chu, J. J. H. Anti-Chikungunya Viral Activities of Aplysiatoxin-Related Compounds from the Marine Cyanobacterium *Trichodesmium erythraeum*. *Mar. Drugs* **2014,** *12*, 115–127.

Hallegraeff_G. M. et al. Perceived Global Increase in Algal Blooms Is Attributable to Intensified Monitoring and Emerging Bloom Impacts. *Commun. Earth Environ.* **2021,** *2*, 117.

Hamada, A. K.; Deglin, S. E.; McLaughlin, J. B.; Deeds, J. R.; Handy, S. M.; Knolhoff, A. M. Suspected Palytoxin Inhalation Exposures Associated With Zoanthid Corals in Aquarium Shops and Homes.—Alaska, 2012–2014. *Morb. Mortal. Wkly. Rep.* **2015,** *64*, 852–855.

Han, B. N.; Liang, T. T.; Keen, L. J.; Fan, T. T.; Zhang, X. D.; Xu, L.; Zhao, Q.; Wang, S. P.; Lin, H. W. Two Marine Cyanobacterial Aplysiatoxin Polyketides, Neo-Debromoaplysiatoxin A and B, with K+Channel Inhibition Activity. *Org. Lett.* **2018,** *20*, 578–581.

Hashimoto, Y.; Oshima, Y. Separation of Grammistins A, B and C from a Soapfish *Pogonoperca punctata. Toxicon* **1972,** *10,* 279–284.

Hieda, M.; Sorada, A.; Kinoshita, M.; Matsumori, N. Amphidinol 3 Preferentially Binds to Cholesterol in Disordered Domains and Disrupts Membrane Phase Separation. *Biochem. Biophys. Rep.* **2021,** *26*, 100941.

Higa, T.; Kuniyoshi, M. Toxins Associated with Medicinal and Edible Seaweeds. *J. Toxicol.* **2000,** *19*, 119–137.

Hoffmann, K.; Hermanns-Clausen, M.; Buhl, C.; Büchler, M. W. A Case of Palytoxin Poisoning Due to Contact with Zoanthid Corals Through a Skin Injury. *Toxicon* **2008,** *51*, 1535–1537.

Honkanen, R. E.; Dukelow, M.; Zwiller, J.; Moore, R. E.; Khatra, B. S.; Boynton, A. L. Cyanobacterial Nodularin Is a Potent Inhibitor of Type 1 and Type 2A Protein Phosphatases. *Mol. Pharmacol.* **1991,** *40*, 577–583.

Honma, T.; Shiomi, K. Peptide Toxins in Sea Anemones: Structural and Functional Aspects. *Mar. Biotechnol.* **2006,** *8*, 1–10.

Hort, V.; Arnich, N.; Guerin, T.; Gwenaëlle Lavison-Bompard, G.; Nicolas, M. First Detection of Tetrodotoxin in Bivalves and Gastropods from the French Mainland Coasts. *Toxin* **2020,** *12*, 599.

Hossen, V.; Silva, N. J.; Guillois-Bécel, Y.; Marchal, J.; Krys, S. Food Poisoning Outbreaks Linked to Mussels Contaminated with Okadaic Acid and Ester Dinophysistoxin-3 in France, June 2009. *Eur. Survelll.* **2011,** *16*, 20020.

https://www.seafish.org/trade-and-regulation/legislation-to-control-contaminants-in-seafood/marine-biotoxins

https://www.marine.ie/site-area/areas-activity/seafood-safety/shellfish-biotoxins

https://www.ncbi.nlm.nih.gov/books/NBK470225/

https://www.dhs.wisconsin.gov/publications/p02198.pdf

https://hero.epa.gov/hero/index.cfm/reference/details/reference_id/1850440

https://en.wikipedia.org/wiki/Brevetoxin

https://en.wikipedia.org/wiki/Holothurin

https://cadivingnews.com/domoic-acid-advisory-issued-for-consumers-of-lobsters-and-rock-crabs

http://extoxnet.orst.edu/tics/domoic.asc;

https://reefkeeping.com/issues/2004–08/rhf/index.php

https://sites.chemengr.ucsb.edu/~ceweb/faculty/scott/Chemical%20SOPs/DiethylEther.pdf).

http://haedat.iode.org

https://wwwn.cdc.gov/DCS/ContactUs/Form

https://reefs.com/magazine/toxins-venoms-and-inhibitory-chemicals-in-marine-organisms/)

https://en.wikipedia.org/wiki/Anatoxin-a

https://wwwnc.cdc.gov/eid/article/19/8/12–1824_article

https://www.whoi.edu/science/B/redtide/notedevents/foreign/DSP_Spain_9–25–02.html
http://www.t3db.ca/toxins/T3D2463
https://www.newfoodmagazine.com/article/112319/marine-biotoxins/
https://www.sciencedirect.com/topics/pharmacology-toxicology-and-pharmaceutical-science/dinophysistoxin-1
https://en.wikipedia.org/wiki/Yessotoxin
https://hero.epa.gov/hero/index.cfm/reference/details/reference_id/7425459
https://cdnsciencepub.com/doi/abs/10.1139/v97–218?journalCode=cjc
https://wchh.onlinelibrary.wiley.com/doi/10.1002/psb.1837
https://dr.lib.iastate.edu/entities/publication/3f35276b-991c-4461–9e8d-3c6efa194b29
https://www.verywellhealth.com/does-jellyfish-protein-work-for-memory-loss-2252421
https://www.whoi.edu/science/B/redtide/notedevents/foreign/DSP_Spain_9–25–02.html
https://www.sciencedirect.com/topics/pharmacology-toxicology-and-pharmaceutical-science/saponin;
https://danboater.org/travel-health-and-safety/sargassum-toxicity-health-hazard-what-you-need-to-know.html
http://www.gov.pe.ca/photos/original/af_domoic_acid.pdf
https://patents.google.com/patent/US8975281B2/en
/https://patents.google.com/patent/EP1664057B1/en
https://www.sciencedirect.com/topics/agricultural-and-biological-sciences/domoic-acid
http://www.gov.pe.ca/photos/original/af_domoic_acid.pdf
https://en.wikipedia.org/wiki/Holothurin
https://www.cdc.gov/mmwr/preview/mmwrhtml/00039077.htm
https://www.forbes.com/health/body/health-benefits-of-seaweed
https://www.forbes.com/health/body/health-benefits-of-seaweed/
https://www.cdc.gov/mmwr/preview/mmwrhtml/00039077.htm
https://www.sciencedirect.com/topics/chemistry/holothurin
https://www.sciencedirect.com/topics/immunology-and-microbiology/pseudo-nitzschia
https://www.eurekalert.org/news-releases/569329
https://oehha.ca.gov/fish/general-info/domoic-acid-marine-biotoxin-fish-and-shellfish
https://www.sciencedirect.com/topics/nursing-and-health-professions/ciguatoxin
https://oehha.ca.gov/fish/general-info/domoic-acid-marine-biotoxin-fish-and-shellfish
https://www.sciencedirect.com/topics/nursing-and-health-professions/ciguatoxin;
https://www.thelancet.com/journals/ebiom/article/PIIS2352–3964(23)00169-X/fulltext
https://www.sciencedirect.com/topics/pharmacology-toxicology-and-pharmaceutical-science/dinophysistoxin-1)
https://coastalscience.noaa.gov/news/common-coastal-seaweeds-release-toxins-harmful-marine-organisms/
https://www.atsdr.cdc.gov/csem/polychlorinated-biphenyls/adverse_health.html
https://www.drugs.com/npp/ciguatera.html
https://www.ncbi.nlm.nih.gov/books/NBK470225/
https://ncceh.ca/content/blog/palytoxin-potent-poorly-understood-marine-toxin-found-aquarium-coral
https://www.news-medical.net/health/Dopamine-Therapeutic-Use.aspx).
https://www.anses.fr/en/system/files/ERCA2013sa0198EN.pdf
https://www.sciencedirect.com/topics/agricultural-and-biological-sciences/domoic-acid

Hu, X.; Tao, N; Wang, X.;Xiao, J.; Wang, M. Marine-derived bioactive compounds with anti-obesity effect: A review. *J. Funct. Foods*, **2016,** *21*, 372–387.

Huang, K. et al. Apratoxin A Shows Novel Pancreas-Targeting Activity through the Binding of Sec 61. *Mol. Cancer Ther.* **2016,** *15*, 1208–1216.

Huang, L.; Gong, L.; Xue, X.; Zhang, Y. Marine Neurotoxins: A DoubleEdged Sword in Food Industry and Brain Research. *Med One* **2017,** *2*, e170012.

Huang, T.; Lee, J.; Chen, J. Pardaxin, an Antimicrobial Peptide, Triggers Caspase-Dependent and ROS-Mediated Apoptosis in HT-1080 Cells. *Mar. Drugs* **2011,** *9*, 1995–2009.

Huc, T. et al. Chronic, low-dose TMAO Treatment Reduces Diastolic Dysfunction and Heart Fibrosis in Hypertensive Rats. *Am. J. Physiol. Heart Circ. Physiol.* **2018,** *315*, H1805–H1820.

Hwang, D. F.; Arakawa, O.; Saito, T.; Noguchi, T.; Simidu, U.; Tsukamoto, K.; Shida, Y.; Hashimoto, K. Tetrodotoxin-Producing Bacteria from the Blue-Ringed Octopus *Octopus maculosus. Mar. Bio.* **1989,** *100*, 327–332.

Iduarte, B.; Juárez, P. Compounds Isolated from Corals: A Potential Therapeutic Strategy in Bone Diseases. *HubLE My Mentor & I.* **2020**. DOI: 10.13140/RG.2.2.35753.95847.

Ito, M. et al. Local Differences in the Toxin Amount and Composition of Tetrodotoxin and Related Compounds in Pufferfish (*Chelonodon patoca*) and Toxic Goby (*Yongeichthys criniger)* Juveniles. *Toxins* (Basel) **2022,** *14*, 150.

Ivanchina, N. V.; Kalinovsky, A. I.; Malyarenko, T. V.; Kicha, A. A.; Dmitrenok, P. S. A Holothurian Triterpene Glycoside Holothurin A2 (= Echinoside A) Isolated from the Starfish *Choriaster granulatus. Nat. Prod. Commun.* **2019,** *14*, 1934578X1985852.

Iwasaki, K. et al. Total Synthesis of Polycavernosides A and B, Two Lethal Toxins from Red Alga. *J. Org. Chem.* **2017,** *82*, 13204–13219.

Jacups, S. P.; Currie, B. J. Blue-Ringed Octopuses: A Brief Review of Their Toxicology. *North. Territ. Nat.* **2008,** *20*, 50–57.

Jayathilake, J. M. N. J.; Gunathilake, K. V. K. Cnidarian Toxins: Recent Evidences for Potential Therapeutic Uses. *Eur. Zool. J.* **2020,** *87*, 708–713.

Kanagarajan, K.; Bragadeeswaran, S.; Venkatashwaran, K. On Some Toxinological Aspects of the Starfish, *Stellaser equestris* (Retzius, 1805). *Venom. Anim. Toxins Incl. Trop. Dis.* **2008,** *14*, 435–449.

Karmakar S. et al. Isolation of a Haemorrhagic Protein Toxin (SA-HT) from the Indina Venomous Butterfish (*Scatophagus argus* Linnn) Sting Extract. *Indian J. Exp. Biol.* **2004,** *42*, 452–460.

Karthigayan, S.; SriBalasubashini, M.; Sengottuvelan, M.; Balasubramanian, T.; Somasundaram, S. T. Anticancer Principles from Salivary Gland Extract of *Octopus ageina. Int. J. Cancer Res.* **2006,** *2*, 242–252.

Karthik, R.; Manigandan, V.; Saravanan, R. Toxicity, Teratogenicity and Antibacterial Activity of Posterior Salivary Gland (PSG) Toxin from the Cuttlefish *Sepia pharaonis* (Ehrenberg, 1831). *J. Chromatogr. B Analyt. Technol. Biomed. Life Sci.* **2017,** *1064*, 28–35.

Kat, M. Diarrhetic Mussel Poisoning in the Netherlands Related to the Dinoflagellate *Dinophysis acuminata. Antonie Van Leeuwenhoek* **1983,** *49*, 417–427.

Katikou, P.; Gokbulut, C.; Kosker, A. R.; Campàs, M.; Ozogul, F. An Updated Review of Tetrodotoxin and Its Peculiarities. *Mar. Drugs* **2022,** *20*, 47.

Kawaguchi, M.; Satake, M.; Zhang, B.; Xiao, Y.; Fukuoka, M.; Uchida, H.; Nagai, H. Neo-Aplysiatoxin A Isolated from Okinawan *Cyanobacterium moorea* Producens. *Molecules* **2020,** *25*, 457.

Kerbrat, A. S. et al. First Evidence of Palytoxin and 42-Hydroxy-palytoxin in the Marine Cyanobacterium *Trichodesmium. Mar Drugs* **2011,** *9*, 543–560.

Khoei, Z. A. Venoms and Medicinal Properties of Cnidarians. *Iran. South Med. J.* **2015,** *18*, 898–916.

Kim, C. H.; Lee, Y. J.; Go, H. J.; Oh, H. Y.; Lee, T. K.; Park, J. B.; Park, N. G. Defensin-Neurotoxin Dyad in a Basally Branching Metazoan Sea Anemone. *FEBS J.* **2017,** *284*, 3320–3338.

Kim, S.; Himaya, S. W. A. Triterpene Glycosides from Sea Cucumbers and Their Biological Activities. *Adv. Food Nutr. Res.* **2012,** *65*, 297–319.

Kim, Y.; An, H.; Kim, J.; Jeon, Y. Current Situation of Palytoxins and Cyclic Imines in Asia-Pacific Countries: Causative Phytoplankton Species and Seafood Poisoning. *Int. J. Environ. Res. Public Health* **2022,** *19*, 4921.

Kirchhoff, K. N.; Billion, A.; Voolstra, C. R.; Kremb, S.; Wilke, T.; Vilcinskas, A. V. Stingray Venom Proteins: Mechanisms of Action Revealed Using a Novel Network Pharmacology Approach. *Mar Drugs* **2022,** *20*, 27.

Kitagawa, I.; Nishino, T.; Matsuno, T.; Akutsu, H.; Kyogoku, Y. Structure of holothurin B a Pharmacologically Active Triterpene-Oligoglycoside from the Sea Cucumber *Holothuria leucospilota* Brandt. *Tetrahedron Lett.* **1978,** *19*, 985–988.

Kobayashi, J.; Ishibashi, M.; Walchli, M. R.; Nakamura, H.; Hirata, Y.; Sasaki, T.; Ohizumi, Y. Amphidinolide C: the first twenty-five membered macrocyclic lactone with potent antineoplastic activity from the cultured dinoflagellate *Amphidinium* sp. *J. Am. Chem. Soc.* **1988,** *110*, 490–494.

Kopfmann, S.; Roesch, S. K.; Hess, W. R.Type II Toxin–Antitoxin Systems in the Unicellular Cyanobacterium *Synechocystis* sp. PCC 6803. *Toxins* **2016,** *8*, 228.

Kumar, K. R.; Vennila, R.; Kanchana, S.; Arumugam, M.; Balasubramaniam, T. Fibrinogenolytic and Anticoagulant Activities in the Tissue Covering the Stingers of Marine Stingrays *Dasyatis sephen* and *Aetobatis narinari*. *J. Thromb. Thrombolysis* **2011,** *31*, 464–471.

Kungsuwan, A.;Noguchi, T.; Arakawa, O.; Simidu, U.;Tsukamoto, K.; Shida, Y.; Hashimoto, K. Tetrodotoxin-Producing Bacteria from the Horseshoe Crab *Carcinoscorpius rotundicauda. Nippon Suisan Gakkaishi* **1988,** *54*, 1799–1802.

Kuwabara, S. Purification and Properties of Peditoxin and the Structure of Its Prosthetic Group, Pedoxin, from the Sea Urchin *Toxopneustes pileolus* (Lamarck) *Biol. Chem.* **1994,** *269*, 26734–26738.

Kwon, Y. S.; Min, S. K.; Yeon, S. J.; Hwang, J. H.; Hong, J.; Shin, H. S. Assessment Of Neuronal Cell-Based Cytotoxicity of Neurotoxins from an Estuarine Nemertean in the Han River Estuary. *J. Microbiol. Biotechnol.* **2017,** *27*, 725–730.

Kvrgić, K.; Lešić, T.; Džafić, N.; Pleadin, J. Occurrence and Seasonal Monitoring of Domoic Acid in Three Shellfish Species from the Northern Adriatic Sea. *Toxins* **2022,** *14*, 33.

Lago, J.; Rodriguez, L.; Blanco, L.; Vieites, J. M.; Cabado, A. G. Tetrodotoxin, an Extremely Potent Marine Neurotoxin: Distribution, Toxicity, Origin and Therapeutical Uses. *Mar. Drugs* **2015,** *13*, 6384–6406

Lassen, S.; Helmholz, H.; Ruhnau, C.; Prange, A. A Novel Proteinaceous Cytotoxin from the Northern Scyphozoa *Cyanea capillata* (L.) with Structural Homology to Cubozoan Haemolysins. *Toxicon* **2011,** *57*, 721–729.

Lawrence, J.; Loreal, H.; Toyofuku, H.; Hess, P.; Karunasagar, I.; Ababouch, L. Assessment and Management of Biotoxin Risks in Bivalve Mollusks. *FAO Fisheries and Aquaculture Technical Paper* No. 551; Rome, FAO, 2011.

Layer, R. T.; McIntosh, J. M. Conotoxins: Therapeutic Potential and Application. *Mar. Drugs* **2006,** *4*, 119–142.

Lazcano-Pérez, F. Hemolytic, Anticancer and Antigiardial Activity of *Palythoa caribaeorum* Venom. *J. Venom Anim. Toxins Incl. Trop. Dis*. **2018,** *17*, 12.

Lee, H. et al. Anticancer Effect of *Nemopilema nomurai* Jellyfish Venom on HepG2 Cells and a Tumor Xenograft Animal Model. *Evid. Based Complement. Alternat. Med.* **2017,** *2017*, 2752716.

Lee, C.; Hsieh, H. J.; Hsieh, C.; Hwang, D. Plancitoxin I from the Venom Of Crown-of-Thorns Starfish (*Acanthaster planci)* Induces Oxidative and Endoplasmic Reticulum Stress Associated Cytotoxicity in A375.S2 Cells. *Exp. Mol. Pathol*. **2015,** *99*, 7–15.

Lim, C. C.; Yoon, J.; Reynolds, K.; Gerald, L. B.; Ault, A. P.; Heo, S.; Bell, M. L. Harmful Algal Bloom Aerosols and Human Health. *eBioMed.* **2023,** *93*, 104604.

Lin, X. Y.; Ishida, M.; Nagashima, Y.; Shiomi, K. A Polypeptide Toxin in the Sea Anemone *Actinia equina* Homologous with Other Sea Anemone Sodium Channel Toxins: Isolation and Amino Acid Sequence. *Toxicon* **1996,** *34*, 57–65.

Lin, Y.; Qi, X.; Liu, H.; Xue, K.; Xu, S.; Tian, Z. The Anti-Cancer Effects of Fucoidan: A Review of Both In Vivo and In Vitro Investigations. *Cancer Cell Int.* **2020,** *20*, 154.

Liu, B. S.; Yi, Y. H.; Li, L.; Sun, P.; Yuan, W. H.; Sun, G. Q.; Han, H.; Xue, M. Argusides B and C, Two New Cytotoxic Triterpene Glycosides from the Sea Cucumber *Bohadschia argus*. *Chem. Biodivers*. **2008,** *5*, 1288–1297.

Liu, C.; Qi, J.; Shan, B.; Ma, Y. Tachyplesin Causes Membrane Instability That Kills Multidrug-Resistant Bacteria by Inhibiting the 3 Ketoacyl Carrier Protein Reductase FabG. *Front. Microbiol.* **2018,** *9*, 825.

Lopes, V. M.; Lopes, A. R.; Costa, P.; Rosa, R. Cephalopods as Vectors of Harmful Algal Bloom Toxins in Marine Food Webs. *Mar. Drugs* **2013,** *11*, 3381–3409.

Luiz, A. P.; Wood, J. N. *Pharmacological Mechanisms and the Modulation of Pain, in Advances in Pharmacology*; Academic Press, 2016.

Luo, Y.-J.; Kanda, M.; Koyanagi, R.; Hisata, K.; Akiyama, T.; Sakamoto, H.; Sakamoto, T.; Satoh, N. Nemertean and Phoronid Genomes Reveal Lophotrochozoan Evolution and the Origin of Bilaterian Heads. *Nat. Ecol. Evol.* **2018,** *2*, 141.

Mackieh, R.; Abou-Nader, R.; Wehbe, R.; Mattei, C.; Legros, C.; Fajloun, Z.; Sabatier, J. M. Voltage-Gated Sodium Channels: A Prominent Target of Marine Toxins. *Mar. Drugs* **2021,** *19*, 562.

Madejska, A.; Osek, J. Occurrence of Marine Biotoxins in Bivalve Mollusks Available in Poland in 2014–**2018,** *J. Vet. Res.* **2021,** *65*, 329–333.

Maeno, Y.; Kotaki, Y.; Terada, R.; Cho, Y.; i Konoki, K.; Yotsu-Yamashita, M. Six Domoic Acid Related Compounds from the Red Alga, *Chondria armata,* and Domoic Acid Biosynthesis by the Diatom, *Pseudo-nitzschia multiseries. Sci. Rep.* **2018,** *8*, 356.

Maier, M. S.; Roccatagliata, A. J.; Kuriss, A.; Chludil, H.; Seldes, A. M.; Pujol, C. A.; Damonte, E. B. Two New Cytotoxic and Virucidal Trisulfated Triterpene Glycosides from the Antarctic Sea Cucumber *Staurocucumis liouvillei*. *J. Nat. Prod.* **2001,** *64*, 732–736.

Majumder, U.; Cox, J. M.; Johnson, H. W. B.; Rainier, J. D. Total Synthesis of Gambierol: The Generation of the A–C and F–H Subunits by Using a C-Glycoside Centered Strategy. *Chem. Eur. J.* **2006,** *12*, 1736–1746.

Malaise, O.; de Seny, D. Therapeutic Advances in Arthritis Diseases. *Biochem. Pharmacol.* **2019,** *165*, 1–3.

Mana, C. N.; Noor, N. M.; Harn, G. L.; Lajis, R.; Mohamad, S. Screening of Tetrodotoxin in Puffers Using Gas Chromatography–Mass Spectrometry. *J. Chromatogr. A* **2010,** *1217*, 7455–7459.

Martinez, K. A. et al. Amphidinol 22, a New Cytotoxic and Antifungal Amphidinol from the Dinoflagellate *Amphidinium carterae*. *Mar Drugs* **2019,** *17*, 385.

Matthew, S.; Schupp, P. J.; Luesch, H.; Apratoxin E, a Cytotoxic Peptolide from a Guamanian Collection of the Marine Cyanobacterium *Lyngbya bouillonii*. *J. Nat. Prod.* **2008,** *71*, 1113–1116.

McLeod, C.; Burrell, S.; Holland, P. Review of the Currently Available Field Methods for Detection of Marine Biotoxins in Shellfish Flesh FS102086, September 2015. https://www.foodstandards.gov.scot/downloads/Review_of_field_testing_methods_for_biotoxins_in_shellfish_-_Final_Report.pdf

Mejía-Camacho, A. L.; Durán-Riveroll, L. M.; Cembella, A. D. Toxicity Bioassay and Cytotoxic Effects of the Benthic Marine Dinoflagellate *Amphidinium operculatum*. *J. Xenobiot.* **2021,** *11*, 33–45.

Meunier, F. A.; Feng, Z. P.; Molgó, J.; Zamponi, G. W.; Shiavo, G. Glycerotoxin from *Glycera convoluta* Stimulates Neurosecretion By Up-Regulating N-Type Ca2+ Channel Activity. *EMBO J.* **2002,** *21*, 6733–6743.

Miller, M. A. et al. Evidence for a Novel Marine Harmful Algal Bloom: Cyanotoxin (Microcystin) Transfer from Land to Sea Otters. *PLoS One* **2010,** *5*, e12576.

Milligan, K. E.; Marquez, B. L.; Williamson, R. T.; Gerwick, W. H. Lyngbyabellin B, a Toxic and Antifungal Secondary Metabolite from the Marine Cyanobacterium *Lyngbya majuscula*. *J. Nat. Prod.* **2000,** *63*, 1440–1443.

Mir, R.; Karim, S.; Kamal, M. A.; Wilson, C.; Mirza, Z. Conotoxins: Structure, Therapeutic Potential and Pharmacological Applications. *Curr. Pharm. Des.* **2016,** *22*, 582–589.

Miyata, T.; Tokunaga, F.; Yoneya, T.; Yoshikawa, K.; Iwanaga, S.; Niwa, M.; Takao, T.; Shimonishi, Y. Antimicrobial Peptides, Isolated from Horseshoe Crab Hemocytes, Tachyplesin II, and Polyphemusins I and II: Chemical Structures and Biological Activity. *J. Biochem.* **1989,** *106*, 663–668.

Molgó, J.; Benoit, E.; Aráoz, R.; Zakarian, A.; Iorga, B. *Spirolides and Cyclic Imines: Toxicological Profile. Marine and Freshwater Toxins*; Springer Netherlands, 2015.

Mollo, E. Chemical Studies of Caribbean Sacoglossans: Dietary Relationships with Green Algae and Ecological Implications. *J. Chem. Ecol.* **2000,** *26*, 1563–1578.

Monastyrnaya, M. M.; Kalina, R. S.; Kozlovskaya, E. P. Pharmacologically Active Peptides of the Sea Anemone *Heteractis crispa* and Their Biological Templates. *Biomed. J. Sci. Tech. Res.* **2019,** *20*, 15115–15120.

Monks, N. R.; Shuqian Liu, S.; Yongsheng Xu, Y.; Hui Yu, H.; Adam S Bendelow, A. S. Potent Cytotoxicity of the Phosphatase Inhibitor Microcystin LR and Microcystin

Analogues in OARP181- and OATP1B3-Expressing HeLa Cells. *Mol. Cancer Ther.* **2007,** *6*, 587–598.

Monks, N. R.; Liu, S.; Moscow, J. A. Microcystins as Agents for Treatment of Cancer. *Pharmaceutical Sciences Faculty Patents*. **2015,** *35.* https://uknowledge.uky.edu/ps_patents/35

Moore, R. E.; Blackman, A. J.; Cheuk, C. E.; Mynderse, J. S.; Matsumoto, G. K.; Clardy, J.; Woodard, R. W.; Craig, J. C. Absolute Stereochemistries of the Aplysiatoxins and Oscillatoxin A. *J. Org. Chem*. **1984,** *49*, 2484–2489.

Moore, R. E.; Niemczura, W. P.; Carmichael, W. W. Anatoxin-a(s)- A Potent Anticholinesterase from *Anabaena flos-aquae*. Shigeki Matsunaga. *J. Am. Chem. Soc.* **1989,** *111*, 8021–8023.

Morabito, S.; Silvestro, S.; Faggio, C. How the Marine Biotoxins Affect Human Health. *Nat. Prod. Res*. **2018,** *32*, 621–631.

Morsy, N. et al. Isolation and Structure Elucidation of a New Amphidinol with a Truncated Polyhydroxyl Chain from *Amphidinium klebsii. Tetrahedron* **2005,** *61*, 8606–8610.

Mudadu, A. G.; Bazzoni, A. M.; Bazzardi, R.; Lorenzoni, G.; Soro, B.; Bardino, N.; Arras, I.; Sanna, G.; Vodret, B.; Virgilio, S. Influence of seasonality on the presence of Okadaic Acid Associated with *Dinophysis* Species: A Four-Year Study in Sardinia (Italy). *Ital. J. Food Saf.* **2021,** *10*, 8947.

Murphy, E.; Barone, M. E.; Campanile, F.; Touzet, N.; Thomas, O. P. Amphidinol C, a Major Polyketide from an Irish Strain of the Dinoflagellate *Amphidinium carterae. Phytochem. Lett*. **2022,** *51*, 104–108.

Murray, J. S. et al. Structural Characterization of Maitotoxins Produced by Toxic *Gambierdiscus* Species. *Mar. Drugs* **2022,** *20,* 453.

Murray, T. F. Marine Neurotoxins. Advances in Neurotoxicology, 2021. https://www.sciencedirect.com/topics/agricultural-and-biological-sciences/brevetoxin

Mynderse, J. S.; Moore, R. E.; Kashiwagi, M.; Norton, T. R. Antileukemia Activity in the Oscillatoriaceae: Isolation of Debromoaplysiatoxin from *Lyngbya. Science* **1977,** *196*, 538–540.

Nagai, H.; Murata, M.; Torigoe, K.; Satake, M.; and Yasumoto, T. Gambieric Acids, New Potent Antifungal Substances with Unprecedented Polyether Structures from a Marine Dinoflagellate *Gambierdiscus toxicus. J. Org. Chem.* **1992,** *57*, 5448–5453.

Nagai, H.; Sato, S.; Iida, K.; Hayashi, K.; Kawaguchi, M.; Uchida, H.; Satake, M. Oscillatoxin I: A New Aplysiatoxin Derivative, from a Marine Cyanobacterium. *Toxins* **2019,** *11*, 366.

Nagasaka, K. et al. A Novel Cytotoxic Protein, Karatoxin, from the Dorsal Spines of the Redfin Velvetfish, *Hypodytes rubripinnis. Toxin Rev*. **2009,** *28*, 260–265.

Nakamura, T.; Furunaka, H.; Miyata, T.; Tokunaga, F.; Muta, T.; Iwanaga, S.; Niwa, M.; Takao, T.; Shimonishi, Y. Tachyplesin, a Class of Antimicrobial Peptide from the Hemocytes of the Horseshoe Crab (*Tachypleus tridentatus*): Isolation and Chemical Structure. *J. Biol. Chem.* **1988,** *263*, 16709–16713

Ndandala, C. B. et al. The Perspective of Fish Venom: An Overview of the Physiology, Evolution, Molecular and Genetics. *Front. Mar. Sci.* **2023,** *10.* doi.org/10.3389/fmars.2023.1085669

Nelsen, D. R.; Nisani, Z.; Cooper, A. M.; Fox, G. A.; Gren, E. C. K.; Corbit, A. G.; Hayes, W. K. Poisons, Toxungens, and Venoms: Redefining and Classifying Toxic Biological

Secretions and the Organisms That Employ Them. *Biol. Rev. Camb. Philos. Soc.* **2014,** *89*, 450–465.

Nielsen, P. G.; Carlé, J. S.; Christophersen, C. Final Structure of Caulerpicin, a Toxin Mixture from the Green Alga *Caulerpa racemosa. Phytochem* **1982,** *21*, 1643–1645.

Nuzzo, G. et al. Potent Cytotoxic Analogs of Amphidinolides from the Atlantic Octocoral *Stragulum bicolor. Mar. Drugs* **2019,** *17*, 58.

Nuzzo, G.; Cutignano, A.; Sardo, A.; Fontana, A. Antifungal Amphidinol 18 and its 7-Sulfate Derivative from the Marine Dinoflagellate *Amphidinium carterae. J. Nat. Prod.* **2014,** *77*, 1524–1527.

Oshiro, N. et al. High Levels of Tetrodotoxin in the Flesh, Usually an Edible Part of the Pufferfish *Takifugu flavipterus*, Caused by Migration from the Skin and the Regional Characteristics of Toxin Accumulation. *J. Mar. Sci. Eng.* **2021,** *9*, 1312.

Osterbauer, P. J.; Dobbs, M. R. Neurobiological Weapons. In: *Clinical Neurotoxicology*; 2009. https://www.sciencedirect.com/topics/medicine-and-dentistry/biological-warfare

Otero, A.; Chapela, M.; Atanassova, M.; Vieites, J. M.; Cabado, A. G. Cyclic Imines: Chemistry and Mechanism of Action: A Review 2. *Chem. Res. Toxicol.* **2011,** *24*, 1817–1829.

Ovchinnikova, T. V.; Aleshina, G. M.; Balandin, S. V.; Krasnosdembskaya, A. D.; Markelov, M. L.; Frolova, E. I.; Leonova, Y. F.; Tagaev, A. A.; Krasnodembsky, E. G.; Kokryakov, V. N. Purification and Primary Structure of Two Isoforms of Arenicin, a Novel Antimicrobial Peptide from Marine Polychaeta *Arenicola marina. FEBS Lett.* **2004,** *577*, 209–214.

Pachaiyappan, A.; Sadhasivam, G.; Arumugam, S.; Muthuvel, A. Bio Prospecting the Enzymatic and Anticancer Potential of Spine Secretions of Marine Catfish (*Plotosus lineatus*). *J. Biol. Act. Prod. Nat.* **2015,** *5*, 406–418.

Papiol, G. G.; Casanova, A.; Fernández-Tejedor, M.; Iglesia, P.; Diogène, J. Management of Domoic Acid Monitoring in Shellfish from the Catalan Coast. *Environ. Monit. Assess.* **2013,** *185*, 6653–6666.

Paul, V. J. Guam Seaweed Poisoning: Seaweed Testing on Guam. *Micronesica.* **1993,** *26*, 45–47.

Paz, B.; Daranas, A. H.; Norte, M.; Riobo, P.; Franco, J. M.; Fernadez, J. J. Yessotoxins, a Group of Marine Polyether Toxins: An Overview. *Mar. Drugs*. **2008,** *6*, 73–102.

Perez, S. et al. Effect of Gambierol and Its Tetracyclic and Heptacyclic Analogues in Cultured Cerebellar Neurons: A Structure−Activity Relationships Study, *Chem. Res. Toxicol.* **2012,** *25*, 1929−1937.

Pinto-Silva, C. C.; Moukha, S.; Matias, W.; Creppy, E. E. Domoic Acid Induces Direct DNA Damage and Apoptosis in Caco-2 Cells: Recent Advances. *Environ Toxicol.* **2008,** *23*, 657–663.

Pope, C. N.; Brimijoin, S. Cholinesterases and the Fine Line Between Poison and Remedy. *Biochem. Pharmacol.* **2018,** *153*, 205–216.

Pradhan, B.; Ki, J. Phytoplankton Toxins and Their Potential Therapeutic Applications: A Journey Toward the Quest for Potent Pharmaceuticals. *Mar. Drugs* **2022,** *20*, 271.

Preece, E. P.; Hardy, F. J.; Moore, B. C.; Brya, M. A Review of Microcystin Detections in Estuarine and Marine Waters: Environmental Implications and Human Health Risk. *Harmful Algae* **2017,** *61*, 31–45.

Prester, L.; Orct, T.; Macan, J.; Vukušić, J.; Kipčić, D. K. Determination of Biogenic Amines and Endotoxin in Squid, Musky Octopus, Norway Lobster, and Mussel Stored at Room Temperature. *Arh. Hig. Rada Toksikol.* **2010,** *61*, 389–397.

Qamar, H.; Hussain, K.; Soni, A.; Khan, A.; Hussain, T.; Chénais, B. Cyanobacteria as Natural Therapeutics and Pharmaceutical Potential: Role in Antitumor Activity and as Nanovectors. *Molecules* **2021,** *26*, 247.

Quan, L. D.; Chun, M. S.; Qing, Y. X.; Hua, F. L.; Ping, L. X. Caulerchlorin, a Novel Chlorinated Bisindole Alkaloid with Antifungal Activity from the Chinese Green Alga *Caulerpa racemosa*. *Heterocycles* **2012,** *85*, 661–666.

Rajeshkumar, R. K.; Vennila, R.; Karthikeyan, S.; Prasad, N. R.; Arumugam, M.; Velpandian, T.; Balasubramaniam, T. Antiproliferative Activity of Marine Stingray *Dasyatis sephen* Venom on Human Cervical Carcinoma Cell Line. *J. Venom Anim. Toxins Incl. Trop. Dis.* **2015,** *21,* 41.

Raksakulthai, R.; Norman, F. Haard, Purification and Characterization of a Carboxypeptidase from Squid Hepatopancreas (*Illex illecebrosus). J. Agric. Food Chem.* **2001,** *49*, 5019–5030.

Ramezani, M.; Samiei, F.; Pourahmad, J. Anti-Glioma Effect of *Pseudosynanceia melanostigma* Venom on Isolated Mitochondria from Glioblastoma Cells. *Asian Pac. J. Cancer Prev.* **2021,** *22*, 2295–2302.

Ramos, D. F.; Matthiensen, A.; Colvara, W.; de Votto, A. P. S.; Trindade, G. S.; da Silva, P. E. A.; Yunes, J. S. Antimycobacterial and Cytotoxicity Activity of Microcystins *J. Venom. Anim. Toxins Incl. Trop. Dis.* **2015,** *21*, 9.

Ramos, V.; Vasconcelos, V. Palytoxin and Analogs: Biological and Ecological Effects. *Mar Drugs* **2010,** *8*, 2021–2037.

Randall, J. E. Review of Clupeotoxism, an Often Fatal Illness from the Consumption of Clupeoid Fishes. *Pac. Sci.* **2005,** *59*, 73–77.

Rao, D. S.; James, D. B.; Vallabhan, K. G. G.; Muthuswamy, S.; Najmuddin, M. Biotoxicity in Echinoderms. *J. Mar. Biol. Ass. India* **1985,** *27*, 88–96.

Ravaonindrina, N.; Andriamaso, T. H.; Rasolofonirina, N. Puffer Fish Poisoning in Madagascar: Four Case Reports. *Arch. Inst. Pasteur Madag.* **2001,** *67*, 61–64.

Réveillon, D. et al. Tetrodotoxins in French Bivalve Mollusks—Analytical Methodology, Environmental Dynamics and Screening of Bacterial Strain Collections. *Toxins* **2021,** *13*, 740.

Reyes J. G. et al. Ch.23: Maitotoxin: An Enigmatic Toxic Molecule with Useful Applications in the Biomedical Sciences. In: Botana, L. M., Ed.; *Seafood and Freshwater Toxins: Pharmacology, Physiology, and Detection*; CRC Press, 2014.

Richter, S. et al. Comparative Analyses of Glycerotoxin Expression Unveil a Novel Structural Organization of the Bloodworm Venom System. *BMC Evol. Biol.* **2017,** *17*, 64.

Rio, G. J.; Stempien Jr.M. F.; Nigrelli, R. F.; Ruggieri, G. D. Echinoderm Toxins—I. Some Biochemical and Physiological Properties of Toxins from Several Species of Asteroidea. *Toxicon* **1965,** *3*, 147–155.

Rodrigo, A. P.; Lopes, A. R.; Baptista, P. V.; Costa, M. H.; Fernandes, A. R.; Costa, P. M. The Great Biotechnological Potential of a Marine Polychaete: An Alliance Between Toxin and Natural Fluorescence. *Front. Mar. Sci. Conf. Abst. 2018.*

Rodrigo, A. P. et al. Specific Antiproliferative Properties of Proteinaceous Toxin Secretions from the Marine Annelid *Eulalia* sp. onto Ovarian Cancer Cells. *Mar. Drugs* **2021,** *19*, 31.

Rodriguez, J.; Castro, R.; Riguera, R. Holothurinosides: New Antitumour Non Sulphated Triterpenoid Glycosides from the Sea Cucumber *Holothuria forskalii*. *Tetrahedron* **1991,** *47*, 4753–4762.

Roos, R. Guam Seaweed Poisoning: Symptoms and Treatment of Cases. *Micronesica* **1993,** *26*, 23–29.

Rossignoli, A. E.; Mariño, C.; Martín, H.; Blanco, J. First Report of Two Gymnodimines and Two Tetrodotoxin Analogues in Invertebrates from the North Atlantic Coast of Spain. *Mar. Drugs* **2023,** *21*, 232.

Rushdi, M. I. et al. A review on the diversity, chemical and pharmacological potential of the green algae genus *Caulerpa. S. Afr. J. Bot.* **2020,** *132*, 226–241.

Sandvang, D.; Kristensen, H.-H.; Neve, S. Arenicin-3: A Novel Antimicrobial Peptide Showing Potent In Vitro Activity Against Gram-Negative Multi-Resistant Clinical Isolates. In: *Proceedings of the 46th Annual Meeting Infectious Disease Society of America*, Washington, DC, 2008.

Sansone, C.; Nuzzo, G.; Galasso, C.; Casotti, R.; Fontana, A.; Romano, G.; Ianora, A. The Marine Dinoflagellate *Alexandrium andersoni* Induces Cell Death in Lung and Colorectal Tumor Cell Lines. *Mar. Biotechnol.* **2018,** *20*, 343–352.

Santhanam, R.; Ramesh, S.; David, S. R. N. *Biology and Ecology of Pharmaceutical Marine Life: Echinoderms*; CRC Press (Taylor &Francis), 2019.

Santhanam, R. *Biology and Ecology of Edible Marine Bivalve Mollusks*; Apple Academic Press, 2018.

Santhanam, R. *Biology and Ecology of Edible Marine Gastropod Mollusks*; Apple Academic Press, 2018.

Santhanam, R.; Ramesh, S.; Shivakumar, G. *Biology and Ecology of Pharmaceutical Marine Cnidarians*; CRC Press (Taylor &Francis), 2019a.

Santhanam, R.; Ramesh, S.; *Biology and Ecology of Pharmaceutical Marine Tunicates*; CRC Press (Taylor &Francis), 2019b.

Santhanam, R. *Biology and Ecology of Venomous Marine Cnidarians*; Springer Nature: Singapore, 2020.

Santhanam, R.; Ramesh, S.; Nivedhitha, S.; Balasundari, S. *Pharmaceuticals and Nutraceuticals from Fish and Fish Wastes*; Apple Academic Press, 2022.

Satake, M.; Shoji, M.; Oshima, Y.; Naoki, H. Gymnocin-A, a Cytotoxic Polyether from the Notorious Red Tide Dinoflagellate, *Gymnodinium mikimotoi. Tetrahedron Lett.* **2002,** *43*, 5829–5832.

Sazdova, I. et al. Review of Cyanotoxicity Studies Based on Cell Cultures. *J Toxicol.* **2022,** *2022*, 5647178.

Scheuer, P. J. Toxins from Fish and other Marine Organisms. *Adv. Food Res.* **1970,** *18*, 141–161.

Schroeder, C. I.; Lewis, R. J. ω-Conotoxins GVIA, MVIIA and CVID: SAR and Clinical Potential. *Mar Drugs* **2006,** *4*, 193–214.

Seymour, B.; Seymour, J. Ch.7. *Cardiovascular Toxicity from Marine Envenomation: Heart and Toxins*; Elsevier Inc., 2014.

Shrestha, S. K.; Min, K. H.; Kim, Sw.; Kim, H.; Gerwick, W. H.; Soh, Y. Kalkitoxin: A Potent Suppressor of Distant Breast Cancer Metastasis. *Int. J. Mol. Sci.* **2023,** *24*, 1207.

Silchenko, A. S. et al. Structures and Bioactivities of Six New Triterpene Glycosides, Psolusosides E, F, G, H, H1, and I and the Corrected Structure of Psolusoside B from the Sea Cucumber *Psolus fabricii. Mar Drugs* **2019,** *17*, 358.

Silchenko, A. S.; Kalinovsky, A. I.; Avilov, S. A.; Andryjaschenko, P. V.; Dmitrenok, P. S.; Martyyas, E. A.; Kalinin, V. I. Triterpene Glycosides from the Sea Cucumber *Eupentacta fraudatrix*. Structure and Biological Action of Cucumariosides A1, A3, A4, A5, A6, A12 and A15, Seven New Minor Non-Sulfated Tetraosides and Unprecedented 25-keto, 27-Norholostane Aglycone. *Nat. Prod. Commun.* **2012,** *7*, 517–525.

Silva, C. O.; Novalis, S. C.; Soares, A. M. V. M.; Barata, C.; Lemos, M. F. L. Impacts of the Invasive Seaweed *Asparagopsis armata* Exudate on Energetic Metabolism of Rock Pool Invertebrates. *Toxins* **2021,** *13*, 15.

Silva, M. R. E.; Leme, J. G. Mediators of the Inflammatory Reaction. In: *Chemical Mediators of the Acute Inflammatory Reaction*; Elsevier Inc., 1972.

Simmons, M. A. Voltage-Gated Sodium Channel Agents., *xPharm: The Comprehensive Pharmacology Reference*, 2007.

Singh, R. K.; Tiwari, S. R.; Rai, A. K.; Mohapatra, T. M. Cyanobacteria: An Emerging Source for Drug Discovery. *J. Antibiot*. **2011,** *64*, 401–412.

Singh, A.; Benjakul, S. Serine Protease Inhibitors from Squid Ovary: Extraction and Its Effect on Proteolysis and Gel Properties of Surimi. *J. Food Sci. Technol.* **2017,** *54*, 267–275.

Singh, N.; Kumar, R.; Gupta, S.; Dube, A.; Lakshmi, V. Antileishmanial Activity In Vitro and In Vivo of Constituents of Sea Cucumber *Actinopyga lecanora. Parasitol. Res.* **2008,** *103*, 351–354.

Sinurat, E.; Marraskuranto, E.; Subaryon, S. Evaluation of Fucoidan from *Sargassum duplicatum* on Performance Immune Responses, Serum Biochemical, and Hematological Testing in Rats. *J. Appl. Pharm. Sci.* **2021,** *11*, 146–153.

Soliño, L.; Sureda, F. X.; Jorge Diogène, J.; Evaluation of Okadaic Acid, Dinophysistoxin-1 and Dinophysistoxin-2 Toxicity on Neuro-2a, NG108–15 and MCF-7 Cell Lines. *Toxicol. In Vitro* **2015,** *29*, 59–62.

Spielmeyer, A.; Loeffler, C. R.; Kappenstein, O. Identical Ciguatoxin-3C Group Profiles in *Lutjanus bohar* from the Pacific and Indian Oceans Indicating the Need to Re-Evaluate Geographical CTX classifications. *Front. Mar. Sci.* **2022,** *99,* 37438.

Stamman, E.; Sagar, D. A.; Davis, P. G. A Prellimnary Epidemiological Assessment of the Potential for Dirrhetic Shellfish Poisoning in the NE United States. *NOAA Technical Memorandum* NOS OMA 34. *1987.*

Steele, T. S.; Brunson, J. K.; Maeno, Y.; Moore, B. S. Domoic Acid Biosynthesis in the Red Alga *Chondria armata* Suggests a Complex Evolutionary History for Toxin Production. *PNAS*, **2022,** 119, e2117407119.

Stewart, J. E.; Marks L. J.; Gilgan, M. W.; Pfeiffer, E.; Zwicker, B. M. Microbial Utilization of the Neurotoxin Domoic Acid: Blue Mussels (*Mytilus edulis*) and Soft Shell Clams (*Mya arenaria*) as Sources of the Microorganisms. *Can. J. Microbiol*. **1998,** *44*, 456–64.

Sudarshan, S.; Krishnaveni.K. N.; Karthik, R.; Aanand, S. Bioprospecting Life Saving Drugs from Jellyfish Venom. *Biotica Res. Today* **2021,** *3*, 920–924.

Sugiyama, N.; Araki, M.; Nagashima, Y.; Shiomi, K. Biological Activities of Synthetic Grammistins and Analogous Peptides. *Fish. Sci*. **2006,** *72*, 1114–1120.

Takata, Y. et al. Occurrence of Domoic Acid in Tropical Bivalves. *Fish. Sci.* **2009,** *75*, 473–480.

Tan, L. T.; Márquez, B. L.; Gerwick, W. H. Lyngbouilloside, a Novel Glycosidic Macrolide from the Marine Cyanobacterium *Lyngbya bouillonii. J. Nat. Prod.* **2002,** *65*, 925–928.

Tan, S. N. et al. Diatom *Nitzschia navis-varingica* (Bacillariophyceae) and Its Domoic Acid Production from the Mangrove Environments of Malaysia. *Harmful Algae* **2016,** *60*, 139–149.

Tang, Y. et al. Chemical and Biological Study of Aplysiatoxin Derivatives Showing Inhibition of Potassium Channel Kv1.5. *RSC Adv.* **2019,** *9*, 7594–7600.

Taniyama, S.; Mahmud, Y.; Terada, M.; Takatani, T.; Arakawa, O.; Noguchi, T. Occurrence of a Food Poisoning Incident by Palytoxin from a Serranid *Epinephelus* sp. in Japan. *J. Nat. Toxins* **2002,** *11*, 277–282.

Tarcha, E. J.; Olsen, C. M.; Probst, P.; Peckham, D.; Muñoz-Elías, E. J.; Kruger, J. G.; Iadonato, S. P. Safety and Pharmacodynamics of Dalazatide, a Kv1.3 Channel Inhibitor, in the Treatment of Plaque Psoriasis: A Randomized Phase 1b Trial. *Plos One* **2017,** *12*, e0180762.

Thornburg, C. J.; Cowley, E. S.; Sikorska, J.; Shaala, L. A.; Ishmael, J. E.; Youssef, D. T. A.; McPhail, K. L. Apratoxin H and Apratoxin A Sulfoxide from the Red Sea Cyanobacterium *Moorea producens*. *J. Nat. Prod.* **2013,** *76*, 1781–1788.

Thuesen, E. V.; Kogure, K. Bacterial Production of Tetrodotoxin in Four Species of Chaetognatha. *Biol. BulL* **1989,** *176*, 191.

Tidgewell, K.; Gerwick, W. H. Natural Products Structural Diversity-II Secondary Metabolites: Sources, Structures and Chemical Biology. In: *Comprehensive Natural Products II*; Elsevier Ltd., 2010.

Tillmann, U.; Chiter, M. E.; Krock, B.; John, U.; Cembella, A. *Azadinium spinosum* Gen. et sp. nov. (Dinophyceae) Identified as a Primary Producer of Azaspiracid Toxins. *Eur. J. Phycol.* **2009,** *44*, 63–79.

Tiwari, A. S.; Tiwari, B. S. Cyanotherapeutics: An Emerging Field for Future Drug Discovery. *J. Appl. Phycol.* **2020,** *1*, 44–57.

Tiwari, N. et al. 8-Leishmaniasis Control: Limitations of Current Drugs and Prospects of Natural Products. *Nat. Prod. Drug Discov.* **2019,** 293–350.

Tobío, A.; Alfonso, A.; Madera-Salcedo, I.; Botana, L. M.; Blank, U. Yessotoxin, a Marine Toxin, Exhibits Anti-Allergic and Anti-Tumoural Activities Inhibiting Melanoma Tumour Growth in a Preclinical Model. *PLoS One* **2016,** *11*, e0167572.

Trainer, V. L. et al. Diarrhetic Shellfish Toxins and Other Lipophilic Toxins of Human Health Concern in Washington State. *Mar Drugs* **2013,** *11*, 1815–1835.

Tsai, Y.; Ho, P.; Hwang, C.; Hwang, P.; Chen, C.; Hwng, D. Tetrodotoxin in Several Species of Xanthid Crabs in Southern Taiwan. *Food Chem.* **2006,** *95*, 205–212.

Turnbull, A.; Seger, A.; Jolley, J.; Hallegraeff, G.; Knowles, G.; Fitzgibbon, Q. Lobster Supply Chains Are Not at Risk from Paralytic Shellfish Toxin Accumulation During Wet Storage. *Toxins* **2021,** *13*, 129.

Twede, V. D.; Miljanich, G.; Olivera, B. M.; Bulaj, G. Neuroprotective and Cardioprotective Conopeptides: An Emerging Class of Drug Leads. *Curr. Opin. Drug Discov. Devel.* **2009,** *12*, 231–239.

Twiner, M. J.; Rehmann, N.; Hess, P.; Doucette, G. J. A Review on the Chemistry, Ecology, and Toxicology with an Emphasis on Human Health Impacts. *Mar Drugs* **2008,** *6*, 39–72.

Ueda, H.; Itoi, S.; Sugita, H. TTX-Bearing Planocerid Flatworm (Platyhelminthes: Acotylea) in the Ryukyu Islands, Japan. *Mar. Drugs* **2018,** *16*, 37.

Ueoka, R.; Hitora, Y.; Ito, A.; Yoshida, M. : Okada, S.; Takada, K.; Matsunaga, S. Curacin E from the Brittle Star *Ophiocoma scolopendrina*. *J. Nat. Prod.* **2016,** *79*, 2754–2757.

Umnyakova, E. S. et al. Modulation of Human Complement System by Antimicrobial Peptide Arenicin-1 from *Arenicola marina. Mar. Drugs* **2018,** *16*, 480.

Valot, G.; Mailhol, D.; Regens, C. S.; O'Malley, D. P.; Godineau, E.; Takikawa, H.; Philipps, P.; Fürstner, A. Concise Total Syntheses of Amphidinolides C and F. *Chemistry* **2015,** *21*, 2398–2408.

Verdes, A.;, D.; Holford, M. Are Fireworms Venomous? Evidence for the Convergent Evolution of Toxin Homologs in Three Species of Fireworms (Annelida, Amphinomidae). *Genome Biol. Evol.* **2018,** *10*, 249–268.

Vest, S. E.; Dawes, C. J.; Romeo, J. T. Distribution of Caulerpin and Caulerpicin in Eight Species of the Green Alga *Caulerpa* (Caulerpales) *Bot. Mar.* **1983,** *26*, 313–316.

Vidal, J. P.; Laurant, D.; Kabore, S. A.; Rechencq, E.; Boucard, M.; Girard, J. P.; Escale, R.; Rossi, J. C. Caulerpin, Caulerpicin, *Caulerpa scalpelliformis*: Comparative Acute Toxicity Study. *Bot. Mar.* **1984,** *27*, 533–537.

Visciano, P.; Schirone, M.; Berti, M.; Milandri, A.; Tofalo, R.; Suzzi, G. Marine Biotoxins: Occurrence, Toxicity, Regulatory Limits and Reference Methods. *Front. Microbiol.* **2016,** *7*, 1051.

Walsh, C. J.; Luer, C. A.; Bodine, A. B.; Smith, C. A.; Cox, H. L.; Noyes, D. R.; and Gasparetto Maura, G. Elasmobranch Immune Cells as a Source of Novel Tumor Cell Inhibitors: Implications for Public Health. *Integr .Comp. Biol.* **2006,** *46*, 1072–1081.

Wang, D.; Xin, Y.; Wang, M. *Gambierdiscus* and Its Associated Toxins: A Minireview. *Toxins* **2022,** *14*, 485.

Wang, Y. et al. Effects of Two Sulfated Triterpene Saponins Echinoside A and Holothurin A on the Inhibition of Dietary Fat Absorption and Obesity Reduction. *Biosci. Biotechnol. Biochem.* **2014,** *78*, 139–146.

Wang, T. et al. Mitochondrial Dysfunction Contributes to the Cytotoxicity Induced by Tentacle Extract from the Jellyfish *Cyanea capillata* in Rat Renal Tubular Epithelial NRK-52E Cells. *Toxicon* **2013,** *74*, 1–7.

Wang, Z.; Zhang, S.; Hao, S.; Yang, W. Neurotoxins and Pore Forming Toxins in Sea Anemones: Potential Candidates for New Drug Development. *Histol. Histopathol* **2023,** *38*, 9–28.

Wargasetia, T. L.; Widodo. Mechanisms of Cancer Cell Killing by Sea Cucumber-Derived Compounds. *Invest. New Drugs* **2017,** *35*, 820–826.

Warrell, D. A. *Poisonous Plants and Aquatic Animals. Hunter's Tropical Medicine and Emerging Infectious Diseases.* Ryan et al., Eds.; Elsevier, 2020.

Watters, D. J. Ascidian Toxins with Potential for Drug Development. *Mar Drugs* **2018,** *16*, 162.

Watters, M. R. Seafood Neurotoxins II: Other Ingestible Marine Biotoxins—Ciguatera, Tetrodotoxin, Cyanotoxins. *Clinic. Neurotoxicol.*, **2009,** DOI:10.1016/B978-032305260-3.50047-2.

Whitelaw, B. L.; Cooke, I. R.; Finn, J.; da Fonseca, R. R.; Ritschard, E. A.; Gilbert, M. T. P.; Simakov, O.; Strugnell, J. M. Adaptive Venom Evolution and Toxicity in Octopods Is Driven by Extensive Novel Gene Formation, Expansion, and Loss. *GigaSci.* **2020,** *9*, 120.

WHO. Biotoxins, Aquatic (Marine and Freshwater) (EHC 37, 1984). World Health Orgnization Geneva, 1984. https://apps.who.int/iris/bitstream/handle/10665/37292/9241540974-eng.pdf

Wiese, M.; D'Agostino, P. M. Mihali, T. K.; Moffitt, M. C.; Neilan, B. A. Neurotoxic Alkaloids: Saxitoxin and Its Analogs. *Mar. Drugs* **2010,** *8*, 2185–2211.

Wijanarko, A.; Firdianna, A.; J Ginting, M. J.; Lischer, K.; Sahlan, M.; Hermansyah, H. Isolation and Anticancer Activity Assay of Stonefish (*Synanceia horrida*) bioactive stonustoxin. *AIP Conf. Proc.* **2022,** 2537, 020002.

Williams, B. L.; Stark, .M. R.; Caldwell, R. L. Microdistribution of Tetrodotoxin in Two Species of Blue-Ringed Octopuses (*Hapalochlaena lunulata* and *Hapalochlaena fasciata)* Detected by Fluorescent Immunolabeling. *Toxicon* **2012,** *60*, 1307–1313.

Wratten, S. J.; Fenical, W.; Faulkner, J. Ptilosarcone, the Toxin from *Ptilosarcus gurneyi. Tetrahedron Lett.* **1977,** *18*, 1959–1962.

Wu, J.; Yi, Y. H.; Tang, H. F.; Zou, Z. R.; Wu, H. M. Structure and Cytotoxicity of a New Lanostane-Type Triterpene Glycoside from the Sea Cucumber *Holothuria hilla. Chem. Biodivers.* **2006,** *3*, 1249–1254.

Wu, S.; Huang, T.; Lin, C.; Hui, C.; Lin, C.; Chen, Y. Pardaxin, a Fish Antimicrobial Peptide, Exhibits Antitumor Activity Toward Murine Fibrosarcoma In Vitro and In Vivo. *Mar. Drugs* **2012,** *10*, 1852–1872.

Wu, Y.; Lin, C.; Chen, C.; Hsieh, C.; Jen, H.; Jian, S.; Hwang, D. Toxin and Species Identification of Toxic Octopus Implicated into Food Poisoning in Taiwan. *Toxicon* **2014,** *91*, 96–102.

Xu, J. et al. The Voltage-Gated Potassium Channel Kv1.3 Regulates Peripheral Insulin Sensitivity. *PNAS*, **2004,** *101*, 3112 - 3117.

Yadav, S. K.; Palanisamy, M. Therapeutic Potentials of Marine Macro Algae of Karnataka. *J. Algal Biomass Utln.* **2019,** *10*, 40–51.

Yotsu-Yamashita, M.; Nagaoka, Y.; Muramoto, K.; Cho, Y.; konoki, K. Pufferfish Saxitoxin and Tetrodotoxin Binding Protein (PSTBP) Analogues in the Blood Plasma of the Pufferfish *Arothron nigropunctatus, A. hispidus, A. manilensis,* and *Chelonodon patoca. Mar. Drugs* **2018,** *16*, 224.

Yang, X. *Occurrence of the Cyanobacterial Neurotoxin, Anatoxin-a, in New York State Waters.* Ph.D. Thesis, State University of New York College of Environmental Science and Forestry Syracuse, New York, 2007.

Yasumoto, T. A. Guam Seaweed Poisoning: Biochemistry of the Gracilaria Toxins. *Micronesica* **1993,** *26*, 53–57.

Yasumoto, Y. Fish Poisoning Due to Toxins of Microalgal Origins in the Pacific. *Toxicon* **1998,** *36*, 515–1518.

Yi, Y. H.; Xu, Q. Z.; Li, L.; Zhang, S. L.; Wu, H. M.; Ding, J.; Tong, Y. G.; Tan, W. F.; Li, M. H.; Tian, F. et al. Philinopsides A and B, Two New Sulfated Triterpene Glycosides from the Sea Cucumber *Pentacta quadrangularis. Helv. Chim. Acta* **2006,** *9*, 54–63.

Yotsu-Yamashita, M.; Yasumoto, T.; Yamada, S.; Bajarias, F. F. A.; Formeloza, M. A.; Romero, M. L.; Fukuyo, Y. Identification of Polycavernoside A as the Causative Agent of the Fatal Food Poisoning Resulting from Ingestion of the Red Alga *Gracilaria edulis* in the Philippines. *Chem. Res. Toxicol.* **2004,** *17*, 1265–1271.

You, J.; Yue, Y.; Xing, F.; Xia, W.; Lai, S. Y.; Zhang, F. Tetrodotoxin Poisoning Caused by Goby Fish Consumption in Southeast China: A Retrospective Case Series Analysis. *Clinics* (Sao Paulo) **2015,** *70*, 24–29.

Young, N.; Sharpe, R. A.; Barciela, R.; Nichols, G.; Davidson, K.; Berdalet, E.; Fleming, L. E. Marine Harmful Algal Blooms and Human Health: A Systematic Scoping Review. *Harmful Algae* **2020,** *98*, 101901.

Zanchett, G.; Oliveira-Filho, E. C. Cyanobacteria and Cyanotoxins: From Impacts on Aquatic Ecosystems and Human Health to Anticarcinogenic Effects. *Toxins* **2013,** *5*, 1896–1917.

Zhang, Y. et al. Geographic Variations in the Toxin Profile of the Xanthid Crab *Zosimus aeneus* in a Single Reef on Ishigaki Island, Okinawa, Japan. *Mar. Drugs* **2021a,** *19*, 670.

Zhang, Y.; Abe, J.; Siddiq, A.; Nakagavwa, H.; Honda, S.; Ichida, S. UT841 Purified from Sea Urchin (*Toxopneustes pileolus*) Venom Inhibits Time-Dependent (45)Ca(2+) Uptake in Crude Synaptosome Fraction from Chick Brain. *Toxicon* **2001b,** *39*, 1223–1229.

Zhang, S.-Y.; Yi, Y.-H.; Tang, H.-F. Bioactive Triterpene Glycosides from the Sea Cucumber *Holothuria fuscocinerea*. *J. Nat. Prod.* **2006a,** *69*, 1492–1495.

Zhang, S. L.; Li, L.; Yi, Y. H.; Sun, P. Philinopsides E and F, Two New Sulfated Triterpene Glycosides from the Sea Cucumber *Pentacta quadrangularis*. *Nat. Prod. Res.* **2006b,** *20*, 399–407.

Zhang, S.; Yi, Y.; Tang, H. Cytotoxic Sulfated Triterpene Glycosides from the Sea Cucumber *Pseudocolochirus violaceus*. *Chem. Biodivers.* **2006c,** *3*, 807–817.

Zhang, S.; Li, L.; Sun, P.; Yi, Y. Lecanorosides A and B, Two New Triterpene Glycosides from the Sea Cucumber *Actinopyga lecanora*. *J. Asian Nat. Prod. Res.* **2008,** *10*, 1097–10103.

Zhu, J. et al. Tetrodotoxin in Asian Horseshoe Crabs *Carcinoscorpius rotundicauda* and *Tachypleus tridentatus* Across Different Life Stages from Northern Beibu Gulf, China. *Mar. Pollut. Bull.* **2022,** *184*, 114195.

Ziegman, R.; Alewood, P. Bioactive Components in Fish Venoms. *Toxins* **2015,** *7*, 1497–1531.

Zou, Z.-R.; Yi, Y.-H.; Wu, H.-M.; Wu, J.-H.; Liaw, C. C.; Lee, K. H. Intercedensides A–C, Three New Cytotoxic Triterpene Glycosides from the Sea Cucumber *Mensamaria intercedens* Lampert. *J. Nat. Prod.* **2003,** *66*, 1055–1060.

Index

E

G

H

I

K

L

M